Vehicular Technology Handbook

Volume I

Vehicular Technology Handbook
Volume I

Edited by **Mackenzie Ruiz**

CLANRYE INTERNATIONAL
New Jersey

Published by Clanrye International,
55 Van Reypen Street,
Jersey City, NJ 07306, USA
www.clanryeinternational.com

Vehicular Technology Handbook: Volume I
Edited by Mackenzie Ruiz

International Standard Book Number: 978-1-63240-513-5 (Hardback)

Printed in the United States of America.

Contents

Preface

The world today has experienced a sea of change in technology. One such field which has undergone rapid changes is that of vehicular technology. The field has various applications in real life that affect us and thus there is a constant flux in the changes being made in such technology. There are many factors that affect changes in vehicular technology. In recent years there have been concerns about the environment and climate changes, combined with growing petroleum demand. This is such a factor that has affected the advances in vehicular technology. The technology has accelerated towards more electrified and sustainable transportation mechanisms and vehicles. Auto manufacturers always keep in mind the needs and demands of the market, especially the economy, in deciding the direction of research that vehicular technology needs to have for updates and changes in efficiency. There are various ground- breaking changes that have happened because of such research in vehicular technology like advanced control, storage, consumption of electricity and signal processing in modern vehicular platforms. There is thus a demand in this field for skilled researchers and engineers who can keep up with the rapidly changing pace of vehicular technology.

This book is an attempt to compile, collate and understand the research being done in the field of vehicular technology and the data found in the advancements in this field. I am grateful to those who put in their hard work and efforts into this field as well as those who supported us in this endeavor. I would also like to thank my family for their endless support at every step of the publication process.

Editor

Modeling and Deployment of Model-Based Decentralized Embedded Diagnosis inside Vehicles: Application to Smart Distance Keeping Function

Othman Nasri,[1] **Hassan Shraim,**[1] **Phillippe Dague,**[1] **Olivier Heron,**[2] **and Michael Cartron**[2]

[1] *LRI, University Paris-Sud 11, CNRS & INRIA Saclay Île-de-France, Bât 490, 91405 Orsay Cedex, France*
[2] *CEA LIST, Saclay, 91191 Gif-sur-Yvette Cedex, France*

Correspondence should be addressed to Hassan Shraim, hassan.shraim@gmail.com

Academic Editor: David Fernández Llorca

The deployment of a fault diagnosis strategy in the Smart Distance Keeping (SDK) system with a decentralized architecture is presented. The SDK system is an advanced Adaptive Cruise Control (ACC) system implemented in a Renault-Volvo Trucks vehicle to increase safety by overcoming some ACC limitations. One of the main differences between this new system and the classical ACC is the choice of the safe distance. This latter is the distance between the vehicle equipped with the ACC or the SDK system and the obstacle-in-front (which may be another vehicle). It is supposed fixed in the case of the ACC, while variable in the case of the SDK. The variation of this distance depends essentially on the relative velocity between the vehicle and the obstacle-in-front. The main goal of this work is to analyze measurements, issued from the SDK elements, in order to detect, to localize, and to identify some faults that may occur. Our main contribution is the proposition of a decentralized approach permitting to carry out an on-line diagnosis without computing the global model and to achieve most of the work locally avoiding huge extra diagnostic information traffic between components. After a detailed description of the SDK system, this paper explains the model-based decentralized solution and its application to the embedded diagnosis of the SDK system inside Renault-Volvo Truck with five control units connected via a CAN-bus using "Hardware in the Loop" (HIL) technique. We also discuss the constraints that must be fulfilled.

1. Introduction

In order to respond to the increasing demands of safety and driving comfort, more and more electronic functions are embedded in the vehicles such as engine control (to optimize fuel economy and to reduce pollution), Antilock Braking System (ABS), and Electronic Stability Program (ESP), Each global safety or comfort system contains one or more functions which may be distributed on several Electronic Control Unit ECUs. Most of these functions are modular and respecting some norms (such as AUTOSAR). They exchange information (e.g., vehicle speed) with other functions via communication interfaces. That means that if the system is not equipped with a certain diagnosis strategy, any fault generated from a function may influence all the functions are related to it. This fact highlights the problem of fault propagation and the need of an on-board (i.e., on running car) fault diagnosis in the vehicle.

An increasing number of vehicles are being equipped with adaptive cruise control (ACC). The ACC adjusts the brake and/or throttle, within limited ranges, to maintain a constant headway from any other vehicle that intrudes upon the path of the driver's vehicle. While ACC provides a potential safety benefit in helping drivers to maintain a constant speed and headway Davis [2], as with other types of automation, there is the potential for misuse and disuse Parasuraman and Riley [3]. It provides assistance to the driver in the task of longitudinal control of his vehicle during motorway driving. For ACC to be effective, drivers need to understand the capabilities of the technology, which include braking and sensor limitations. Based on this understanding, they must be able to intervene when a given situation exceeds ACC capabilities. However, drivers have difficulties in understanding how ACC functions Stanton and Marsden [4]. As a result, they tend to rely on the system inappropriately. For instance, Nilsson [5] showed that drivers failed to intervene when approaching a stopped queue of vehicles because they believed that the ACC could effectively respond to the situation. Stanton et al. [6] introduced an unexpected acceleration into the ACC system during routine driving conditions, which resulted in a collision 33 percent of the time. Whether or not drivers can respond effectively when automation fails depends on their understanding of the type of failure that occurs and the context in which it occurs Lee et al. [7].

To ensure safe and effective use, ACC limits of operation should be identifiable and interpretable Goodrich et al. [8]. One approach to help drivers to detect and to respond to these limits is to match the limits of the ACC algorithm to the natural boundaries drivers use to switch between car-following and active braking behaviors, as defined by environmental cues (e.g., time headway (THW) and time to collision (TTC)). In order to avoid several problems that may be produced because of the misuse or the disuse of the ACC, the Smart Distance Keeping (SDK) system is proposed. (The SDK system is an advanced Adaptive Cruise Control (ACC) system implemented in a Renault-Volvo Trucks vehicle to increase safety by overcoming some ACC limitations.) This system must be understood as a function to enhance the driver's capability to manage his longitudinal environment and is dedicated to a use on highways or expressways (straight line, low curvatures, oneway roads). The SDK is based on the immediate front environment sensing on one hand, and on the automated management of the truck longitudinal actuators (brakes, engine, gearbox) on the other hand, all this being monitored and controllable at any time by the driver through the in-cabin human machine interface and the conventional driving commands (pedals, switches). This system is composed of many subsystems (micro controllers, cables, CAN or FlexRay bus, sensors, actuators, etc.) coming from different suppliers. That is why its diagnostic is a challenging task.

In this work, we will focus on the modeling and the fault diagnosis of the SDK system. Faults study is limited to those that may be produced on sensors measurements due to their direct influence on the SDK system decision. So, they should be always checked to ensure that they are within their expected operating range. Simple checks on the recent rate of change or variance of the output can also be incorporated. Faults which cause the sensors to have an offset or altered gain will affect the control system but may not be detected by this first level approach. The traditional approach to sensor fault checking is to include hardware redundancy for sensors. If two sensors measure the same quantity disagree, there is likely to be a fault in one of them, and if three or more measurements are available, the fault is likely to be in the sensor which disagrees most. But due to the high cost of providing direct hardware redundancy for sensors, the analytic redundancy techniques were proposed. Conceptually, this equates to creating virtual sensors from other available measurements, to compare with the one being monitored. Analytic redundancy is used in available passenger car control systems.

Considering the size and the complexity of the SDK system, a centralized on-board diagnosis is not adequate because it requires the establishment of a global model of the system and too much communication and memory resources and prevents to act immediately each time a diagnosis could be found at local level. To detect and isolate possible faults in the SDK system and manage its architectural complexity, we have chosen to apply the model-based fault diagnosis approach (FDI) Patten et al. [9] in a decentralized manner. A local diagnoser is associated with each component of the SDK system based on a modular modeling of the plant elements. All local diagnosis decisions are transmitted via CAN-bus and merged by a dedicated supervisor in order to obtain a global decision and carry out any recovery action. A strategy for applying this merging operation was developed in order to be efficient.

The paper is organized as follows. In Section 2, we present a general description and modeling of the SDK controller with its environment system. The modeling in this work includes a simplified mathematical model of the wheels and the engine. Then, in Section 3, decentralized algorithm and strategy for the detection and the isolation of sensors faults that may affect the overall system are developed. Section 4 explains how to deploy this approach in order to achieve on-line diagnosis of the SDK system. Then, we evaluate the proposed diagnosis approach in Section 5. Finally, we conclude with a discussion on the related work.

2. System Description and Modeling

2.1. Smart Distance Keeping System. Smart Distance Keeping (SDK) or "enhanced Adaptive Cruise Control (ACC)" is a system which automatically controls the vehicle's longitudinal velocity, by acting on the engine, gearbox, retarder, and braking system. This requires the vehicle to be equipped with a radar system connected to a dedicated control unit as shown in Figure 1.

The global SDK system may be decomposed into two main parts, the SDK controller and the SDK physical system displayed in Figure 2.

The main functions of the SDK controller (the block "SDK Function") are to

(i) receive the distance between the truck and the object-in-front,

(ii) find the deceleration (acceleration) needed to realize the correct functioning of the SDK system (maintaining a minimal safety distance with the vehicle-in-front),

(iii) use a control algorithm for acting through engine, braking system, and so forth in order to adjust the velocity of the truck.

A realistic representation of the SDK controller is given by the Input Output Symbolic Transition System (IOSTS) model as shown in Figure 3. This model describes exhaustively the relevant driving situations where the distance control intervenes. Apart from the "Refresh (R)" mode (initial state), three categories of "basic" modes can be distinguished: the "Cruise Control (C)" mode, the "Approch (A)" mode, and the "Follow (F)" mode. The variables of this

Modeling and Deployment of Model-Based Decentralized Embedded Diagnosis inside Vehicles: Application to Smart Distance Keeping Function

3

FIGURE 1: Radar installed on the Renault Magnum truck.

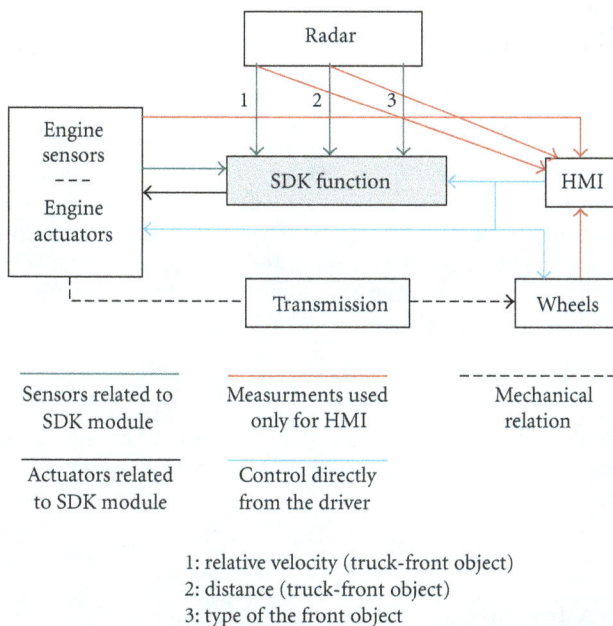

1: relative velocity (truck-front object)
2: distance (truck-front object)
3: type of the front object

FIGURE 2: Environment of the SDK system.

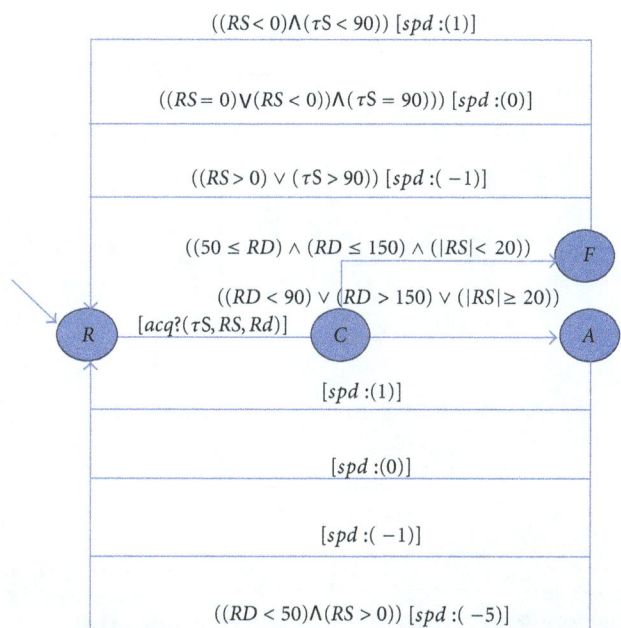

FIGURE 3: IOSTS system of the SDK controller.

IOSTS system are $TS(= \text{TruckSpeed})$, $RS(= \text{RelativeSpeed})$, and $RD(= \text{RelativeDistance})$, while the communication channels are $acq(= acquisition)$ and $spd(= speed)$ where $acq?(TS, RS, RD)$ and $spd!(-5, -1, 0, 1)$.

The decision of the SDK controller depends essentially on the data issued from some sensors: truck velocity sensor, wheels angular velocity sensors, radar, and engine sensors.

2.2. The Radar. The SDK needs to be informed about the object-in-front presence, and about its relative position and velocity. Within this work, the sensor is a 3-beam Doppler effect ARS100 Radar. This radar monitors the traffic in front of the vehicle using three stationary independent millimeter waves.

Moving and stationary objects are detected and their distance and relative velocity are measured and processed sixteen times per second.

Due to its physical nature, the radar sensor is offering excellent performance characteristics even in adverse weather conditions.

Since the data issued from the radar depend on the external object, then in order to realize any simulation, several scenarios should be prepared for the movements of the SDK vehicle and of the object-in-front. In this work, we suppose that the distance and the relative velocity between the SDK vehicle and the vehicle-in-front are the inputs to the system (depending on the scenarios that we are choosing).

2.3. The Wheels. The linear truck velocity, which is one of the important inputs to the SDK controller, is calculated based on the angular velocities of the six wheels.

The wheel rotational dynamics is given in (1) by applying Newton's Law:

$$I_{w_i}\dot{w}_i = R_i F_{X_i} + \text{Torque}(m_i) - \text{Torque}(b_i), \qquad (1)$$

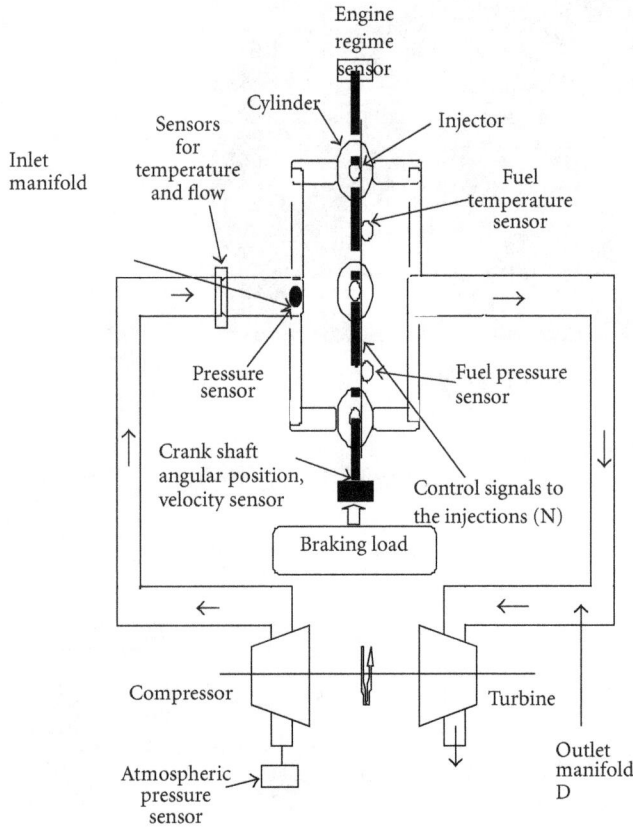

FIGURE 4: Engine Architecture Peysson et al. [1].

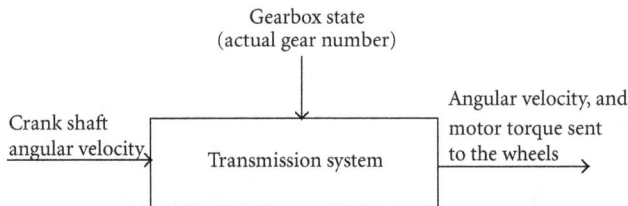

FIGURE 5: Transmission block.

where I_{w_i} is the moment of inertia of the wheel number i, R_i its effective radius, w_i is the angular velocity of the wheel, F_{X_i} is the friction force, $\mathrm{Torque}(m_i)$ is the applied tractive torque, and $\mathrm{Torque}(b_i)$ is the braking torque.

2.4. Motor and Power Train.
Modern diesel engines are essentially made up of the following subsystems.

We present in this section the simplified model for the diesel engine (see Figure 4) that we have developed, in order to be used by the SDK controller. Based on Peysson et al. [1], the dynamics of motor rotation is given by.

$$J_e \dot{w}_e(t) = M_{\mathrm{ind}}(t - \tau_i) - M_f(t) - M_{\mathrm{load}}(t), \qquad (2)$$

where w_e is the crank shaft angular velocity, M_{ind} is the indicated torque, τ_i is the delay, M_f is the friction torque, M_{load} is the torque due to the load, and J_e is the effective inertia of the engine.

In this work, the transmission system is represented as in the Figure 5. For simplification purposes, the "Transmission System" block is composed only of several constants, depending on the gearbox state.

2.4.1. The Admission and Intake Manifold. The temperature T_{im} of the intake manifold is assumed to remain constant due to the intercooler. Therefore, the analysis will be based essentially on the variation of the pressure P_{im}. In this study, the input flow is characterized by the output flow $\dot{m}_c$ of the compressor (see (3)):

$$\dot{P}_{\mathrm{im}} + \frac{\tau_v V_d N_e}{120 V_{\mathrm{im}}} P_{\mathrm{im}} = \dot{m}_c \frac{R T_{\mathrm{im}}}{V_{\mathrm{im}}}, \qquad (3)$$

where V_{im} is the volume of the intake manifold, τ_v is the volumic efficiency, V_d is the exchange volume in the engine, and N_e is the rotational velocity of the engine in rpm ($N_e = 60 w_e / 2\pi$).

2.4.2. The Indicated Torque M_{ind}. In order to calculate the indicated torque, we should calculate firstly the indicated efficiency. Normally, this efficiency is a specific characteristic to the engine and it is found from empirical data. This efficiency is higher when the mixture is light, and it may be approximated by (4)

$$\mu_{\mathrm{ind}} = a + b\lambda + c\lambda^2. \qquad (4)$$

The coefficients (a, b, c) are found by identification (three different tests have been made) and λ is defined by (5)

$$\lambda = \frac{(P_{\mathrm{im}} w_e V_d / 4\pi R)\tau_v}{T_{\mathrm{im}} \dot{m}_f}. \qquad (5)$$

Then the indicated torque M_{ind} can be found by (6)

$$M_{\mathrm{ind}} = \dot{m}_f p_{c_i} \mu_{\mathrm{ind}}, \qquad (6)$$

where $\dot{m}_f$ is the flow of fuel and p_{c_i} is a characteristic for the diesel ($40000000 \, \mathrm{JKg}^{-1}$).

2.4.3. The Injection. The injection system controls the quantity of fuel that will be introduced into the combustion chamber. The mixture Air/fuel should be capable to auto ignite by the effect of temperature and the high pressure. The calorific power of combustion is related to the quantity of fuel injected. The following model gives the flow of fuel $\dot{m}_f$ in function of the position x_p of the accelerator pedal and the engine speed (see (7)) Peysson et al. [1]:

$$\begin{aligned} \dot{m}_f &= i_0 + \Delta\dot{m}_{(\mathrm{SDK})} + \Delta\dot{m}_f, \\ \Delta\dot{m}_f &= w_e\left(i_1 + i_2 x_p + i_3 x_p^2 + i_4 w_e\right), \end{aligned} \qquad (7)$$

where i_0 characterizes the minimal injection flow (greater than zero, when the engine is idle) and $\Delta\dot{m}_f$ models the variations of the flow around i_0. $i_1, \ldots, i_4$ are constants obtained by identification following simplification on the flow control model (regulator slide).

Modeling and Deployment of Model-Based Decentralized Embedded Diagnosis inside Vehicles: Application to Smart
Distance Keeping Function

5

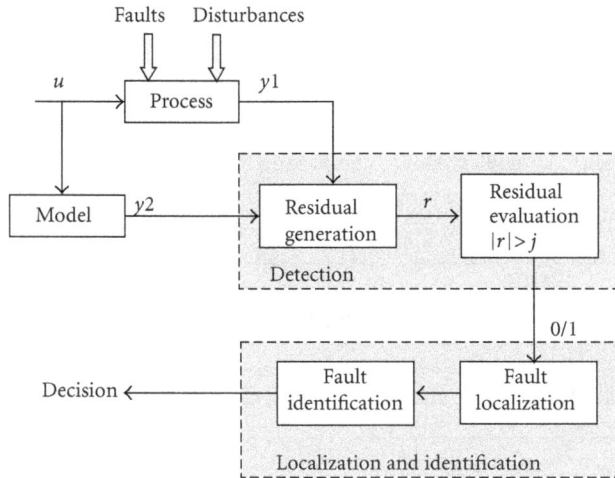

FIGURE 6: Model-based fault diagnosis using residual approach.

2.4.4. The Friction Torque M_f.

The friction torque may be calculated by

$$M_f = \frac{(c_0 + c_1 w_e + c_1 w_e^2) V_d}{2\pi nr}. \tag{8}$$

2.4.5. The Load Torque M_{load}.

The torque M_{load} depends on the type of the road, the vehicle velocity, the turnings, and so forth. In this work we will suppose that this torque is an input to the system and has a constant value.

3. Decentralized Diagnosis

In most automated systems, the command part (which implements the control of the operative part) is generally represented through a model to be applied to the operative part (mechanical components which should be controlled by means of actuators, such as engines). Realizing a diagnosis requires also to be able to represent the state of the operative part using a model that can be integrated to the one of the command part, separated or mixed. Thus, when a fault occurs, it is possible to get information regarding the process and to compare model and process. This is called model-based diagnosis, more particularly Fault Detection and Isolation (FDI) Darkhovski [10].

The method of FDI is based on the use of model-based analytical redundancy, that is, relations among the measured variables (see Figure 6). It can be divided into several steps Patten et al. [9].

(i) *First*, data containing information about the process states are transmitted to the residual generation module. This module generates a vector that carries information about symptoms and particular possible faults.

(ii) *Second*, the successive generated vectors are evaluated and filtered in order to extract the primary cause of the observed evolution, that is, to achieve fault localization and identification,. The structured residual

approach Chow et al. [11] has been chosen in order to perform this stage.

3.1. Motivations. A decision-making structure for fault diagnosis must be defined to face combinatorial explosion and real-time problems and/or communication problems between various components of a process. The choice of a structure depends on the distribution of the available information (model and observation): centralized or distributed, and on the nature of the process: simple with only one control unit or complex with several local control units. There are therefore three main structures of decision-making methods for diagnosis: centralized, decentralized, and distributed.

The centralized structure consists in associating to one global model of the process a single diagnostic module (called diagnoser). It collects the different process information before making its final decision about the operating status of the process Sampath et al. [12]. Although successful in terms of diagnosis for simple systems, the centralized structure is difficult to use for large systems. Indeed, the acquisition of a global model of the process rises difficulties and often leads to combinatorial explosion problem, when it is not just impossible due to the presence of several manufacturers for the different parts of the process and privacy rules.

The decentralized structure is based on several local independent diagnosers that are associated to one global model of the process. Each diagnoser receives the observations which are specific to it and takes a local decision based on its local observations. However, this structure involves problems of indecision when some global specifications require consistency checking between decisions of local diagnosers. To solve these cases, each diagnoser sends its local decision to a coordinator (or supervisor) which will manage the different problems of ambiguity between these diagnosers and will take the final decision Wang et al. [13].

In the distributed structure, the process is modeled through its components (or subsystems) by several local models. Each one is equipped with a local diagnoser responsible of it. In the case of global specifications, a communication protocol allows directly the communication between the different diagnosers to manage conflict decision Qiu [14]. Each diagnoser makes its decision based on its own local observation and that reported by other local diagnosers as answers to its queries. This structure permits to throw off the construction of a coordinator but implies the definition of a protocol for decision making through communication between diagnosers, often impractical because without guarantee of convergence in bounded time and generating more important communication traffic and delays.

In this paper, we adapt the decentralized/distributed structure. In other words, a local diagnoser is associated with each component of the process based on a modular modelling of the plant elements. All local diagnosis decisions are transmitted via a communication environment and merged by a dedicated supervisor in order to obtain a global decision and carry out any recovery action (see Figure 7).

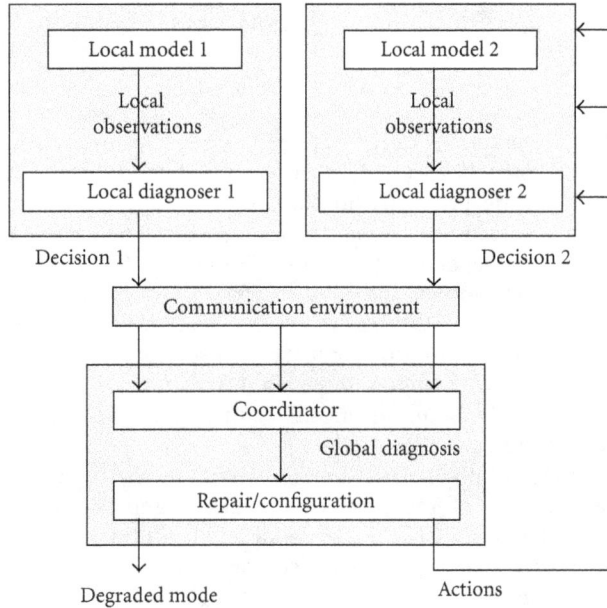

FIGURE 7: Model-based decentralized principle with 2 local diagnosers.

TABLE 1: Angular velocity comparison for the wheels 1, 3, and 5.

Difference	$w_1 - w_3$	$w_1 - w_5$	$w_3 - w_5$
$f_{\text{odd } i}$	f_{13}	f_{15}	f_{35}

TABLE 2: Angular velocity comparison for the wheels 2, 4, and 6.

Difference	$w_2 - w_4$	$w_2 - w_6$	$w_4 - w_6$
$f_{\text{even } i}$	f_{24}	f_{26}	f_{46}

TABLE 3: Comparison of the angular velocity of the wheels: 1, 2, 4, and 6.

Difference	$w_1 - w_2$	$w_1 - w_4$	$w_1 - w_6$
f_{1i}	f_{12}	f_{14}	f_{16}

TABLE 4: Comparison of the angular velocity of the wheels: 3, and 2, 4, 6.

Difference	$w_3 - w_2$	$w_3 - w_4$	$w_3 - w_6$
f_{3i}	f_{32}	f_{34}	f_{36}

TABLE 5: Comparison of the angular velocity of the wheels: 5, and 2, 4, 6.

Difference	$w_5 - w_2$	$w_5 - w_4$	$w_5 - w_6$
f_{5i}	f_{52}	f_{54}	f_{56}

This fusion can be realized by a coordinator based on a set of rules. The goal of this coordinator is to solve the problem of decision conflict and/or ambiguity among local diagnosers in order to obtain a diagnosis performance equivalent to that of a centralized diagnoser. This approach allows carrying out on-line diagnosis without computing the global model and overcoming both the combinatorial and communication traffic explosion problems.

3.2. SDK Algorithm. As shown previously, the decision of the SDK controller depends essentially on the data issued from some sensors (wheels angular velocity sensors, radar, and transmission sensor), which means that any faulty data will influence the SDK system decision. That is why one local diagnoser is associated with each one of these sensors in order to diagnose an SDK fault by using the decentralized model-based approach (see Figure 7).

3.3. Decentralized Diagnosis of the SDK. A model of the SDK environment, that is, the part of the vehicle required to close the loop, is necessary to perform diagnosis (see Figure 7). So, by applying the laws of dynamics, a simplified model of the diesel engine has been developed. It permits to identify the angular velocities of the six wheels and the crank shaft angular velocity in response to an action (on braking system or accelerator pedal or gearbox).

3.3.1. Wheels Diagnoser. The six wheels angular velocities w_i are the inputs of a local algorithm able to detect and isolate any fault that occurs in the wheels velocity sensors. Indeed, it is possible by using these redundant measurements to generate a set of structured residuals and afterwards detect and isolate single or multiple fault.

Two cases are considered based on steering angle.

(i) *Case of Straight Line Motion.* In this case, we suppose that the angular velocities of the six wheels should be approximately equal. Then in order to apply this strategy, we suppose that we have two groups: group 1 (for the wheels: 1, 3, 5), and group 2 (for the wheels: 2, 4, 6), and we calculate the differences in the angular velocities as shown in Tables 1 and 2. Then if $(w_i - w_j) < \epsilon$, we suppose there is no fault, and $f_{ij} = 0$, else we have a fault and $f_{ij} = 1$. To localise the fault in the case of $f_{ij} = 1$, a small algorithm is realized. This algorithm is able to localise from 1 to 4 faults. In the case of more than four, it gives a signal that all the wheels are faulty The realization of this algorithm is based on Tables 1, 2, 3, 4, 5. By completing these tables, the localization of the fault will be evident.

(ii) *Case of a Curve Motion.* in this case, we follow the same strategy proposed in the previous case, with the five tables, but the main difference here, in the case of the curve, is when we compare a wheel in group 1 (for the wheels: 1, 3, 5) to a wheel in group 2 (for the wheels: 2, 4, 6), we should take in consideration a small difference that can be calculated geometrically based on the Ackerman angle theory and based essentially on the steering angle. So we should replace ϵ with ϵ'.

The outputs of this algorithm are the state (normal/abnormal) s_i of the wheels sensors. In addition, it

Modeling and Deployment of Model-Based Decentralized Embedded Diagnosis inside Vehicles: Application to Smart
Distance Keeping Function

7

computes the longitudinal speed TS_w of the truck which is approximated based on the nonfaulty sensors:

$$(s_1, s_2, s_3, s_4, s_5, s_6, TS_w) = \text{wheels}(w_1, w_2, w_3, w_4, w_5, w_6). \tag{9}$$

3.3.2. Transmission Diagnoser.
Since there is no redundancy measurement, the algorithm just computes the longitudinal truck speed TS_t by using the value of the crank shaft angular velocity w_e (the gear box output) and the transmission rate number n:

$$TS_t = \text{transmission}(w_e, n). \tag{10}$$

3.3.3. Radar Diagnoser.
The basic data detection requirement is to measure distance, relative speed, and reflection signal amplitude of moving and stationary objects in three beams. Angular position is calculated by the interpolation algorithms based on signal levels in adjacent beams.

Several scenarios for the radar fault detection analysis ar as follows:

(i) *First*, if the radar is faulty and does not detect any object. So, without the help of another device, we can do nothing,

(ii) *Second*, if the radar works but gives incorrect distances (with a certain shift of x meters): for example $(150\,\text{m} \rightarrow (150 - x)\text{m})$ where x is a constant term, then, we cannot detect this fault.

(iii) *Third*, if the relative velocity and the distance between the vehicles are measured separately (two different measurement tools), then it is important to check at each period (e.g., 2 seconds) if the variation in the distance corresponds to the variation in the relative velocity. If there is a difference then we say that there is a fault.

Exemple: suppose that we initially have the relative velocity RS and the distance RD between the SDK truck and the vehicle-in-front (see Figure 9), so if we consider that the period (that we choose for checking) is equal to 2 seconds, then we should obtain

$$d(t) - d(t - 2) = 2 * \text{Avera}(Rs), \tag{11}$$

where $\text{Avera}(RS)$ is the average value of RS during the period of 2 seconds.

(iv) *Fourth*, suppose that the radar was detecting a vehicle-in-front (see Figure 9).

As we have shown before, the relative speed RS is a measurement given by the radar. And also, the SDK vehicle velocity TS is measured from other sensors (wheels angular velocities or vehicle velocity); then we can find the velocity of the vehicle-in-front: $FS = RS + TS$.

Getting the velocity of the front vehicle, we can analyze as follows: if there is a strong sudden variation (and then its acceleration (deceleration) is not realistic), then we have one of the three following cases:

(i) there is a fault in the radar sensor;

(ii) the value of the SDK vehicle velocity is faulty;

(iii) an intruder vehicle comes in front of the SDK vehicle (see Figure 10).

In all the aforementioned cases, it is important to observe the velocity of the front vehicle for several points before taking any decision.

In some of the previous cases, the calculation of the front vehicle acceleration (deceleration) may give a nonrealistic values. A study about the maximum (minimum) possible acceleration (deceleration) can be given as follows:

$$|a_x|_{\max} = \max\left|\frac{\sum F_{x_{ij}}}{m}\right| = \max\left|\frac{\mu \cdot N_z}{m}\right| = \max\left|\frac{\mu \cdot m \cdot g}{m}\right|$$
$$\leq \mu_{\max}.g. \tag{12}$$

A maximum friction coefficient ($\mu_{\max}$) determines maximum acceleration or deceleration. In order to estimate $\mu_{\max}$, sliding mode observers can be applied. A hierarchical observer is needed for this estimation. In the first step, an observer based on the dynamical equation of the wheels should be developed. This observer takes as an input the applied motor torque (which is estimated statically (existing maps)) and the braking torque, which can be easily found based on the hydraulic pressure sent to the wheels shraim [15]. Then, in parallel to this observer, a sliding modes observer is used to estimate the vertical forces. This observer is based on the suspension system modeling. Then by calculating the longitudinal force and the vertical force, we apply the following formula to estimate the adherence coefficient:

$$\mu_{\max} > \frac{\max(\sum_{i=1}^{n} F_{x_i})}{\min(\sum_{i=1}^{n} F_{z_i})}, \tag{13}$$

where n is the number of the wheels (equals 6), and F_{x_i} and F_{z_i} are, respectively, the longitudinal and the vertical forces applied at the wheel i.

If a nonrealistic acceleration (deceleration) value is found, then if there is no intruder vehicle, we can suppose that there is radar fault. Thus, the outputs of the local algorithm are the radar state (normal/abnormal) s_r and the velocity of the object-in-front:

$$(s_r, FS) = \text{radar}(RD, RV, TV, s_{\inf}, \mu_{\max}), \tag{14}$$

where $s_{\inf}$ is a signal precising if the truck velocity is faulty or not.

3.3.4. Global Diagnoser (or Supervisor).
It takes as inputs the outputs of the local diagnosers and its goal is to do some global consistency checking and to merge the local diagnosis decisions in order to obtain one global diagnosis decision and carry out the appropriate recovery action:

$$(c, TS) = \text{global}(s_1, s_2, s_3, s_4, s_5, s_6, s_r, TS_w, TS_t). \tag{15}$$

c and TS represent the control (recovery action) and the truck speed, respectively.

Several fault scenarios and recovery actions have been analyzed. The first class of scenarios is composed of wheel sensors failures and/or sensor failure on the rotation speed of the shaft engine (transmission fault). The global diagnosis and recovery actions are described as follows.

(a) If 1 to 3 wheels sensors are faulty out of the 6 (determined by the wheels diagnoser), the recovery action consists for the SDK function in using the truck speed TS_w calculated as average values provided by the correct sensors (between 3 and 5). The global diagnoser compares TS_t to TS_w and, in case of discrepancy, concludes also to a faulty transmission sensor.

(b) If 4 wheels sensors are faulty (determined by the wheels diagnoser), a comparison between the speed TS_w computed from the two ones assumed to be correct and TS_t is performed by the global diagnoser. In case of consistency, the recovery action decided by the global diagnoser consists in using the truck speed calculated as average values provided by the correct wheel sensors and the transmission sensor (3 out of 7).

(c) In all other cases, that is, when at most 2 sensors out of 7 provide consistent measurements, the recovery action decided by the global diagnoser consists in disabling the SDK function.

The second scenario category includes the radar failure. It may pass unnoticed, in particular if RD and RS measurements are not independent, because we cannot provide analytical redundancies between the truck and the front vehicle. In absence of such redundancy, the only check we can do is to verify that the behavior of the front vehicle, in terms of velocity and acceleration, as deduced by the global diagnoser from the truck's behavior (TS) and the radar measurement, is not physically impossible, that is, does not violate the laws of dynamics. In case of physical impossibility detected, then the recovery action taken by the global diagnoser consists in disabling the SDK controller.

Obviously both scenarios can combine, allowing multiple fault diagnosis of the wheels and transmission sensors and of the radar.

4. Diagnosis Deployment

In this section, we propose a deployment solution of the decentralized model-based fault diagnosis strategy (see Figure 8) in the electronic architecture of a Renault Truck's vehicle. The SDK electronic architecture part is composed of three Electronic Control Units (ECUs) that communicate between them through a bus topology. The ECUs exchange messages that follow the Control Area Network (CAN) protocol, which is low cost and very wide spread in the road transport domain. The three ECUs (ECU1, 2, and 3) are, respectively, linked to wheels, transmission and radar (+SDK algorithm) control functions. We also assume that

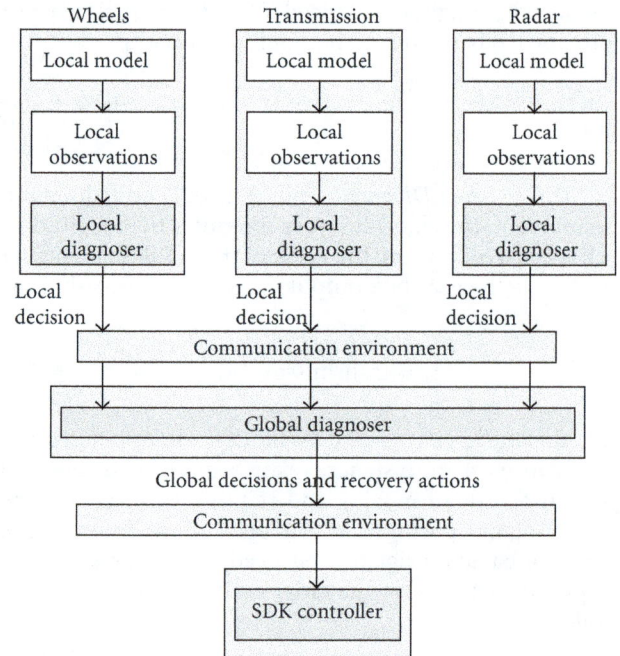

FIGURE 8: Decentralized model-based fault diagnosis of the SDK system.

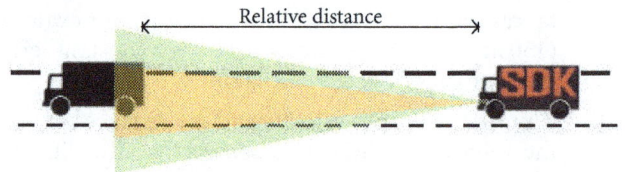

FIGURE 9: SDK Vehicle and the vehicle-in-front.

there is at least another ECU (ECU4) in the vehicle for other high level control functions, which is a very likely situation in modern vehicles. ECU4 will be used for embedding the global diagnoser and recovery functions.

We first list the manufacturing constraints that motivate the adopted deployment solution. We next present the classical on-board diagnosis techniques in vehicles. Finally, we describe the principle of our on-board diagnosis strategy, based on the previous electronic architecture. The next section will present an integration and validation platform.

4.1. Motivations. From an end users point of view, breakdowns and malfunctioning can lead to a loss of safety for the driver and his environment—a major breakdown which necessitates an emergency stop and repair—or a company's performance penalty, due to the need of a vehicle maintenance. From the vehicle manufacturer point's of view, these risks must be avoided because they can mostly cause a loss of corporate image and in-field yield.

Beside that, the road transport industry is a very competitive market and the integration of innovative features, such as diagnosis, is highly driven by economical considerations. That is why most of control organs of modern fuel vehicles

Modeling and Deployment of Model-Based Decentralized Embedded Diagnosis inside Vehicles: Application to Smart Distance Keeping Function

9

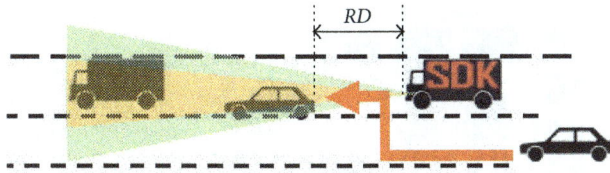

FIGURE 10: SDK Vehicle and the front vehicle AV with intruders (vehicle).

are embedded and architectured around a limited number of ECUs and communication buses. This evolution towards a very wide use of ECUs was also pushed by the need to reach new challenges such as environmental, performance, security, and driving assistance requirements.

As a result, the following four major requirements have driven our diagnosis deployment strategy:

  (i) the number of ECUs has to remain unchanged;

  (ii) the number of buses has to remain unchanged and the same communication protocol should be used if possible;

  (iii) the additional bus load related to diagnosis information must not affect the current real time performance;

  (iv) the diagnosis procedure must be achieved within a bounded time.

Note that, here, we do not address the problem of ECUs load sharing and balancing between the control functions already embedded in them and the diagnosis control ones that will be embedded. Nevertheless, the response time of a control function mainly depends on the total communication delay along the bus lines.

Due to the robustness of the CAN-based bus technology, it is a good candidate for enabling the deployment of a decentralized diagnosis strategy. The electronic architecture remains unchanged (no additional ECU and bus). In order to achieve the two last requirements, we developed an *SW*-based diagnosis service that is an *SW* middleware built on the top of any ECU operating system. It enables the communication of diagnosis information between the local diagnosers and the global diagnoser (as shown in Figure 8) over a loaded CAN bus. It also processes any alert with a bounded time whatever the bus load. It enables the message exchanging with a bounded time while the real-time requirements are verified (no more than 2% of delay time) whatever the bus load.

4.2. Classical Diagnosis Approaches. On-board vehicle diagnosis (OBD) refers to vehicle self-detection, localization, identification, and reporting capabilities (O'Reilly [16] and Greening [17]). Early OBD versions for fuel vehicle managed by electronic simply switched on a malfunction indicator light in the vehicle if any problem was detected. The diagnosis was next performed by an operator in a garage with the aid of a terminal connected to the vehicle electronic.

Efficient diagnosis tools were developed for tracking the problem such as Ressencourt [18].

Modern OBD provides real-time diagnosis data in addition to standardized diagnosis trouble codes (DTCs) which rapidly allow the vehicle to self-identify and, possibly, self-repair by itself the problem during the driving. Otherwise, some vehicles activate a downgraded mode that allows the vehicle driving in safe conditions even in the presence of problems Fromion [19]. In parallel, the vehicle switches on a driver indicator light that points the need of an emergency maintenance (the driver must reach the closest garage) or stop.

Current SAE (Society of Automobile Engineers) and ISO (International Standardization Organization) standards specify the hardware (connector, network) and communication protocol (Open System Interconnection model) for exchanging diagnostic data over the ECUs and external terminals. Some engineering companies propose tools that allow the ECU original equipment manufacturers (OEMs) implement the standardized DTC and customer-specific diagnosis requirements in their ECU, such as Frank et al. [20] (diagnostic-oriented process flow).

4.3. Deployment of the Decentralized Diagnosis. The decentralized diagnosis system, described in Section 3, is deployed in an electronic architecture of four ECUs which communicate over a single CAN-based bus.

The local diagnosers only transmit boolean signals, which take two states: *normal* or *abnormal*, to the global diagnoser. Note that the local diagnosers do not exchange directly diagnosis information with each other. A local diagnoser outputs a *normal* state whenever no fault is detected. During this *normal* state, no additional bus load is due to the diagnosis protocol. Conversely, an *abnormal* state indicates that a problem has likely occurred in a sensor, and when it appears, the local diagnoser must immediately send diagnosis information to the remote global diagnoser in order to compute a global diagnosis and, if the need arises, to perform a recovery procedure.

For enabling this event-based procedure, we insert a Local Diagnosis Service (LDS) in every ECU that embeds a local diagnoser. An LDS reads the outputs of the local diagnoser (*normal/abnormal*). When an event "*normal →abnormal*" occurs, it virtually creates a high-priority communication channel between the local diagnoser and the global diagnoser, so that the former can immediately transmit the diagnosis information to the latter within a bounded time. The diagnosis information is embedded in CAN messages. In addition, a CAN message contains a control header that especially defines the transmitter and receiver identifiers and the priority level of the communication. On the opposite side, a Supervision Diagnosis Service (SDS) is inserted in the global diagnoser ECU. It monitors the bus load and gathers the diagnosis information sent by any LDS. It next triggers the global diagnoser. Note that the LDS and SDS can periodically check if, respectively, the SDS or any LDS is safe and ready.

Figure 11 illustrates the different behavior phases of both LDS and SDS before and after the occurrence of a problem.

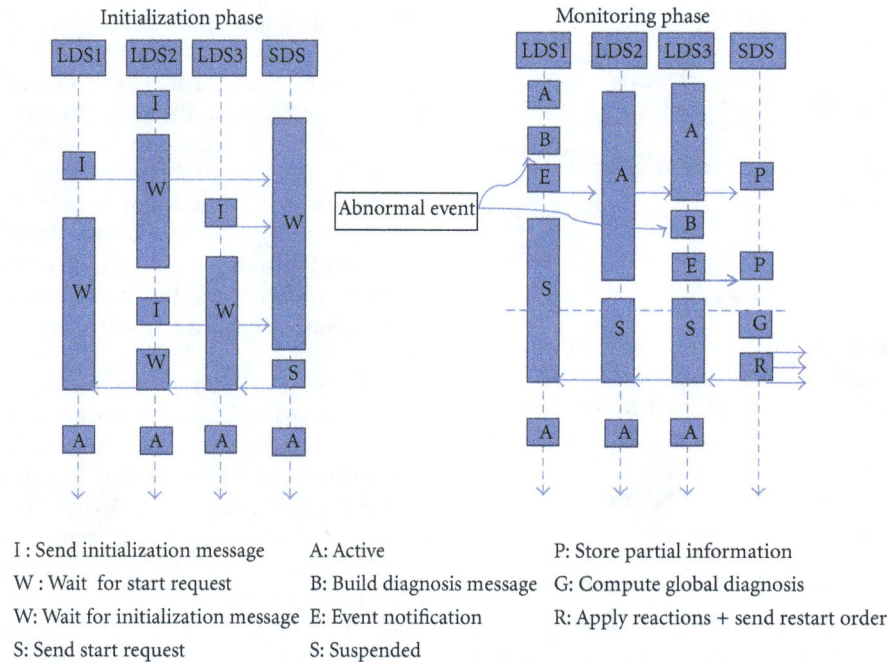

Figure 11: Decentralized diagnosis protocol over a CAN bus.

I : Send initialization message   A: Active   P: Store partial information
W : Wait for start request   B: Build diagnosis message   G: Compute global diagnosis
W: Wait for initialization message   E: Event notification   R: Apply reactions + send restart order
S: Send start request   S: Suspended

Initialisation Phase. The first phase consists in the establishment of the list of available LDSs. Each LDS sends to the SDS a specific CAN message for initialization which includes its identification number. When the SDS has received the initialization message from all LDSs, the protocol enters the monitoring phase. This instant is materialized by the broadcasting of a specific message from the SDS to all LDSs.

Monitoring Phase. During this phase, no messages are exchanged between the LDSs and the SDS. This phase corresponds to a situation where the system is operating normally, without local diagnosis event from the local diagnosers. This implies that there is not any overtraffic due to the diagnosis during the normal operation of the system.

Alert Phase. This phase begins when an abnormal event is detected by a local diagnoser. At this moment, the LDS of the concerned ECU sends a specific alert event message to the SDS. The instant of the first alert event message emission materializes the beginning of a period when the SDS is waiting for other alert event messages that would complete the information for the global diagnosis. At the end of this period, the SDS gives the order to the global diagnoser to compute the global diagnosis. When this is done, the SDS broadcasts a specific message to all LDSs for entering again in the monitoring phase.

Vivacity Check. While the protocol is in monitoring phase, the SDS can initiate a vivacity check: the SDS broadcasts a specific CAN message to all LDSs. When the LDSs receive this message during their monitoring phase, they send an

acknowledgement message for proving that the on-board diagnosis is correctly running.

5. Evaluation of the Diagnosis

5.1. Evaluation of the Diagnosis Algorithms

5.1.1. Presentation of the Scenarios. Several fault scenarios were chosen to valid the diagnostic approach for this SDK system. The selection was done according to

(i) relevance of fault during SDK system operation,

(ii) ease of inducing the fault,

(iii) exclusion of danger possibly caused by the induced fault.

Such faults in the SDK system are failures of the wheels, the transmission, and the radar. All these failures can occur intermittently. Moreover, single or multiple faults may be considered.

5.1.2. Evaluation of the Scenarios. We developed a physical simulation model of the SDK environment in MATLAB/Simulink in order to evaluate our model-based diagnosis approach. The simulation serves as a virtual test bed where we can easily study a large number of fault scenarios to develop our diagnosis models and test our algorithms. We adopted a component-based modeling paradigm, where parameterized simulation models of generic components (SDK controller, radar, wheels, transmission, engine, and supervisor) were developed within a component library. The different local models are constructed by instantiating different components from the library, specifying their

Modeling and Deployment of Model-Based Decentralized Embedded Diagnosis inside Vehicles: Application to Smart
Distance Keeping Function

11

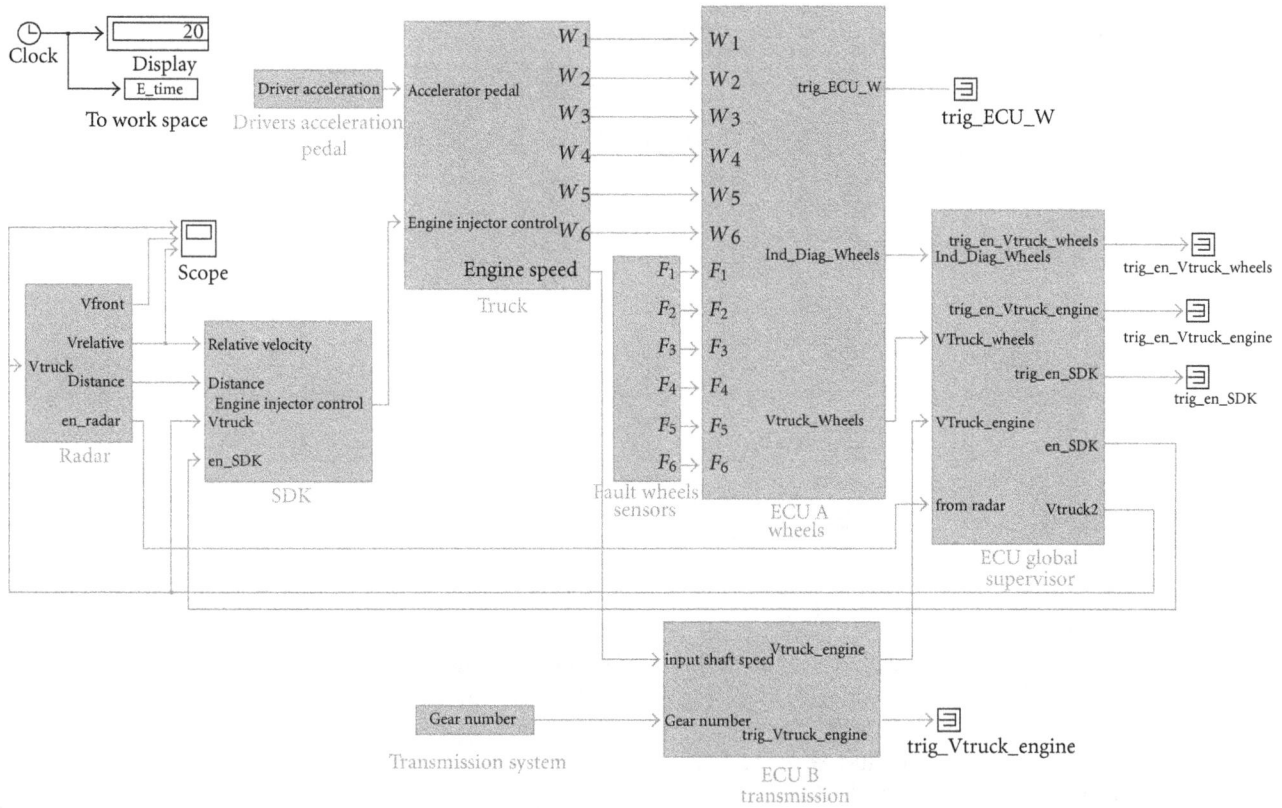

FIGURE 12: Simulink block diagram for detection of SDK sensor faults.

parameters, and connecting the components to each other in the appropriate fashion. The top level of the Simulink block diagram is illustrated in Figure 12.

Except the supervisor, each component model includes its associated fault modes. The fault mode, time of fault injection, and fault magnitude (where applicable) can all be specified. In general, each fault mode is mapped to a change in component mode and a fault-dependent magnitude parameter. Because each fault mode is parameterized within the Simulink model, a fault can be injected programmatically (i.e., the fault mode, injection time, and magnitude are specified) either at the beginning of the simulation or while the simulation is running.

As mentioned previously, a fault will be detected by observing residual values which should be close to zero in the nominal behavior of the process, otherwise, significantly different from zero. Ideally the residual signal should carry only information about faults but, practically, it also contains disturbances, which is the effect of model uncertainty. It is necessary in this case to establish thresholds on residuals to avoid false alarms. The fault occurs only when the residual values exceed the prescribed threshold. For this, we have analyzed the residual values in the absence of fault, which helped determine the residual magnitude in functioning healthy state. The resulting thresholds are then used to the fault detection mechanism.

By using this simulator, the fault scenarios and recovery actions presented in Section 3.3 have been performed. Then,

the injected fault has been detected and localized. Finally, the appropriate recovery action has been applied.

5.2. Performance Evaluation of the CAN Integration. A prototype of the diagnosis system has been realized, according to the diagnosis organization presented in Section 3 and using the deployment scheme described in Section 4. The objective of this prototype is to validate the functionality of the diagnosis implementation and the performance of the deployment architecture. More specifically, we want to check the real-time property provided by the proposed deployment scheme. For that, we developed a validation environment that allows the emulation of both LDS and SDS modules with a high accuracy (real time simulation). In the following, a description of the experiment will be done; then we will present and analyze the results.

5.2.1. Prototype Presentation. As represented in Figure 13, some parts of the prototype are emulated with a simulation tool while others are implemented in a real hardware ECU. The electromechanical truck subsystem models and the local diagnosis algorithms are modeled with the Matlab Simulink simulation tool. Every LDS module, presented in Section 4, is modeled in the CANoe tool Frank and Schmitds [20], which is a suitable emulator of CAN-based systems. This tool simulates the behavior of a CAN-based system with a high accuracy (real time). In addition, this tool allows

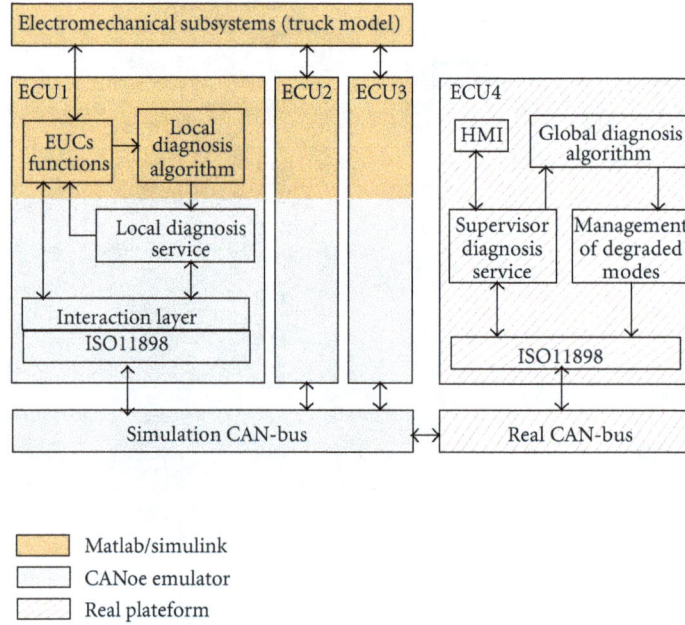

FIGURE 13: General diagram of the prototype of embedded diagnosis of the SDK function.

a cosimulation with a real CAN-bus and real ECU. For performance evaluation, a CAN based traffic generator was included for generating different reference loads that are adjusted for evaluating under which condition the LDS-SDS protocol under study verifies the real-time condition. In addition, the priority of the bus load can be configured as *lower* or *higher*, compared to the messages' priority of our diagnosis protocol. The global diagnosis algorithm (Section 4), the SDS module (Section 4), and a human-machine interface (HMI) are implemented in a hardware ECU which is the Freescale board MAC7100EVB, equipped with an ARM MAC7111 microprocessor, which includes 4 CAN peripherals, one of which is used by our application. The CAN bus is configured at a bit rate of 250 kbit/s.

For stimulating the diagnosis protocol between the LDSs and the SDS, a fault is injected in the electromechanical model of the truck. The fault generates a divergence between the model and the golden model of the local diagnosis. We considered 8 scenarios that correspond to various configurations of load values and priorities. For each scenario, a series of 20 faults is applied. The different scenarios are given in Table 6.

5.2.2. Performance Results. The experimental results are presented in Table 7. We observe that if the bus load does not exceed 60%, no particular impact can be noticed on the processing latency of the diagnosis, for higher and lower load priorities. But when the load priority is high and the load is over 60%, an impact on the latency can be measured, which is growing with the bus load level. These configurations must be avoided for ensuring a reliable use of this decentralized diagnosis algorithm.

In practice, the CAN bus load of vehicles never exceeds 40% by design, for avoiding congestion problems, so considering this statement, our decentralized diagnosis algorithm

TABLE 6: Scenarios for the evaluation of the CAN integration.

Configuration	Load priority	Load level (%)
Config0	No Load	0
Config1	Lower	2.67
Config2	Lower	8.11
Config3	Lower	62.44
Config4	Higher	2.65
Config5	Higher	8.06
Config6	Higher	62.02
Config7	Higher	97.18

TABLE 7: Performance evaluation of the CAN integration.

Configuration	Processing latency of the diagnosis (s)			
	Min	Max	Average	Std. dev.
Config0	20.72	21.09	20.91	0.37
Config1	20.82	22.22	21.17	0.52
Config2	20.75	22.24	21.34	0.65
Config3	20.60	21.02	20.87	0.13
Config4	20.75	22.00	21.11	0.41
Config5	20.67	21.05	20.93	0.12
Config6	20.68	28.09	21.81	2.26
Config7	21.06	101.76	32.90	30.37

verifies the real-time condition, that is, the delay between the LDS and SDS over the network remains quite constant whatever the bus traffic conditions.

Modeling and Deployment of Model-Based Decentralized Embedded Diagnosis inside Vehicles: Application to Smart Distance Keeping Function

13

6. Conclusions

In this paper, we have developed a diagnosis algorithm of the SDK system and a software architecture in order to board it inside truck vehicle in a decentralized manner. The model-based diagnosis approach has been used because it presents the advantage that no prior knowledge of possible faults or symptoms is needed. It relies only on a given model of the correct functioning of the system and proceeds by comparing the behaviors of the model and of the actual system (as known through the observations given by sensors).

The originality of the accomplished work is based on two contributions. The first one is the distribution of diagnosis algorithms on several ECUs by using the decentralized diagnosis approach (the method has only been applied, in the practical context of industrial applications, in a centralized manner). This approach uses a set of diagnosers. Each diagnoser observes a part of the SDK system and takes a local decision about the occurrence of a fault and its localization. The construction of the local diagnosers is based on a modular modeling of the plant elements. All local diagnosers decisions must be merged by a dedicated supervisor in order to obtain one global diagnosis decision and to take also any recovery action. This fusion can be realized by a coordinator. The second contribution is the design and deployment of the decentralized model-based fault diagnosis approach for the SDK system. The attention was paid to minimize the additional traffic generated by the diagnosis function and to respect the real application constraints (performance, diagnosis latency, etc.). An on-board diagnosis using Hardware-in-the-Loop scheme under specifications (constrains) near to a truck "Renault Trucks" has been performed.

The decentralized embedded diagnosis for the SDK system inside real vehicles is mature from the point of view of research and of feasibility. However, we should extend and develop algorithms robustness, take into account the protocols and details of the hardware architecture and existing software, conduct tests on many scenarios, and measure in situ the quality (correctness, precision, time delay) of the diagnosis and its compatibility with existing functions (induced traffic on the CAN-bus, transparency). Moreover, in order to verify a priori that the set of local diagnosers and supervisor is capable of diagnosing a given set of faults within a bounded delay, a notion of diagnosability must be studied.

Acknowledgments

The work reported in this paper was supported by the project DIAFORE (Diagnosis for Distributed Functions) which is funded by the French National Research Agency (ANR) and SYSTEM@TIC PARIS-REGION Cluster, with the support of French Environment and Energy Management Agency (ADEME) and French Programme of Research, Experimentation and Innovation in Land transport (PREDIT). Many thanks to industrial partners (Renault Trucks/Volvo SAS and SERMA INGENIERIE) in this project for their support upon completion of the simulation platform.

References

[1] F. Peysson, H. Noura, and R. Younes, "Diagnostic de défauts sur un moteur diesel," in *Proceedings of the Conférence Internationale Francophone d'Automatique (CIFA '06)*, Bordeaux, France, 2006.

[2] L. C. Davis, "Effect of adaptive cruise control systems on traffic flow," *Physical Review E*, vol. 69, no. 6, Article ID 066110, 8 pages, 2004.

[3] R. Parasuraman and V. Riley, "Humans and automation: use, misuse, disuse, abuse," *Human Factors*, vol. 39, no. 2, pp. 230–253, 1997.

[4] N. A. Stanton and P. Marsden, "From fly-by-wire to drive-by-wire: safety implications of automation in vehicles," *Safety Science*, vol. 24, no. 1, pp. 35–49, 1996.

[5] L. Nilsson, "Safety effects of adaptive cruise control in critical traffic situations," in *Proceedings of the nd World Congress on Intelligent Transport Systems: Steps Forward*, vol. 3, pp. 1254–1259, VERTIS, 1995.

[6] N. A. Stanton, M. Young, and B. McCaulder, "Drive-by-wire: the case of driver workload and reclaiming control with adaptive cruise control," *Safety Science*, vol. 27, no. 2-3, pp. 149–159, 1997.

[7] J. D. Lee and K. A. See, "Trust in automation: designing for appropriate reliance," *Human Factors*, vol. 46, no. 1, pp. 50–80, 2004.

[8] M. A. Goodrich and E. R. Boer, "Model-based human-centered task automation: a case study in ACC system design," *IEEE Transactions on Systems, Man, and Cybernetics Part A*, vol. 33, no. 3, pp. 325–336, 2003.

[9] R. J. Patten, P. M. Frank, and R. Clark, *Issues of Fault Diagnosis for Dynamic Systems*, Springer, London, UK, 2000.

[10] B. Darkhovski and M. Staroswiecki, "A game-theoretic approach to decision in FDI," *IEEE Transactions on Automatic Control*, vol. 48, no. 5, pp. 853–858, 2003.

[11] E. Chow and A. Willsky, "Analytic redundancy and the design of robust failure detection systems," *IEEE Transactions on Automatic Control*, vol. 29, pp. 603–614, 1984.

[12] M. Sampath, R. Sungupta, S. Lafortune, K. Sinnamohideen, and D. Teneketzis, "Diagnosability of discrete event systems," in *Proceedings of the 11th International Conference Analysis Optimization of Systems: Discrete Event Systems*, Sophia-Antipolis, France, 1994.

[13] Y. Wang, T. S. Yoo, and S. Lafortune, "New results on decentralized diagnosis of discrete event systems," in *Annual Allerton Conference on Communication, Control and Computing*, 2004.

[14] W. Qiu, *Decentralized/distributed failure diagnosis and supervisory control of discrete event systems*, Ph.D. thesis, Iowa State University, 2005.

[15] H. Shraim, *Modeling, Estimation and control for vehicle dynamics*, Ph.D. thesis, Université Paul Cézanne, Aix Marseille III, 2007.

[16] P. O'Reilly, "An overview of the potential contribution of diagnostics to improving vehicle safety and reducing vehicle emissions," in *IEEE Colloquium on Vehicle Diagnostics in Europe*, pp. 1–12, February 1994.

[17] P. Greening, "On-board diagnostics for control of vehicle emissions," in *IEEE Colloquium on Vehicle Diagnostics in Europe*, pp. 1–6, February 1994.

[18] H. Ressencourt, *Diagnostic hors-ligne à base de modèles: approche multi-modèle pour la génération automatique de séquences de tests. Application au domaine de l'utomobile*, Ph.D. thesis, 2008.

[19] A. Fromion, Apparatus for the differential transmission of information between at least two devices in a vehicle. European patent EP19960401324, 2003.

[20] H. Frank and U. Schmidts, Vehicle Diagnostics—The whole Story. Vector Informatik, Article press, 2007, http://www.vector.com/.

2

Vehicle Dynamics Approach to Driver Warning

Youssef A. Ghoneim

Research and Development Center, General Motors Corporation, 30500 Mound Road, Warren, MI 48090, USA

Correspondence should be addressed to Youssef A. Ghoneim; youssef.ghoneim@gm.com

Academic Editor: Martin Reisslein

This paper discusses a concept for enhanced active safety by introducing a driver warning system based on vehicle dynamics that predicts a potential loss of control condition prior to stability control activation. This real-time warning algorithm builds on available technologies such as the Electronic Stability Control (ESC). The driver warning system computes several indices based on yaw rate, side-slip velocity, and vehicle understeer using ESC sensor suite. An arbitrator block arbitrates between the different indices and determines the status index of the driving vehicle. The status index is compared to predetermined stability levels which correspond to high and low stability levels. If the index exceeds the high stability level, a warning signal (haptic, acoustic, or visual) is issued to alert the driver of a potential loss of control and ESC activation. This alert will remain in effect until the index is less than the low stability level at which time the warning signal will be terminated. A vehicle speed advisory algorithm is integrated with the warning algorithm to provide a desired vehicle speed of a vehicle traveling on a curve. Simulation results and vehicle tests were conducted to illustrate the effectiveness of the warning algorithm.

1. Introduction

Freeway entrance and exit ramp interchanges are the sites of far more crashes per mile driven than other segments of interstate highways. Crashes most common on exit ramps—run-off-road crashes—frequently occurred when vehicles were exiting interstates at night, in bad weather, or on curved portions of ramps. When the vehicle is driving under these conditions at a higher speed than the surface can allow, the understeer gradient of the vehicle can increase causing the vehicle to plow or decrease and becomes negative causing the vehicle to spinout.

In recent years, electronic stability control systems for motor vehicles have become increasingly popular [1–12]. Conventionally, such systems monitor vehicle stability-related quantities such as a yaw rate error, that is, a deviation between yaw rates expected based on vehicle speed and steering wheel angle and an observed yaw rate, and selectively brake the wheels at one side of the vehicle in order to assist cornering, that is, to decrease the deviation between expected and observed yaw rates.

Electronic Stability Control (ESC) helps keep the vehicle on its steered path during a turn, to avoid sliding or skidding. It uses a computer linked to a series of sensors—detecting wheel speed, steering angle, yaw rate and lateral acceleration of the vehicle. During normal driving, ESC works in the background and continuously monitors steering and vehicle direction. It compares the driver's intended direction (determined through the measured steering wheel angle) to the vehicle's actual direction (determined through measured lateral acceleration, vehicle rotation (yaw), and individual road wheel speeds). ESC intervenes only when it detects a probable loss of steering control, that is, when the vehicle is not going where the driver is steering. This may happen, for example, when skidding during emergency evasive swerves, understeer or oversteer during poorly judged turns on slippery roads, or hydroplaning. If the vehicle starts to drift, the system momentarily brakes one or more wheels and, depending on the system, reduces engine power to keep the car on the steered course. However, Electronic Stability Control cannot override the laws of physics. If a driver exceeds the friction capabilities of the road surface, ESC cannot prevent a crash. It is a tool to help the driver maintain control.

In this paper, we introduce a driver warning algorithm integrated with the Electronic Stability Control system. The warning system is designed to further assist a driver by warning of an impending ESC activation so that the driver will reduce the vehicle's speed prior to the need for ESC

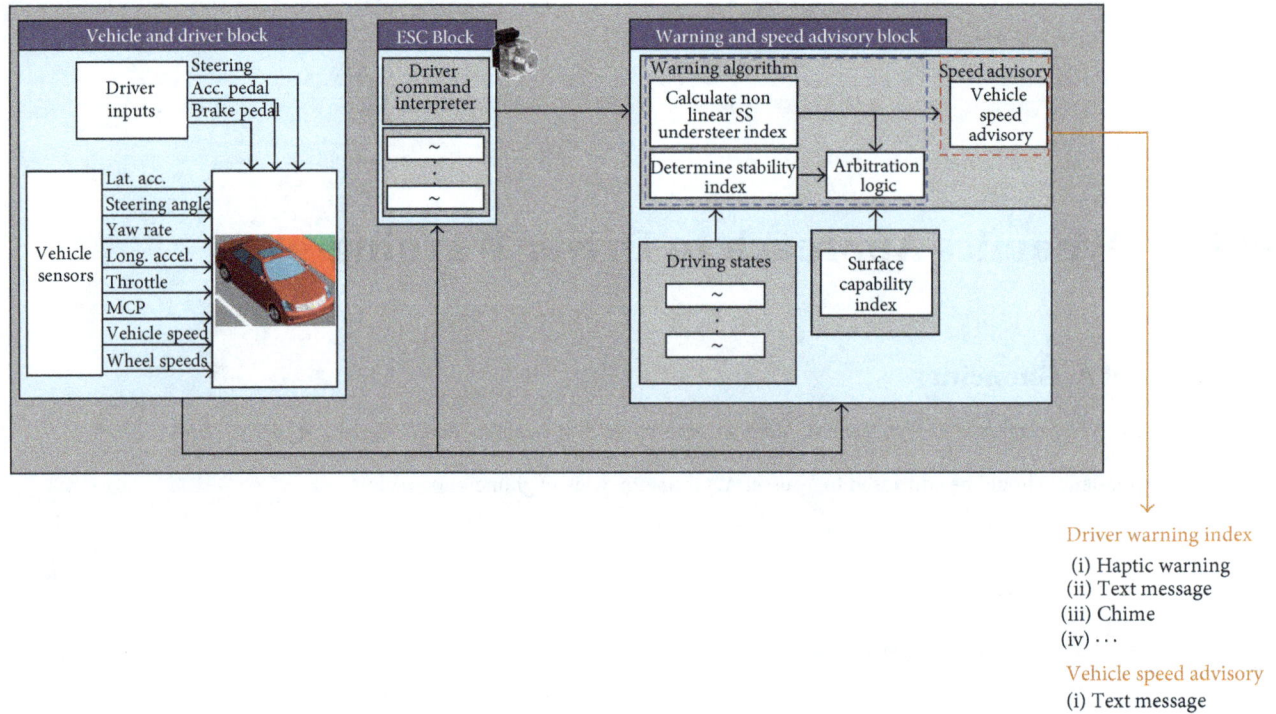

Driver warning index
(i) Haptic warning
(ii) Text message
(iii) Chime
(iv) ···

Vehicle speed advisory
(i) Text message

FIGURE 1: Schematic diagram of the warning algorithm.

intervention. It is hoped that the warning system will assist the driver in recognizing and avoiding instances of potential loss of vehicle control and also reduce usage of ESC.

The body of the paper begins first with system architecture including a brief description of the different blocks used in the warning algorithm. Second, we develop the warning signal command for both the transient and the steady-state modes of the vehicle. Third, we calculate a vehicle advisory speed in a curve based on vehicle understeer gradient. Finally, we present simulation results and vehicle tests.

2. Warning Algorithm Architecture

Figure 1 illustrates the schematic diagram of the warning algorithm which consists of three major parts:

(1) the vehicle and driver block which contains the following:

    (i) driver inputs (steering, accelerator pedal, and brake pedal),

    (ii) vehicle sensors (lateral acceleration, yaw rate, wheel speeds to estimate the vehicle longitudinal speed, throttle position, and master cylinder pressure sensor);

(2) the ESC block:

    (i) command Interpreter block;

(3) the warning and speed advisory block:

    (i) the driving states,

(ii) the Lateral Surface Capability Index,

(iii) the warning Algorithm:

    (a) the understeer index,
    (b) the stability index,
    (c) the arbitration logic,

(iv) the vehicle speed advisory.

2.1. The ESC Block

2.1.1. The Command Interpreter Block. Vehicle Yaw-Plane Dynamics. In this section, we describe the equations of dynamics for the yaw-plane motion of a vehicle. While a vehicle is undergoing handling maneuvers, it not only incurs a yaw motion, it also experiences a side-slip motion at the same time. The yaw-plane dynamics determine the performance of vehicle yaw motion characterized by vehicle yaw rate, as well as the lateral motion characterized by side-slip velocity.

The vehicle yaw-plane dynamics can be described by a second-order state [7]

$$\begin{bmatrix} \dot{v}_y \\ \ddot{\psi} \end{bmatrix} = \begin{bmatrix} a_{11} & a_{12} \\ a_{21} & a_{22} \end{bmatrix} \begin{bmatrix} v_y \\ \dot{\psi} \end{bmatrix} + \begin{bmatrix} b_1 \\ b_2 \end{bmatrix} \delta, \qquad (1)$$

where v_y and $\dot{\psi}$ are the vehicle side-slip velocity (defined as the component of the vehicle velocity vector in the y direction [13]) and yaw rate (defined as vehicle's angular velocity around its vertical axis), respectively, and the system coefficients, a_{ij}'s and b_i's, are functions of vehicle mass, vehicle speed, vehicle inertia, and the front and rear cornering stiffness, and inevitably, the location of vehicle center of gravity are described by the parameters a and b:

$$a_{11} = -\frac{C_f + C_r}{M_v v_x} \qquad a_{12} = \frac{-aC_f + bC_r}{M_v v_x} - v_x,$$

$$a_{21} = \frac{-aC_f + bC_r}{I_z v_x} \qquad a_{22} = -\frac{a^2 C_f + b^2 C_r}{I_z v_x}, \qquad (2)$$

$$b_1 = \frac{C_f}{M_v} \qquad b_2 = \frac{aC_f}{I_z}.$$

The transfer functions from the steering input to the vehicle yaw rate and side-slip velocity can be derived from the state

$$\frac{\dot{\psi}(s)}{\delta(s)} = \frac{b_2 s + (a_{21} b_1 - a_{11} b_2)}{s^2 - (a_{11} + a_{22}) s + (a_{11} a_{22} - a_{12} a_{21})}, \qquad (3)$$

$$\frac{v_y(s)}{\delta(s)} = \frac{b_1 s + (a_{12} b_2 - a_{22} b_1)}{s^2 - (a_{11} + a_{22}) s + (a_{11} a_{22} - a_{12} a_{21})}. \qquad (4)$$

Equations (4) and (5) can be used to design a closed-loop control system for vehicle stability enhancement by regulating the measured vehicle performance to its desired value determined in the following section.

Desired Vehicle Response. The desired yaw rate and the side-slip velocity commands are determined by the desired vehicle response to the driver's steering input. There are mainly two approaches of implementation: (a) using the state in (1) to perform real-time integration or (b) using transfer functions such as (4) and (5) which consist of the steady-state value of the desired yaw rate and side-slip velocity and a dynamic filter representing the desired vehicle dynamics.

When the state approach is employed to generate the commands, the desired vehicle side-slip velocity and desired yaw rate are computed based on the system differentials using nominal values of system parameters defined in (1)–(5).

When the state equation approach is employed to generate the commands, the desired vehicle side-slip velocity and desired yaw rate are computed based on the system differential equations using nominal values of system parameters:

$$\begin{bmatrix} \dot{v}_{yd} \\ \ddot{\psi}_d \end{bmatrix} = \begin{bmatrix} a_{11} & a_{12} \\ a_{21} & a_{22} \end{bmatrix} \begin{bmatrix} v_{yd} \\ \dot{\psi}_d \end{bmatrix} + \begin{bmatrix} b_1 \\ b_2 \end{bmatrix} \delta. \qquad (5)$$

In this reference model, the C_f and C_r are replaced with constants C_{f0} and C_{r0} representing the values of a high-coefficient condition on dry surface. Integration of (1) results in a desired time trace of vehicle side-slip velocity and yaw rate.

The transfer-function approach for obtaining the desired vehicle response is based on the structure of the system input-output transfer function derived in (4) and (5), but

substituting the system coefficients with the nominal values forms the first-stage command:

$$\frac{v_{yd}(s)}{\delta(s)} = \frac{b_1 s + (a_{12} b_2 - a_{22} b_1)}{s^2 - (a_{11} + a_{22}) s + (a_{11} a_{22} - a_{12} a_{21})},$$

$$\frac{\dot{\psi}_d(s)}{\delta(s)} = \frac{b_2 s + (a_{21} b_1 - a_{11} b_2)}{s^2 - (a_{11} + a_{22}) s + (a_{11} a_{22} - a_{12} a_{21})}. \qquad (6)$$

Equations (6) are mathematically equivalent to (5). The practical difference is that the desired natural frequency and damping ratio determined by (6) can be specified without regard to their original formation derived from the vehicle parameters.

Therefore, rewriting (6) in terms of system natural frequency and damping ratio yields

$$\frac{v_{yd}(s)}{\delta(s)} = \frac{(s/z_v + 1)\omega_n^2}{s^2 + 2\zeta\omega_n s + \omega_n^2} V_{y\,dss\_gain},$$

$$\frac{\dot{\psi}_d(s)}{\delta(s)} = \frac{(s/z_\psi + 1)\omega_n^2}{s^2 + 2\zeta\omega_n s + \omega_n^2} \dot{\psi}_{dss\_gain}, \qquad (7)$$

where

$$V_{y\,dss\_gain} = \frac{(a_{12} b_2 - v_x) - a_{22} b_1}{a_{12} a_{22} - a_{12} a_{21}}, \qquad (8)$$

$$\dot{\psi}_{dss\_gain} = \frac{a_{21} b_1 - a_{11} b_2}{a_{12} a_{22} - a_{12} a_{21}} \qquad (9)$$

are the steady-state gains of desired steady-state side-slip velocity and yaw rate. The damping ratio and natural frequency of the desired vehicle performance can be expressed in terms of system parameters

$$\omega_n = \sqrt{a_{11} a_{22} - a_{12} a_{21}},$$

$$\zeta = -\frac{a_{11} + a_{22}}{2\omega_n} \qquad (10)$$

and tabulated as functions of vehicle speed in control software. The variables z_v and z_ψ are the negative of the system zeroes for both side slip velocity and yaw rate and can be represented by

$$z_v = (a_{12} - v_x)\frac{b_2}{b_1} - a_{22}, \qquad (11)$$

$$z_\psi = a_{21}\frac{b_2}{b_1} - a_{22}. \qquad (12)$$

The steady-state value of the vehicle side-slip velocity can be obtained by multiplying the gain with the steering angle; that is,

$$V_{y\,dss} = V_{y\,dss\_gain}\delta. \qquad (13)$$

Substituting expressions in (3) and (8) into (13) yields

$$V_{y\,dss} = \frac{v_x\left((a+b)\,bC_fC_r - aC_fM_vv_x^2\right)}{(a+b)^2C_fC_r + M_vv_x^2\left(-aC_f + bC_r\right)}\delta$$

$$= \left(\frac{v_x\delta}{(a+b)+K_uv_x^2}\right)\left(b - \frac{a}{a+b}\frac{M_vv_x^2}{C_r}\right), \qquad (14)$$

where

$$K_u = \frac{M_v}{a+b}\left(\frac{b}{C_f} - \frac{a}{C_r}\right). \qquad (15)$$

Equation (14) is composed of a product of two terms where the first term represents the steady-state of desired yaw rate. From this fact, the steady-state value of desired side-slip velocity can be implemented in terms of steady-state desired yaw rate

$$V_{y\,dss} = \dot{\psi}_{dss}\left(b - \frac{a}{a+b}\frac{M_vv_x^2}{C_r}\right). \qquad (16)$$

The steady-state desired yaw rate $\dot{\psi}_{dss}$ during control computation can be obtained using the following:

$$\dot{\psi}_{dss} = \frac{v_x\delta}{(a+b)+K_uv_x^2}, \qquad (17)$$

where K_u is obtained from computation of (15).

Therefore, (14), (15), and (16) can be used to obtain the value of $V_{y\,dss}$.

Given the steady-state side-slip velocity, the next step is to process such value through a dynamic filter with desired damping ratio and natural frequency representative of the system dynamics described in (7) and (8). But one question remains regarding to the significance of the non-minimum-phase zero in the implementation of the side slip command.

The dynamic filter with desired damping ratio, natural frequency, and zero can be implemented using a set of two first-order differentials

$$\dot{x}_1 = V_{y\,dss} - 2\zeta\omega_nx_1 - \omega_n^2x_2,$$

$$\dot{x}_2 = x_1, \qquad (18)$$

$$v_{yd} = \omega_n^2\left(x_2 + \frac{x_1}{z_v}\right),$$

where z_v is expressed in (11) and implemented in a lookup table as a function of vehicle speed. A similar second-order dynamic filter can be implemented in control computation to provide the desired yaw-rate command. Consider

$$\dot{x}_1 = \dot{\psi}_{y\,dss} - 2\zeta\omega_nx_1 - \omega_n^2x_2,$$

$$\dot{x}_2 = x_1, \qquad (19)$$

$$v_{yd} = \omega_n^2\left(x_2 + \frac{x_1}{z_{\dot{\psi}}}\right).$$

Equations (18) and (19), which take separate desired steady-state values as the inputs, can be combined into one set of dynamic filters with a common input. Dynamics are imparted to steering wheel angle through a second-order filter with two states. Desired yaw rate and lateral velocity are computed based on linear combinations of these states. Steady-state yaw gain, as mentioned before, can be implemented in a table look-up to account for variation of understeer coefficient when the vehicle is operated at the limit of tire adhesion. Steady state lateral gain is computed from the steady-state yaw gain through kinematics relationships. Both filter natural frequency and damping ratio are functions of vehicle velocity and are implemented as look-up tables.

The dynamic relationship between the steering angle and desired side-slip velocity can be expressed into the following form:

$$\frac{v_{yd}(s)}{\delta(s)} = \frac{b_1s + a_{12}b_2 - a_{22}b_1}{s^2 + 2\zeta\omega_ns + \omega_n^2}. \qquad (20)$$

Using this relationship, a dynamic filter can be established with steering angle as input:

$$\dot{x}_1 = \delta - 2\zeta\omega_nx_1 - \omega_n^2x_2,$$

$$\dot{x}_2 = x_1, \qquad (21)$$

where δ = steering wheel angle, ζ = damping ratio, function of vehicle speed, and ω_n = natural frequency, function of vehicle speed.

Defining

$$V_{y\,gain} = \Omega_{gain}g_r\left(b - \frac{aM_v}{a+b}\frac{V_x^2}{C_r}\right),$$

$$\Omega_{gain} = \text{the yaw gain from 3-d table,} \qquad (22)$$

$$g_r = \text{steering ratio,}$$

we can compute for both the dynamic desired side-slip velocity and yaw rate using the two states x_1 and x_2 of a single dynamic filter described in (21):

$$\dot{\psi}_{des} = \left[\frac{a}{I_z}\right]C_fx_1 + \Omega_{gain}g_r\omega_n^2x_2, \qquad (23)$$

$$V_{y\,des} = \left[\frac{1}{M_v}\right]C_fx_1 + V_{y\,gain}\omega_n^2x_2. \qquad (24)$$

A block diagram of the command interpreter block is illustrated in Figure 2.

2.2. The Warning and Speed Advisory Block

2.2.1. The Driving States.
This block reads all available sensor signals, namely, the lateral and longitudinal accelerations, steering wheel angle, and yaw rate to detect the current driving situation. This block distinguishes 11 different driving modes such as cruising, braking, cornering, transient, low speed maneuvering, and reversing.

$$\dot{\psi}_{\mathrm{des\_ss}} = \frac{(a_{21}b_1/v_x) - (a_{11}b_2/v_x)}{(a_{12}a_{22}/v_x^2) - (a_{12}a_{21}/v_x^2)}\delta$$

Steady-state
desired yaw rate

$$\dot{x}_1 = \delta - 2\zeta\omega_n x_1 - \omega_n^2 x_2$$
$$\dot{x}_2 = x_1$$
$$\Downarrow$$
$$\dot{\psi}_{\mathrm{des}} = \left[\frac{a}{I_z}\right]C_f x_1 + \Omega_{\mathrm{gain}}\, g_r \omega_n^2 x_2$$
$$\dot{V}_{y\,\mathrm{des}} = \frac{\Delta}{dt}\left(\left[\frac{1}{M_v}\right]C_f x_1 + V_{y\,\mathrm{gain}}\,\omega_n^2 x_2\right)$$

Dynamic filter

δ_f
V_x

$$V_{y\,\mathrm{des\_ss}} = \frac{((a_{12}b_2/v_x) - v_x) - (a_{22}/v_x)b_1}{(a_{12}a_{22}/v_x^2) - (a_{12}a_{21}/v_x^2)}\delta$$

Steady-state
desired lateral velocity

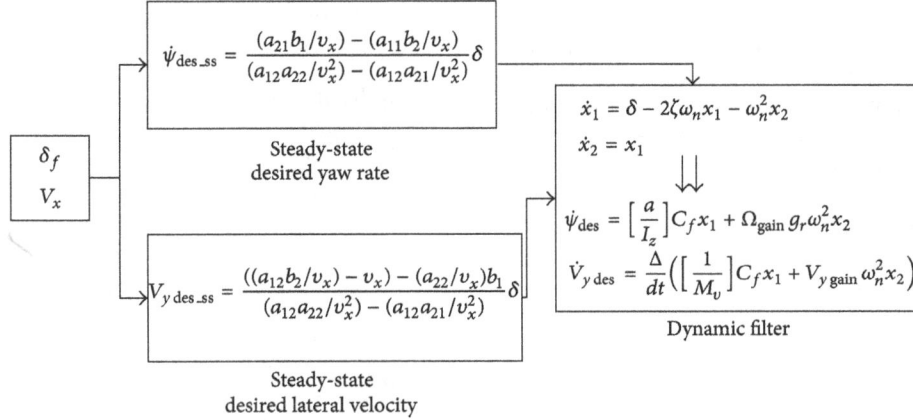

FIGURE 2: Block diagram of the command interpreter algorithm.

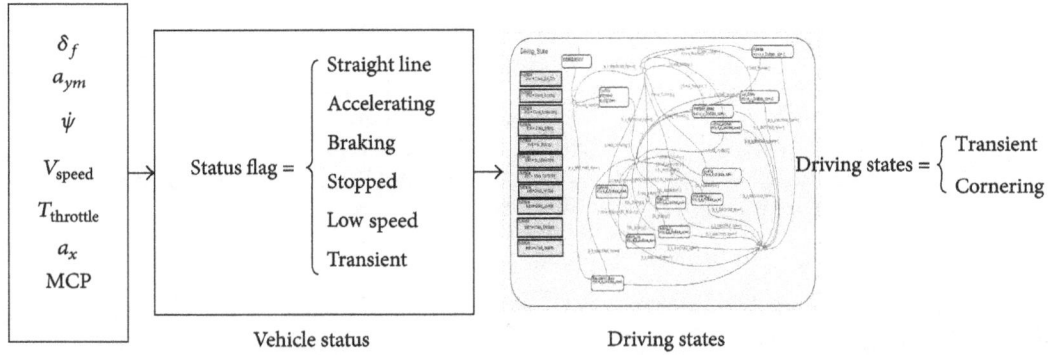

δ_f
a_{ym}
$\dot{\psi}$
V_{speed}
T_{throttle}
a_x
MCP

Status flag = { Straight line, Accelerating, Braking, Stopped, Low speed, Transient }

Driving states = { Transient, Cornering }

Vehicle status   Driving states

FIGURE 3: Block diagram of the driving states algorithm.

For the purpose of this study, it is sufficient to distinguish between a transient mode and a steady turn for both linearized and nonlinear relations between lateral tire forces and slip angles. We define four intermediate flags $F_{i_{\mathrm{tr}}}$, $i = 1, 4$.

The transient mode is determined as follows:

$$\ddot{\psi}\begin{cases} |\ddot{\psi}| > \ddot{\psi}_{\mathrm{trans\,2\_th}}(V_x) \longrightarrow F_{1\_\mathrm{tr}} = \mathrm{true}, \\ |\ddot{\psi}| \le \ddot{\psi}_{\mathrm{trans\_th\,min}}(V_x) \longrightarrow F_{1\_\mathrm{tr}} = \mathrm{false}, \end{cases}$$

$$\dot{V}_y\begin{cases} |\dot{V}_y| > \dot{V}_{y_{\mathrm{trans\_th}}}(V_x) \longrightarrow F_{2\_\mathrm{tr}} = \mathrm{true}, \\ |\dot{V}_y| \le \dot{V}_{y_{\mathrm{trans\_th\,min}}}(V_x) \longrightarrow F_{1\_\mathrm{tr}} = \mathrm{false}, \end{cases}$$

$$\dot{\delta}\begin{cases} |\dot{\delta}| * f(V_x) > \dot{\delta}_{\mathrm{trans\_th}}(V_x) \longrightarrow F_{3\_\mathrm{tr}} = \mathrm{true}, \\ |\dot{\delta}| * f(V_x) \le \dot{\delta}_{\mathrm{trans\_th\,min}}(V_x) \longrightarrow F_{3\_\mathrm{tr}} = \mathrm{false}, \end{cases}$$

$$\delta\begin{cases} [\delta] > \delta_{\mathrm{trans\_th}}(V_x) \longrightarrow F_{4\_\mathrm{tr}} = \mathrm{true}, \\ |\delta| \le \delta_{\mathrm{trans\_th\,min}}(V_x) \longrightarrow F_{4\_\mathrm{tr}} = \mathrm{false}. \end{cases}$$
(25)

Transient mode

$$= \mathrm{delay\_sample}\left\{\left(F_{1\_\mathrm{tr}} \lor F_{2\_\mathrm{tr}} \land F_{3\_\mathrm{tr}} \land F_{4\_\mathrm{tr}}\right), n\right\}.$$
(26)

The delay mode starts when the input signal becomes false. The variable count is incremented by one as long as the old value of the variable is less than n, where n is the number of samples. The output in this case is only true if the old value is less than n. Thus, the falling edge is delayed by n samples. The reason for introducing the delay sample function is to avoid transient mode switching during signals' zero crossing.

We define three intermediate flags F_{i_cr}, $i = 1, 3$. The steady turn is determined as follows:

$$\dot{\psi}\begin{cases} |\dot{\psi}| > \dot{\psi}_{\mathrm{cr\_th}} \longrightarrow F_{1\_\mathrm{cr}} = \mathrm{true}, \\ |\dot{\psi}| \le \dot{\psi}_{\mathrm{cr\_th\,min}} \longrightarrow F_{1\_\mathrm{cr}} = \mathrm{false}, \end{cases}$$

$$\delta\begin{cases} |\delta| > \delta_{\mathrm{cr\_th}} \longrightarrow F_{2\_\mathrm{cr}} = \mathrm{true}, \\ |\delta| \le \delta_{\mathrm{cr\_th\,min}} \longrightarrow F_{2\_\mathrm{cr}} = \mathrm{false}, \end{cases}$$

$$a_y\begin{cases} |a_y| > a_{y_{\mathrm{cr\_th}}} \longrightarrow F_{3\_\mathrm{cr}} = \mathrm{true}, \\ |a_y| \le a_{y_{\mathrm{cr_{th}\,min}}} \longrightarrow F_{3\_\mathrm{cr}} = \mathrm{false}, \end{cases}$$
(27)

Steady turn = $\left\{F_{1\_\mathrm{cr}} \lor F_{2\_\mathrm{cr}} \land F_{3\_\mathrm{cr}}\right\}$.

Figure 3 illustrates the driving states algorithm described earlier.

2.2.2. The Lateral Surface Capability Index. The lateral surface
capability index is determined based on the comparison
between actual vehicle motion obtained from sensor inputs
and vehicle behavior obtained from a linear vehicle motion
model. When the vehicle is in the linear range of operation,
the vehicle motion model is close to the actual vehicle motion
obtained from the sensor inputs. In this case, the lateral
surface index is set to the maximum lateral acceleration that
the vehicle can sustain on dry surface. When the vehicle
lateral motion approaches the limit of adhesion, the vehicle
motion model is substantially different from the actual
vehicle motion, and it can be concluded that the lateral
surface index must at least equal a_y/g, wherein g denotes the
gravity acceleration.

The first step in this block is to detect whether the vehicle
is in a linear mode of motion or not. To this effect, the linear
mode detection evaluates the following three conditions:

$$\left|\dot{\psi}_d v_x + \dot{v}_{yd}\right| - \left|a_y\right| < a_{y_{\text{Thr1}}}, \tag{28}$$

$$\left(\dot{\psi}_d v_x + \dot{v}_{yd}\right) a_y > -a_{y\_\text{Thr2}}, \tag{29}$$

$$\dot{\psi}_{e\_\text{min}} \le \left|\dot{\psi}_d - \dot{\psi}\right| \le \dot{\psi}_{e\_\text{max}}. \tag{30}$$

The desired yaw rate $\dot{\psi}_d$ is obtained from (23). $\dot{v}_{yd}$ is
obtained by differentiating (18) as follows:

$$\dot{v}_{yd} = \omega_n^2\left(x_1 + \frac{1}{z_v}\left[V_{yd\text{ss}} - 2\xi\omega_n x_1 - \omega_n^2 x_2\right]\right). \tag{31}$$

The difference between measured and expected lateral
accelerations in (28) is due to the fact that the assumed linear
relationship between lateral force and sideslip angle does not
hold exactly. In fact, the direct proportionality between lateral
force and sideslip angle is a good approximation as long as
the lateral forces are moderate. When the difference between
measured and desired lateral accelerations exceeds a certain
threshold, the sideslip angle increases much more strongly,
indicating that the tire is moving towards its lateral limit of
adhesion.

Equation (29) compares the product of signed desired and
observed lateral accelerations, to a small negative number
$-a_{y\_\text{Thr2}}$. An excessive negative value is symptomatic of a
situation where the state of motion of the vehicle cannot
follow the driver intended steering. Equation (30) compares
the difference between a desired and observed yaw rate to
upper and lower thresholds. Obviously, if this difference is
above the upper threshold $\dot{\psi}_{e\_\text{max}}$, control of the vehicle is not
exact indicating that the vehicle has departed from its linear
range of operation. On the other hand, if it is below the lower
threshold $\dot{\psi}_{e\_\text{min}}$, it is likely that the vehicle is going straight,
and that no information about the lateral friction properties
of the road surface can be inferred from the data of the various
sensors.

We define four intermediate flags $f_{i\,\text{lin}}$ $i = 1, 4$. The linear
mode detection is determined as follows:

$$\begin{aligned}
&\text{if } \left|\dot{\psi}_d v_x + \dot{v}_{yd}\right| - \left|a_y\right| < a_{y\_\text{Thr1}} \longrightarrow f_{1\,\text{lin}} = \text{true},\\
&\text{else } f_{1\,\text{lin}} = \text{false},\\
&\text{if } \left(\dot{\psi}_d v_x + \dot{v}_{yd}\right) a_y > -a_{y\_\text{Thr2}} \longrightarrow f_{2\,\text{lin}} = \text{true},\\
&\text{else } f_{2\,\text{lin}} = \text{false},\\
&\text{if } \left|\dot{\psi}_d - \dot{\psi}\right| \le \dot{\psi}_{e\_\text{max}} \longrightarrow f_{3\,\text{lin}} = \text{true},\\
&\text{else } f_{3\,\text{lin}} = \text{false},\\
&\text{if } \dot{\psi}_{e\_\text{min}} \le \left|\dot{\psi}_d - \dot{\psi}\right| \longrightarrow f_{4\,\text{lin}} = \text{true},\\
&\text{else } f_{4\,\text{lin}} = \text{false}.
\end{aligned} \tag{32}$$

The linear flag is given by

$$f_{\text{lin}}(t) = f_{\text{lin}}(t-1) \vee f_{4\,\text{lin}} \wedge \left\{f_{1\,\text{lin}} \wedge f_{2\,\text{lin}} \wedge f_{3\,\text{lin}}\right\}. \tag{33}$$

The second step in the lateral surface capability index block is
the straight driving mode detection algorithm. We define the
following intermediate flags $f_{\dot{\psi}\,\text{max}}$, $f_{\dot{\psi}\,\text{min}}$, $f_{\delta\,\text{max}}$, and $f_{\delta\,\text{min}}$.
The straight line flag is determined as follows:

$$\begin{aligned}
&\text{if } \left|\dot{\psi}\right| \ge \Omega_{\text{th max}} \longrightarrow f_{\dot{\psi}\,\text{max}} = \text{false},\\
&\text{else } f_{\dot{\psi}\,\text{max}} = \text{true},\\
&\text{if } \left|\delta\right| \ge \delta_{\text{th max}} \longrightarrow f_{\delta\,\text{max}} = \text{false},\\
&\text{else } f_{\delta\,\text{max}} = \text{true},\\
&\text{if } \left|\dot{\psi}\right| \le \dot{\Omega}_{\text{th min}} \longrightarrow f_{\dot{\psi}\,\text{min}} = \text{true},\\
&\text{else } f_{\dot{\psi}\,\text{min}} = \text{false},\\
&\text{if } \left|\delta\right| \le \delta_{\text{th min}} \longrightarrow f_{\delta\,\text{min}} = \text{true},\\
&\text{else } f_{\delta\,\text{min}} = \text{false}.
\end{aligned} \tag{34}$$

The straight driving flag is given by

$$f_{\text{sl}}(t) = f_{\text{sl}}(t-1) \vee \left\{f_{\dot{\psi}\,\text{min}} \wedge f_{\dot{\psi}\,\text{max}} \wedge f_{\delta\,\text{min}} \wedge f_{\delta\,\text{max}}\right\}. \tag{35}$$

The lateral surface capability index is determined based
on input data from the sensors and the flags $f_{\text{sl}}(t)$ and
$f_{\text{lin}}(t)$. The operation is described referring to the flowchart
of Figure 4. During the initialization step, the index μ is
set to a predetermined default value μ_0, which may be
a typical friction coefficient of a dry, solid road surface.
The lateral acceleration a_y is read from lateral acceleration
sensor. The straight line flag $f_{\text{sl}}(t)$ is then verified. If it is
"false," that is, if the vehicle is going through curves and is
subject to a substantial lateral acceleration, a timer is reset
to zero. Next, if the linear flag $f_{\text{lin}}(t)$ is "true," it can be
concluded that the lateral surface capability index must at
least equal a_y/g, wherein g denotes the gravity acceleration.
The lateral surface capability index is therefore updated to be
$\mu(t) = \max(\mu(t-1), a_y(t)/g)$ the maximum of a_y/g and an
estimate obtained from a previous iteration $\mu(t-1)$. In this
way, if this step is executed repeatedly in subsequent iterations

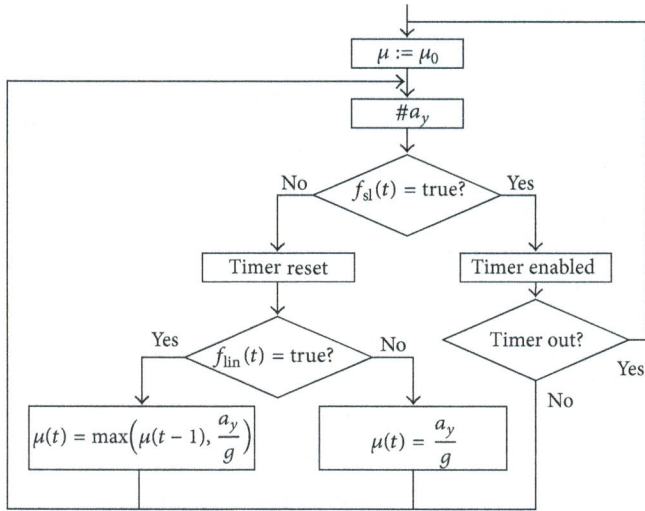

FIGURE 4: Schematic diagram of the lateral surface capability index.

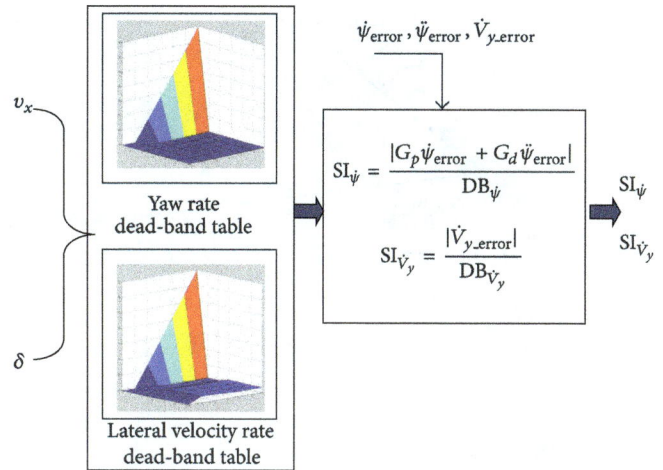

FIGURE 5: Schematic diagram of stability index implementation.

of the procedure of Figure 4, $\mu(t)$ will grow and converge towards the true friction coefficient of the road surface.

On the other hand, if the linear flag is found to be "false," this may be due to the fact that the quality of the road surface has deteriorated and its friction coefficient has decreased, or that the vehicle is going at the stability limit. In that case, a_y/g is set as the new the lateral surface capability index.

If the straight driving flag is found to be "true," no estimation of lateral surface capability index is possible. In this case, the timer mentioned is enabled; that is, the timer starts to run if the straight flag has just switched to "true," or it simply continues to run if the straight flag was "true" already in the previous iteration of the procedure. The value of the timer is thus representative of the time in which the vehicle has been going straight. If the timer has exceeded a predetermined limit, the algorithm resets the friction coefficient to μ_0. In this way, if the vehicle has been going straight for such a long time such that the previously acquired estimate of the lateral surface capability index is no longer reliable, the estimate is reset to μ_0, and the process of iteratively approximating its true value restarts when the straight driving flag $f_{sl}(t)$ becomes "false."

2.2.3. The Warning Algorithm. The warning algorithm computes indices based on yaw-rate error, yaw rate dead-band, side slip velocity error, and side slip velocity dead-band, an understeer error, and an understeer dead-band, using various sensors. It consists of 3 major blocks:

   (i) stability index block;

   (ii) steady state linear/nonlinear understeer index block;

   (iii) arbitration block.

Stability Index Block. The stability index block is mainly used during transient driving situations, in which a driver has to turn the steering wheel of the vehicle quickly and/or in alternating directions, so that a reliable estimation of

the vehicle understeer is difficult. Figure 5 illustrates the schematic diagram of the stability index implementation.

The stability index block requires the yaw rate error $\dot{\psi}_{error}$, the rate of the yaw error $\ddot{\psi}_{error}$, and the lateral velocity error rate $\dot{V}_{y\_error}$ for the computation of the stability index. These quantities are defined as follows:

$$\dot{\psi}_{error} = \dot{\psi}_d - \dot{\psi}, \qquad \ddot{\psi}_{error} = \frac{\left(\dot{\psi}_{error}(t) - \dot{\psi}_{error}(t-T)\right)}{T},$$

$$\dot{V}_{y_{error}} = \dot{v}_{yd} - \dot{v}_y = \dot{v}_{yd} - \left(\dot{\psi}v_x - a_y\right).$$

$$(36)$$

The yaw rate and the lateral velocity rate dead-band lookup tables store a yaw rate dead-band $DB_{\dot{\psi}}$ and a lateral velocity rate dead band $DB_{\dot{V}_y}$, as a function of longitudinal velocity v_x and steering wheel angle δ. These dead bands represent (sometimes called a neutral zone) an area of a signal range or band where no action occurs (the system is dead), The purpose is of the dead band to prevent oscillation or repeated activation-deactivation cycles (called "hunting" in control systems). a predetermined percentage of a yaw rate or a lateral velocity derivative above which, for a given vehicle speed and steering wheel angle, control over the vehicle is lost. The warning algorithm is used in combination with an electronic stability control (ESC system), and the ESC system may use associated dead bands of the yaw rate and the lateral velocity derivative for deciding whether to intervene or not. If, for example, the dead bands of the ESC system are at 70% of a value at which control of the vehicle is lost for a given vehicle speed and steering wheel angle, the dead bands used in the stability index block may be at 50% of such a value, ensuring that the warning algorithm will issue a warning signal prior to any intervention of the ESC.

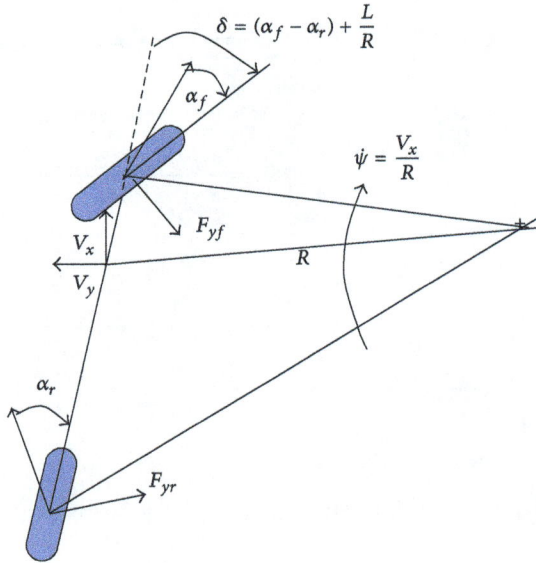

$$\delta = (\alpha_f - \alpha_r) + \frac{L}{R}$$

FIGURE 6: Vehicle in steady turn.

From the dead bands $\mathrm{DB}_{\dot\psi}$, $\mathrm{DB}_{\dot{V}_y}$ and the error signals $\dot\psi_{\mathrm{error}}$, $\ddot\psi_{\mathrm{error}}$, and $\dot{V}_{y\text{-}\mathrm{error}}$, the yaw rate and lateral velocity rate stability indices $\mathrm{SI}_{\dot\psi}$, $\mathrm{SI}_{\dot{V}_y}$ as follows:

$$\mathrm{SI}_{\dot\psi} = \frac{\left| G_p \dot\psi_{\mathrm{error}} + G_d \ddot\psi_{\mathrm{error}} \right|}{\mathrm{DB}_{\dot\psi}},$$

$$\mathrm{SI}_{\dot{V}_y} = \frac{\left| \dot{V}_{y\text{-}\mathrm{error}} \right|}{\mathrm{DB}_{\dot{V}_y}}. \tag{37}$$

The decision whether the vehicle is in a situation approaching ESC activation or not during a transient maneuver can be based on the yaw rate and lateral velocity derivative indices defined in (37), and a general stability index SI is set equal to

$$\mathrm{SI} = \max\left(\mathrm{SI}_{\dot\psi}, \mathrm{SI}_{\dot{V}_y} \right). \tag{38}$$

Steady State Linear/Nonlinear Understeer Index Block. In this section, we will discuss the structure and operation of the understeer index. For estimating the understeer of a vehicle, it is important to know whether the vehicle is moving in a linear regime, in which the sideslip angle of the vehicle is approximately directly proportional to the lateral forces. Figure 6 illustrates the vehicle in a steady turn; to keep the slip angle small, we assume that the turn radius is large. This is normally the case for high-speed turns, in which the vehicle is not skidding.

For steady state mode, the dynamic equation of motion in the body-centered coordinate system is given by

$$\frac{Mv_x^2}{R} = F_{yf} + F_{yr}, \tag{39}$$

$$0 = aF_{yf} + bF_{yr}.$$

This allows solving for the forces

$$F_{yf} = \frac{aMv_x^2}{LR},$$

$$F_{yr} = \frac{bMv_x^2}{LR}. \tag{40}$$

Using the relation $\dot\psi = v_x/R$, the steer angle equation can be written as

$$\delta = -\alpha_f + \alpha_r + \frac{L}{R}. \tag{41}$$

The linearized lateral forces are expressed in terms of the tire slip angles

$$F_{yf} = C_f \alpha_f,$$

$$F_{yr} = C_r \alpha_r. \tag{42}$$

From (40) and (42), the steer angle relation becomes

$$\delta = \left(\frac{M\left(bC_r - aC_f \right)}{LC_f C_r} \right) \frac{v_x^2}{R} + \frac{L}{R}. \tag{43}$$

Equation (43) relates the steer angle to the speed via the understeer coefficient K_{und}

$$\delta = K_{\mathrm{und}} \frac{v_x^2}{R} + \frac{L}{R}, \tag{44}$$

where

$$K_{\mathrm{und}} = \left(\frac{M\left(bC_r - aC_f \right)}{LC_f C_r} \right)$$

$$= \frac{M}{L}\left(\frac{b}{C_f} - \frac{a}{C_R} \right). \tag{45}$$

From (44), we can calculate the change in the steer angle when the L/R is changed by differentiating (44).

In the linear case

$$\frac{\partial \delta}{\partial (L/R)} = K_{\mathrm{und\_linear}} \frac{V_x^2}{L} + 1,$$

$$K_{\mathrm{und\_linear}} = \frac{L}{V_x^2}\left(\frac{\partial \delta}{\partial (L/R)} - 1 \right). \tag{46}$$

Figure 7 shows the steering wheel angle δ versus L/R for constant speed. For the linear tire force assumption, the slopes of the plot are constants since the understeer and vehicle speed are constant.

For the nonlinear tire characteristics, in which the tire force-sideslip angle becomes significantly nonlinear, the slopes are not constant and vary with the turn radius when the vehicle speed is constant as illustrated in Figure 8. In this case, the understeer relates to the local slope of the steer angle curve.

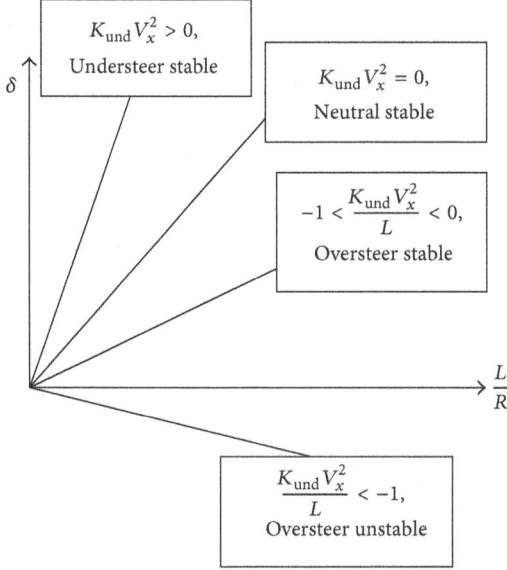

FIGURE 7: Steer angle versus L/R for linear case [14].

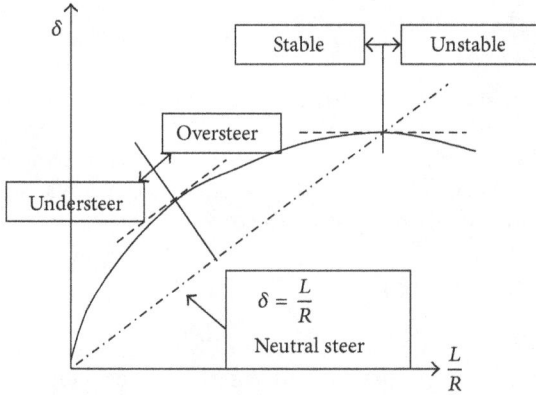

FIGURE 8: Steer angle versus L/R for nonlinear case [14].

Since in the nonlinear case there is no constant understeer coefficient, we define a variable coefficient that expresses how the slip angle difference changes as the lateral acceleration changes

$$K_{\text{und\_nl}} = \frac{d\left(\alpha_f - \alpha_r\right)}{d\left(V_x^2/Rg\right)} = \frac{d\left(\alpha_f - \alpha_r\right)}{d\left(a_y\right)}, \qquad (47)$$

where

$$\left(\alpha_f - \alpha_r\right) = \delta - \frac{a+b}{v_x}\dot{\psi} = \delta - \frac{La_y}{v_x^2}. \qquad (48)$$

By applying Kalman filter technique, it is possible to estimate the linear understeer and nonlinear understeer variable.

Estimation of the Linear and Nonlinear Understeer. In the last section, we developed linear and nonlinear expression for the understeer tendency of the vehicle under steady cornering.

In this section, an identification algorithm using the Kalman filter is developed to estimate the linear understeer and nonlinear understeer variable. The Kalman filter is a set of mathematical equations that provide an efficient computational (recursive) solution of the least-squares method. The filter is very powerful in several aspects; it supports estimations of past, present, and even future states, and it can do so even when the precise nature of the modeled system is unknown.

The Kalman filter addresses the general problem of trying to estimate the state of a discrete-time controlled process that is governed by the linear stochastic difference equation

$$x(t+1) = x(t) + \upsilon(t) \qquad (49)$$

with a measurement $y \in \mathfrak{R}$ that is

$$y(t) = H(t)x(t) + n(t). \qquad (50)$$

We will be using two Kalman filters for the linear and nonlinear operations. Equations (44) and (47) can be recast as

$$\begin{aligned} K_{\text{und}}a_y &= \delta - L\frac{a_y}{v_x^2}, \\ K_{\text{und}_{\text{nl}}}\frac{da_y}{dt} &= \frac{d\left(\alpha_f - \alpha_r\right)}{dt}. \end{aligned} \qquad (51)$$

In the linear range we define

$$x(t) = K_{\text{und}}(t), \quad y(t) = \delta - L\frac{a_y}{v_x^2}, \quad H(t) = a_y. \qquad (52)$$

In the nonlinear range we define

$$x(t) = K_{\text{und\_nl}}(t), \quad y(t) = \frac{d\left(\alpha_f - \alpha_r\right)}{dt}, \quad H(t) = \frac{da_y}{dt}. \qquad (53)$$

The random variables $\upsilon(t)$ and $n(t)$ represent the process and measurement noise, respectively.

They are assumed to be independent (of each other), white, and with normal probability distributions

$$\begin{aligned} p(\upsilon) &\sim N(0, Q), \\ p(n) &\sim N(0, \Gamma). \end{aligned} \qquad (54)$$

Q is the process noise covariance and, Γ is measurement noise covariance. $H(t)$ in the measurement equations (52)-(53) relates the state to the measurement $y(t)$. In practice, it will change with each time step or measurement.

Consider the signal model of (49)–(54), and assume that the initial state and noise sequences are jointly Gaussian. Let $\hat{x}(t+1)$ denote the conditional mean of $x(t+1)$ given

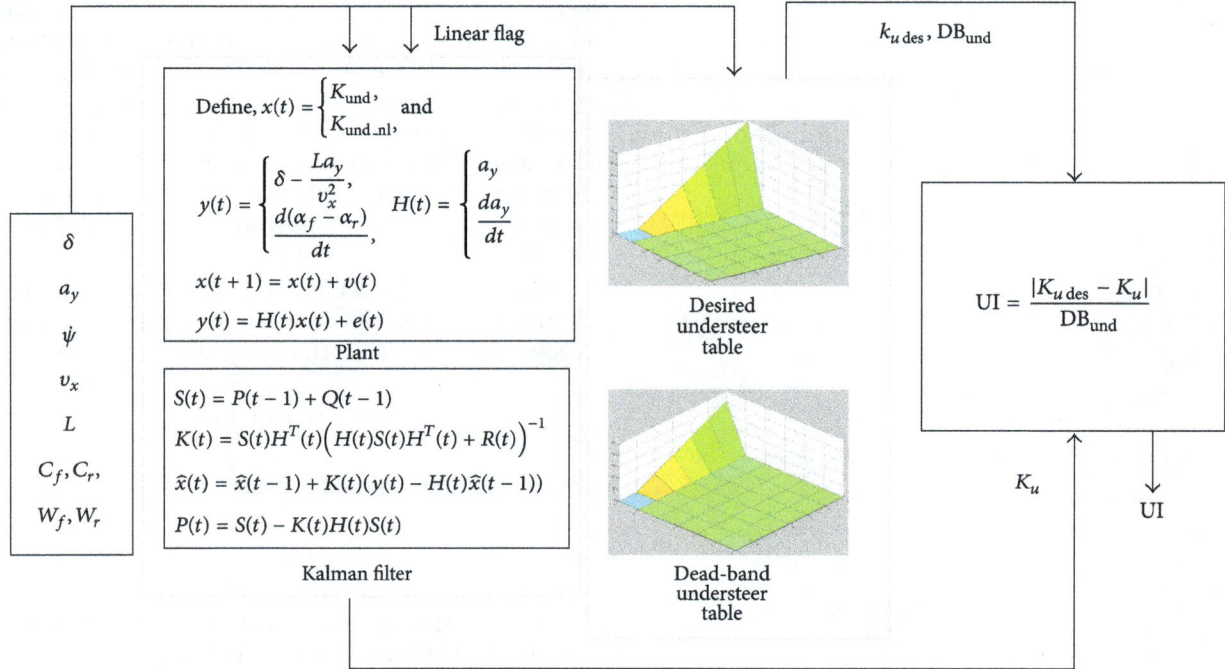

FIGURE 9: Schematic diagram of the understeer index implementation.

observation $\{y(t)\}$ up to and including time t; then, $\hat{x}(t+1)$ satisfies the following recursion:

$$\hat{x}(t+1) = \hat{x}(t) + L(t)\left(y(t) - H(t)\,\hat{x}(t)\right),$$

$$\hat{x}(t_0) = \overline{x}_0,$$

$$S(t+1) = P(t) + Q(t),$$

$$L(t) = S(t)\,H^T(t)\left(H(t)\,S(t)\,H^T(t) + \Gamma(t)\right)^{-1}, \quad (55)$$

$$P(t+1) = S(t+1) - L(t)\,H(t)\,S(t+1),$$

$$P(t_0) = \overline{P}_0,$$

where $L(t)$ is the filter gain and $P(t)$ is the state error covariance.

The linear flag defined in (35) controls the operation of the two Kalman filters (52) and (53) for estimating the understeer of the vehicle in linear and nonlinear regimes, respectively.

When the vehicle is determined to be in non-linear regime, the linear flag becomes zero, filter (52) stops, and filter (53) starts to operate, initialized with the most recent understeer value from filter (52). Similarly, when the linear flag changes back to 1, filter (52) becomes operative again and is initialized with a nominal understeer based on the front and rear lateral tire stiffness and the front and rear vehicle weight distribution and with initial covariance values.

In principle, the two filters might be regarded as a single Kalman filter which swaps $y(t)$ and $H(t)$ according to the value of the linear flag.

In analogy to what was described earlier for the yaw rate and lateral velocity rate stability indices $SI_{\dot{\psi}}$, $SI_{\dot{V}_y}$, lookup tables provided values of a desired understeer $K_{u\,des}$ and a dead band of the understeer DB_{und}. As in case of the yaw rate and the lateral velocity derivative, desired understeer values can be predetermined empirically by measuring the understeer of a test vehicle at given speeds and steering wheel angles. Alternatively, they may be calculated in advance or in real time, for example, using the following formula:

$$K_{u\,des} = \max\left(\frac{1}{v_x^2}\left(\frac{v_x\delta}{\dot{\psi}_{des}} - L\right), \frac{W_f}{C_f} - \frac{W_r}{C_r}\right). \quad (56)$$

Since $K_{u\,des}$ may take impractically high values according to (56), at high speeds and steering wheel angles, it is preferred to define an upper limit of the desired understeer $K_{u\,des}$ as follows:

$K_{u\,des}$
$$= \min\left(\max\left(\frac{1}{v_x^2}\left(\frac{v_x\delta}{\dot{\psi}_{des}} - L\right), \frac{W_f}{C_f} - \frac{W_r}{C_r}\right), K_{und\_max}\right). \quad (57)$$

The upper limit K_{und_max} may be set to, for example, $8°/g$ or $5°/g$, g denoting the gravity acceleration.

Similar to the yaw rate dead band $DB_{\dot{\psi}}$ and a lateral velocity rate dead band $DB_{\dot{V}_y}$, the understeer dead band DB_{und} gives values of the understeer which can be regarded as safe as function of longitudinal velocity v_x and steering wheel angle δ. The understeer index UI based on the effective vehicle understeer K_{und} or K_{und_nl}, defined as K_u, estimated by filter (52) or (53) the desired understeer $K_{u\,des}$ from (57) and the understeer dead band DB_{und}:

$$UI = \frac{|K_{u\,des} - K_u|}{DB_{und}}. \quad (58)$$

The structure and operation of the understeer index are depicted in Figure 9.

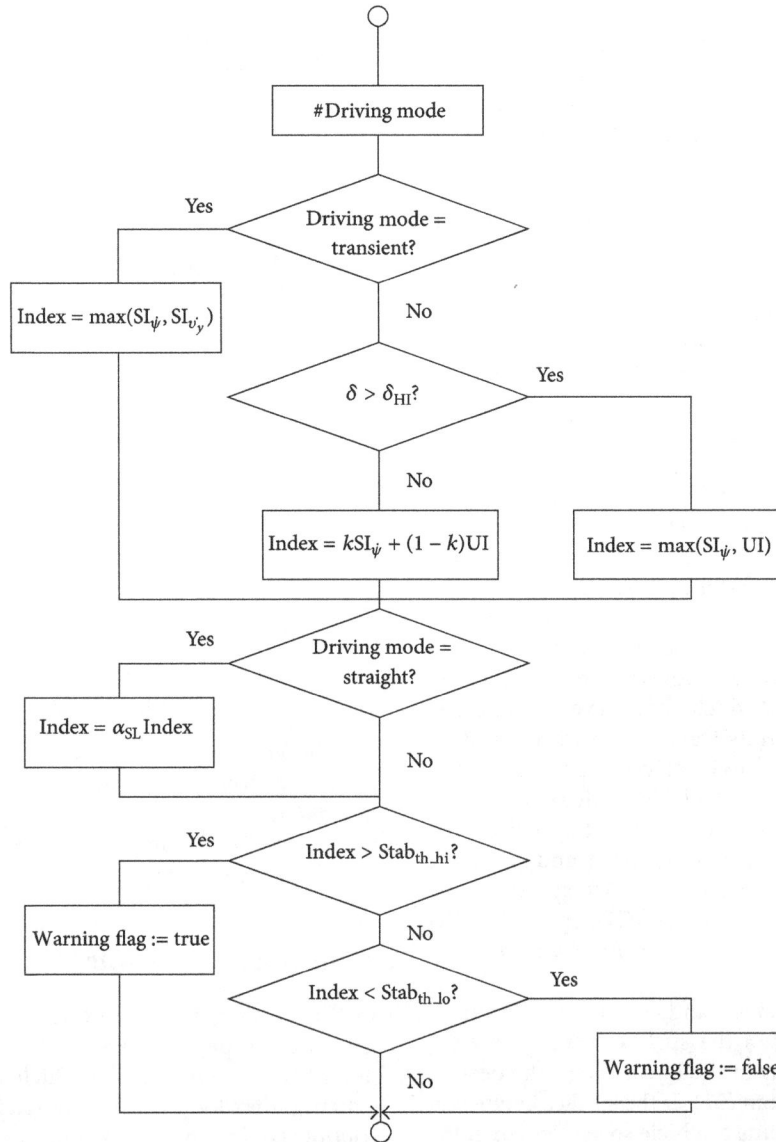

FIGURE 10: Flow chart of the arbitrator logic.

The Arbitrator Block. Referring again to Figures 5 and 9, three indices $SI_{\dot{\psi}}$, $SI_{\dot{v}_y}$, and UI are supplied to the arbitrator. Albeit of different origin, the three indices are comparable in that they are dimensionless and that a value above 1 indicates a critical driving situation.

The operation of the arbitrator is explained referring to the flowchart of Figure 10. The first step is that we distinguish between a transient mode, and a steady state mode. In the transient mode the direction of the vehicle is changing so rapidly that the understeer cannot be relied upon. Therefore, if the driving mode is found to be the transient mode, the decision whether the vehicle is approaching ESC activation or not can only be based on the yaw rate and lateral velocity derivative indices $SI_{\dot{\psi}}$ and $SI_{\dot{v}_y}$.

A general stability index is set equal to the $\max(SI_{\dot{\psi}}, SI_{\dot{v}_y})$. If the vehicle is not in the transient mode, the steering wheel angle δ is compared to a predetermined upper threshold δ_{HI}.

If this threshold is exceeded, there is a considerable risk of the vehicle being unstable, and the system should be rather liberal in issuing a warning. In that case, the general stability index is set equal to the $\max(SI_{\dot{\psi}}, UI)$. Otherwise, a weighted $\text{sum}(kSI_{\dot{\psi}} + (1 - k)UI)$ of the yaw rate stability index and understeer index is calculated. The weighting factor k is tuned to a value between 0 and 1. This weighting factor may be set dependent on the vehicle speed v_x and decreases with the vehicle speed, giving increasing importance to the understeer at high speeds.

If the vehicle is in a straight line driving mode, the general stability index determined prior to this step is multiplied by positive factor α_{SL}, which is smaller than 1, reflecting the fact that the vehicle is least susceptible to a loss of control if it is driving a longer straight line.

Finally, the general stability index is compared to an upper threshold $Stab_{th\_hi}$, for example, 0.8. If it exceeds this

26   Vehicular Technology Handbook

upper threshold, a warning flag is set to TRUE. Since the dead bands for warning algorithms are set lower than those of an ESC system, the warning flag will become TRUE prior to any intervention of the ESC system. If the stability index was found not to exceed the upper threshold, the index is compared to a lower threshold $Stab_{th\_lo}$, for example, 0.2. If it is below this lower threshold, the warning flag is set to FALSE; if not, the warning flag is left as it is until the procedure is repeated.

2.2.4. The Speed Advisory Block. Very often the road signs indicate the safe speed in a curve. However, on low μ surfaces, the suggested posted speed might not be adequate for the road condition.

When the vehicle is driving in a curve at a higher speed than the surface can allow, the understeer gradient of the vehicle increases causing the vehicle to plow or decreases and becomes negative causing the vehicle to spinout. The warning algorithm described earlier will issue a warning to alert the driver that he/she is traveling faster than the road surface can allow. In this section, we develop an advisory speed algorithm in a curve based on vehicle dynamics in conjunction with the driver warning algorithm. The advisory speed algorithm computes the advisory speed which allows a vehicle to travel around the turn or curve in its travel lane without causing an uncomfortable "side force" to its driver or passengers and helps maintain control of the vehicle. The advisory speed is based on the maximum lateral capability of the surface, the driver steering input, the actual vehicle speed, and the actual understeer of the vehicle. A visual advisory speed can be displayed, for example, in the DCI or a HUD display when the warning signal is issued. The visual advisory HMI is outside the scope of this paper.

Based on the steering input δ and vehicle speed v_x, if we assume that the coefficient of the surface would allow the vehicle (theoretically) to stay in the linear range, the desired understeer is determined from (57). If the surface coefficient is much lower, we can compute a vehicle speed (less than the actual speed of the vehicle) such that the understeer gradient is kept within a small deviation from the desired understeer of the vehicle. Assume that ΔK_{und} is the understeer deviation from the linear performance. Thus, the stable vehicle speed limit can be determined from (44) as follows:

$$v_{lim} = \sqrt{\frac{L\mu g}{|\delta| - \left[\min\left(\max\left(\frac{1}{v_x^2}\left(\frac{v_x\delta}{\dot{\psi}_{des}} - L\right), \frac{W_f}{C_f} - \frac{W_r}{C_r}\right), K_{und\_max}\right) + \Delta K_{und}\right]\mu g}},$$
(59)

where μ is the maximum capability of the surface and is determined as described in Section 2.2.2.

Finally, the advisory speed is calculated as follows:

$$V_{adv} = \begin{cases} \min(v_{lim}, v_x) \text{ if the warning flag is set} \\ \text{speed limit if available, when warning flag is not set.} \end{cases}$$
(60)

Under straight line condition v_{lim} is set to ∞.

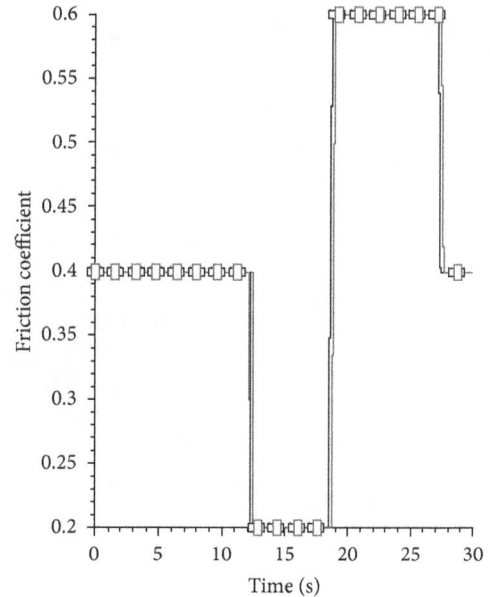

FIGURE 11: Time trace of the road friction coefficient.

The advisory speed is communicated to the driver to allow the driver to make a choice on what action should be taken, or through an intervention system where the engine and/or braking systems are controlled automatically to reduce the vehicle's speed.

3. Simulation Results

In this section, we present some typical simulation results showing the performance of the driver warning algorithm described in Section 2. The vehicle was traveling at 80 kph on a 40 m radius loop with 200 m straight section with variable friction coefficient to simulate the exit ramp of a freeway. A driver model with a driver preview time of 1 second and driver lag of 0.12 second was used in the simulation. Figure 11 shows the time trace of the road friction coefficient as the vehicle travels. In this simulation, the driver did not react to the warning at the time when the warning was issued.

Figure 12 illustrates the time trace of the warning signal and the ESC activation flag. The warning was issued at around 5 seconds when the driver entered the curved section of the road at 80 kph. In order to make ESC system as unintrusive as possible, its activation threshold will have to be set rather high. If the vehicle starts to drift, the ESC system is activated to keep the vehicle back on course. Figure 13 shows the vehicle actual path as compared to the target path. It is noticed that although the ESC intervened to maintain the vehicle stability, the vehicle had started sliding towards the edge of the road and did not follow closely the target path. This is also confirmed in Figure 14, where the vehicle trajectory is illustrated as the vehicle entered the curved section of the road.

To illustrate the performance of the speed advisory algorithm, the simulation was repeated this time with the

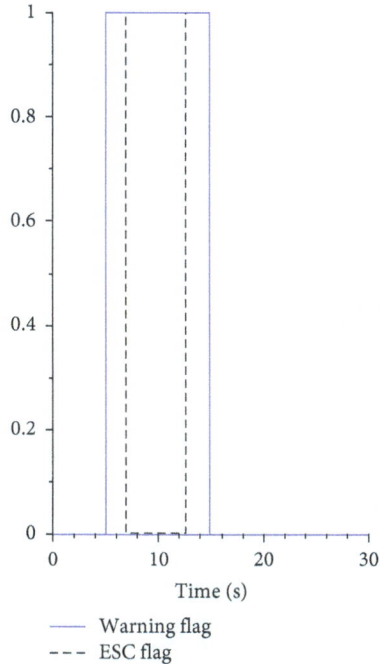

Figure 12: Warning and ESC flags time trace, driver not reacting to the warning at the time the warning is set.

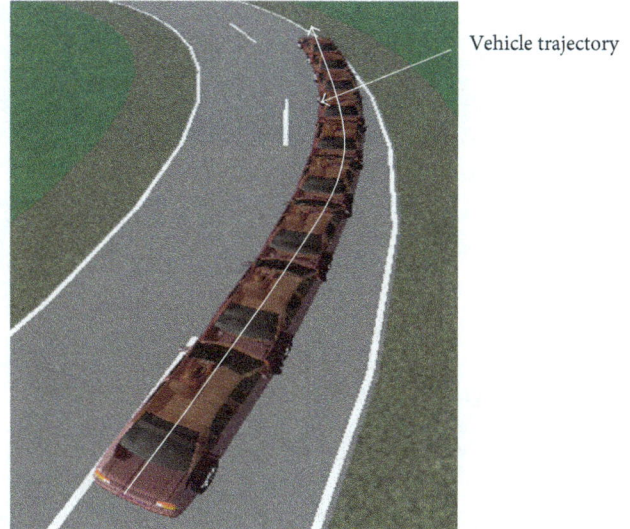

Figure 14: Vehicle trajectory in the curved section of the road, driver not reacting to the warning at the time the warning is set.

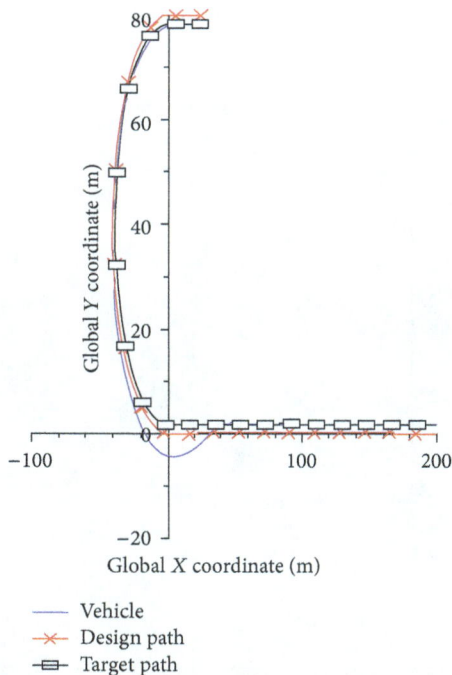

Figure 13: Vehicle path compared to the target path, driver not reacting to the warning at the time the warning is set.

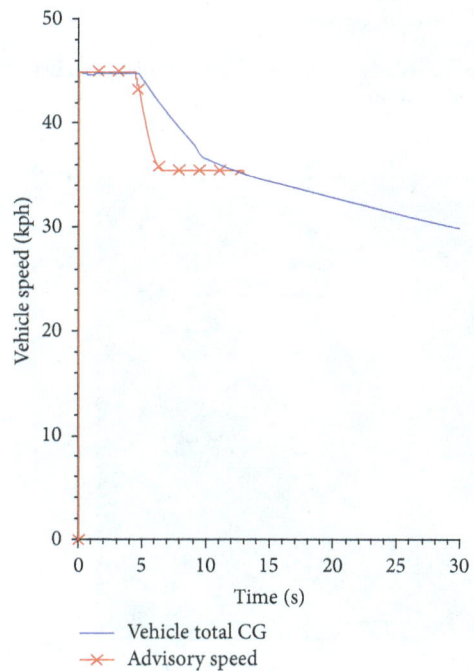

Figure 15: Vehicle advisory speed.

driver reacting to the warning signal by reducing the vehicle speed based on the speed advisory algorithm.

Figure 15 shows the vehicle advisory speed and the vehicle speed as function of time. The advisory speed is initially set equal to the speed of the vehicle before the activation of the warning signal. This speed will not be displayed until

the warning is activated. At the end of the warning signal, the advisory speed will no longer be displayed to the driver. Instead if the vehicle is equipped with GPS receiver and a conventional navigation system, the GPS receiver enables the speed limit detection unit to find out the exact geographic location of the vehicle, to identify, based on map data of the navigation system, a road on which the vehicle is currently moving and to retrieve from the navigation system data on an eventual speed limit on this road. Figure 16 shows the vehicle trajectory; in this case, the vehicle is following more precisely the target path as opposed to the first case where the driver

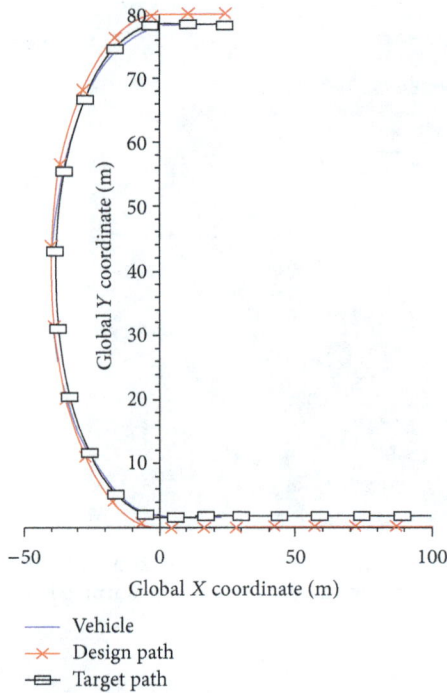

FIGURE 16: Vehicle path compared to the target path, driver reacting to the warning at the time the warning is set.

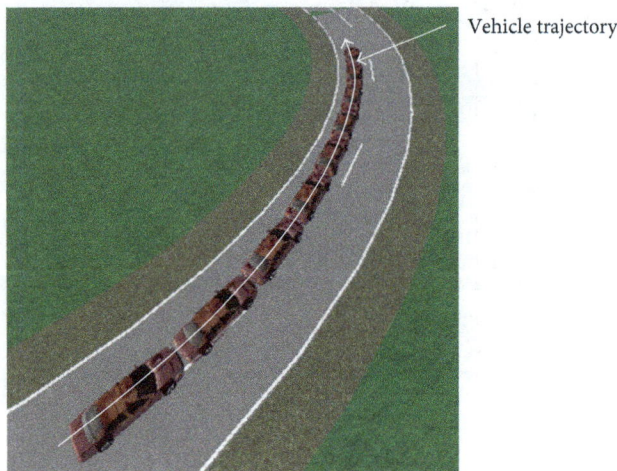

FIGURE 17: Vehicle trajectory in the curved section of the road, driver reacting to the warning at the time the warning is set.

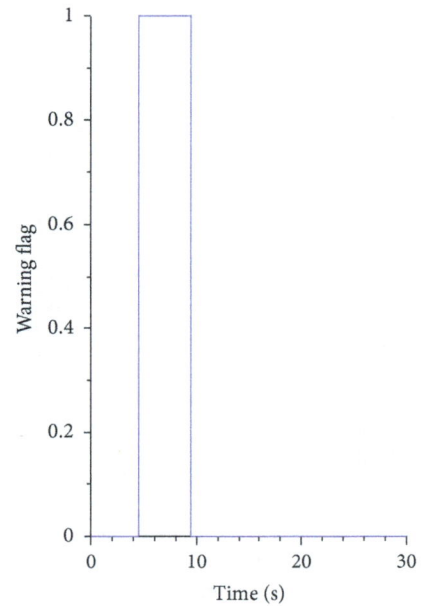

FIGURE 18: Warning and ESC flags time trace, driver reacting to the warning at the time the warning is set.

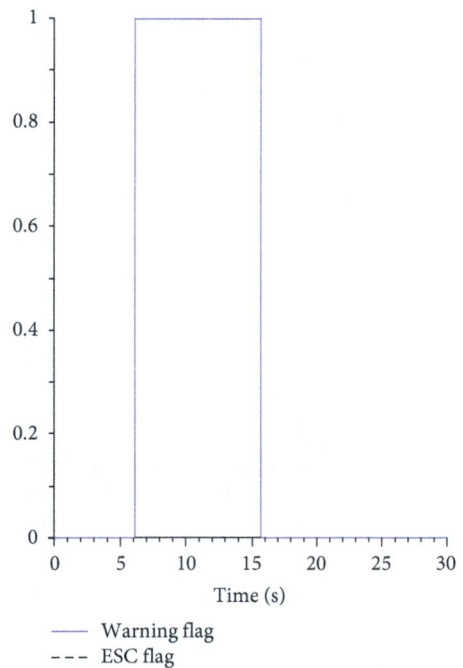

FIGURE 19: False warning issued on a banked road with uncompensated lateral acceleration.

did not react at the time the warning was set. Figure 17 shows a stable trajectory of the vehicle. It is also noticed that the duration of the warning signal has been reduced, and the ESC did not activate as shown in Figure 18.

To study the effect of the vehicle being on a banked road on the warning algorithm, the previous simulation was repeated with the vehicle traveling at 80 kph on a 40 m radius loop with 200 m straight section of dry surface and variable bank angle. When the vehicle operator is driving on a banked road, the measured lateral acceleration will include the effect of the banked road and therefore affecting the straight driving

detection algorithm. In addition, the operator introduces a correction to the steering angle to maintain the vehicle on the road, and therefore the desired yaw rate and understeer commands will indicate that the driver wishes to travel on the bank and not across it. In this case, the warning signal might be triggered unnecessarily especially when the vehicle is in the linear range of operation. Figure 19 shows that algorithm issued a warning signal even though the vehicle was stable

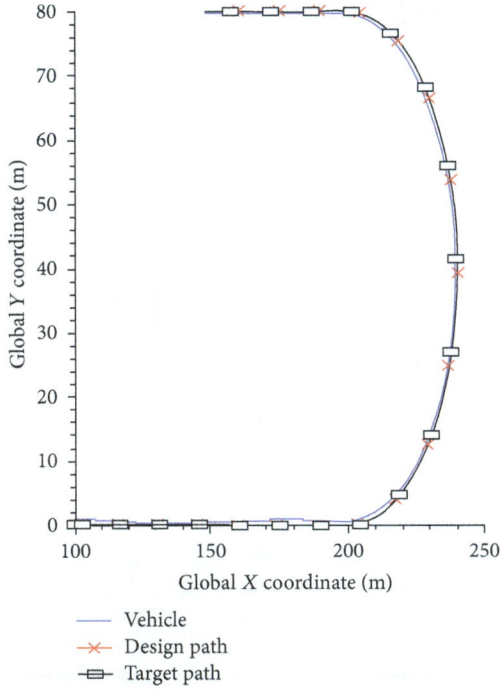

FIGURE 20: Vehicle path compared to the target path on a banked road.

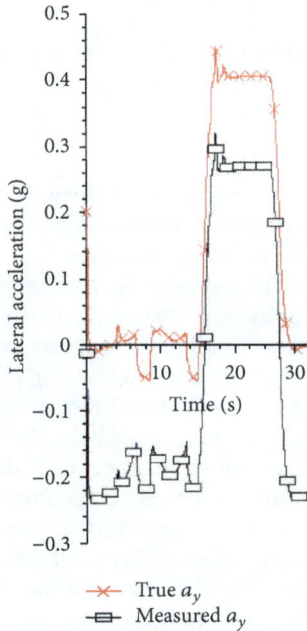

FIGURE 21: True and measured uncompensated lateral acceleration on a banked road.

and followed the target path as shown in Figure 20. Figure 21 compares the measured acceleration including the effect of the bank component to the true lateral acceleration of the vehicle.

Therefore, it is important to compensate for the effect of the bank in the lateral acceleration measurement.

When the vehicle is driven on a banked road, the measured lateral acceleration is corrupted by the bank angle of the road given by the following equation:

$$a_{ym} = a_y + g \sin \phi. \tag{61}$$

The kinematics relationship between the lateral acceleration a_y and the yaw rate of the vehicle $\dot{\psi}$ is given by the following equation:

$$\dot{v}_y = a_y - \dot{\psi} v_x. \tag{62}$$

Under steady state equation, $\dot{v}_y = 0$, and therefore (62) becomes

$$a_y = \dot{\psi} v_x. \tag{63}$$

Define $\varepsilon(k)$ as

$$\varepsilon(k) = a_{ym}(k) - a_y(k) = g \sin \phi$$
$$= a_{ym}(k) - \dot{\psi} v_x(k). \tag{64}$$

Next, we will develop a Kalman filter to estimate $\varepsilon(k)$. Define the following state vector

$$x(k) = \begin{bmatrix} 1 \\ \varepsilon(k) \end{bmatrix}; \tag{65}$$

the state vector is governed by the linear stochastic difference equation

$$x(k+1) = x(k) + \omega(k) \tag{66}$$

with a measurement $y \in \Re$ that is

$$y(k) = \dot{\varphi} v_x(k) = \lfloor a_{ym}(k) - 1 \rfloor x(k) + \eta(k). \tag{67}$$

The random variables $\omega(k)$ and $\eta(k)$ represent the process and measurement noise, respectively.

They are assumed to be independent (of each other), white, and with normal probability distributions

$$p(\omega) \sim N(0, Q),$$
$$p(\eta) \sim N(0, \Gamma). \tag{68}$$

The Kalman filter developed in (55) is applied to obtain an estimate $\hat{\varepsilon}(k)$ for the difference between the measured and the actual lateral acceleration.

The previous simulation was repeated using the compensated lateral acceleration

$$a_{y\,comp} = a_y + \hat{\varepsilon}(k). \tag{69}$$

In Figure 22, the compensated measured lateral acceleration shows a good agreement with the true lateral acceleration of the vehicle. Referring to Figure 23, the simulation result represents an example of a correct warning (no warning issued) on a banked curve in the linear range of the vehicle when the lateral acceleration is compensated.

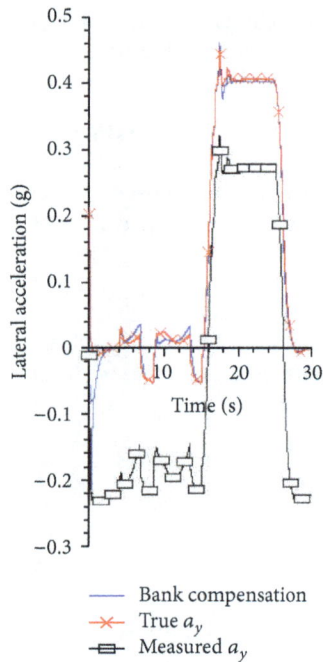

FIGURE 22: True lateral acceleration and measured uncompensated and compensated lateral acceleration on a banked road.

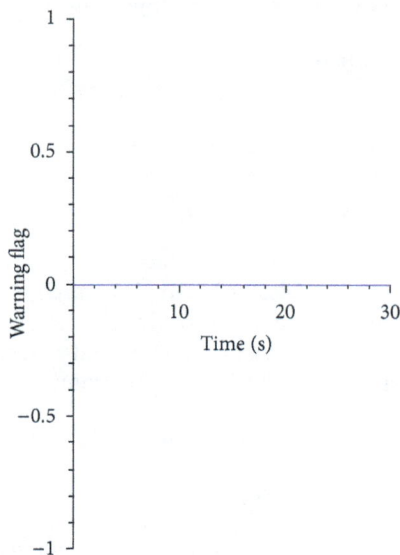

(a)

(b)

(c)

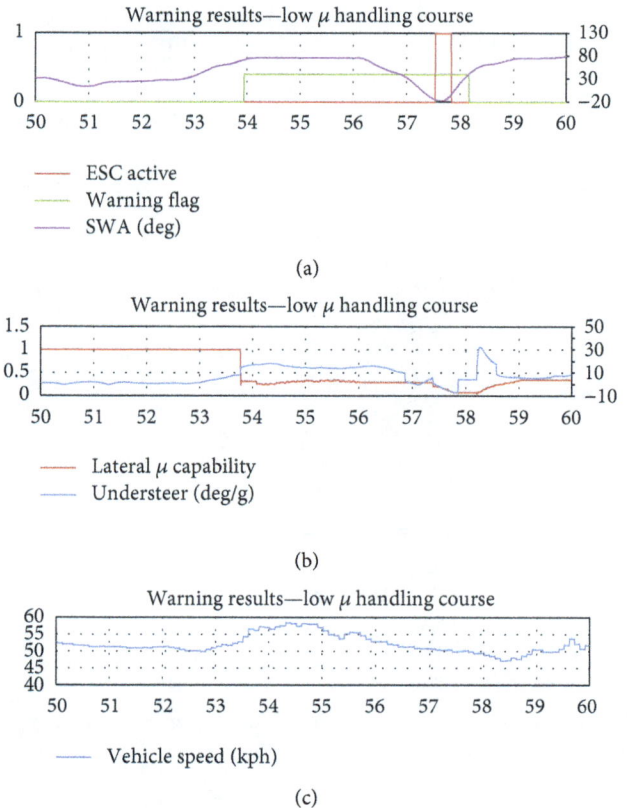

FIGURE 24: Vehicle simulating a freeway exit on low μ surface.

FIGURE 23: Warning signal on a banked road with compensated lateral acceleration.

4. Vehicle Test Results

To evaluate the warning algorithm performance, the following results are based on experimental data obtained using an Opel Omega vehicle equipped with ESC sensors. The tests were conducted on low-mu handling track with straight and curved sections to simulate a freeway exit. Time slices from different driving sessions are zoomed in to illustrate the performance of the warning algorithm.

A time slice of 10 seconds (50–60 seconds) is shown in Figure 24. The graphs show the performance of the warning system as the vehicle simulates a free way exit. In the first 4 seconds, the driver enters a curved section of the course.

At time $t = 54$ seconds, the driver maintains an 80 deg steering wheel angle at speed of 58 kph. The vehicle starts to understeer and reaches an understeer of 20 deg/g. The lateral mu capability algorithm identifies a maximum lateral surface capability of 0.3. At this moment, the warning signal was issued, and the driver released his accelerator pedal but did not apply his brakes. The ESC was activated 3.5 seconds after the warning signal and slowed down the car. The sudden change in the understeer at 58 seconds is due to the fact that vehicle was not in a steady state mode and the understeer is not defined outside the steady state behavior of the vehicle.

Figure 25 represents a series of transient maneuvers on the handling course. As seen from the warning signal and the ESC active flag, the warning signal was issued 3 to 4 seconds prior to the ESC activation which allows the driver to react to the warning signal and reduce the vehicle speed.

Figure 26 shows a large increasing steering angle maneuver at relatively low speed. It is noticed that under this scenario the warning is issued almost at the same time as the ESC activation. The warning signal is not effective in this case

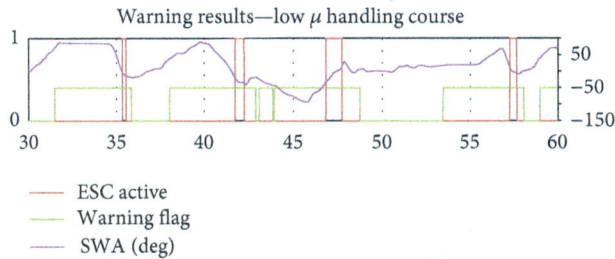

FIGURE 25: Warning and ESC signals under transient maneuvers.

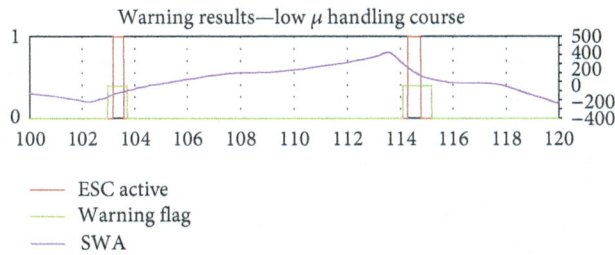

FIGURE 26: Warning and ESC signals with increasing steering angle at low speed.

since the driver does not have time to react to the warning signal before the ESC activation.

Figure 27 shows the vehicle going into a steady state maneuver. In this case, the speed advisory algorithm was triggered as the driver enters the curved section of the road. The vehicle starts to understeer and the warning signal is set. The driver releases the accelerator pedal and reduces the steering input. The warning signal was then terminated and the speed advisory ended. The ESC did not activate in this case since the driver reduced his steering input and the understeer gradient of the vehicle is reduced. The vehicle did not a have an automatic brake control to control the vehicle speed to the advisory speed. The warning was set until the driver reduced the steering input.

5. Conclusions

(1) In some of the conditions evaluated, the warning signal is issued 3 to 4 seconds prior to ESC activation to allow the driver to react to the warning.

(2) If the driver reacts to the warning signal by reducing the vehicle speed in the curve, the vehicle will follow more precisely the road curves with minimum or no ESC intervention.

(3) For large and increasing steering angle maneuver at relatively low speed, the warning is issued almost at the same time as the ESC activation. The warning signal is not effective in this case since the driver does not have time to react to the warning signal before the ESC activation. However, at these low speeds, the ESC is very effective and can stabilize the vehicle without heavy brake intervention.

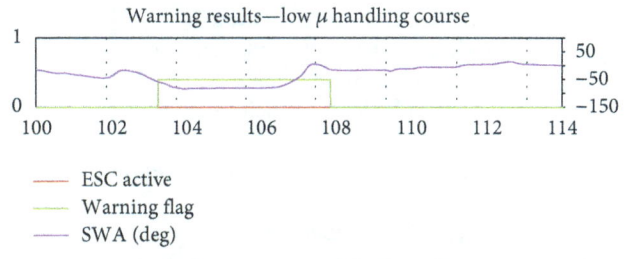

(a)

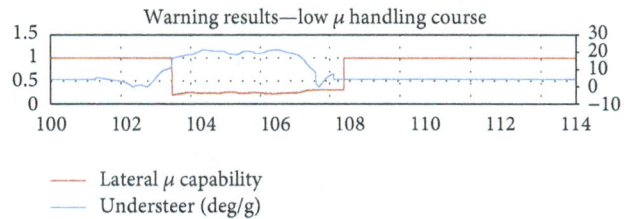

(b)

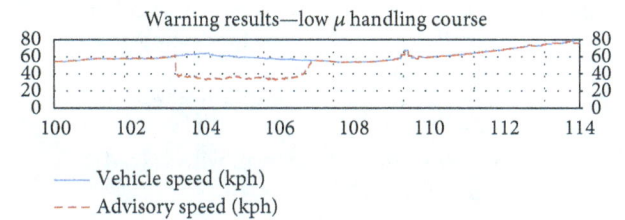

(c)

FIGURE 27: Vehicle simulating a freeway exit on low μ surface with speed advisory.

(4) When the vehicle is driven on a banked road, the uncompensated lateral acceleration measurement can false trigger the warning algorithm. Therefore, it is important to compensate for the effect of the bank in the lateral acceleration measurement.

(5) This warning system should be evaluated in a wider range of road and vehicle conditions to more fully evaluate its usefulness.

Nomenclature

a : Distance from the center of gravity of vehicle to the front axle (m)

b: Distance from the center of gravity of vehicle to the rear axle (m)

C_f: Cornering stiffness of both tires of front axle (N/rad)

C_r: Cornering stiffness of both tires of rear axle (N/rad)

g: Acceleration of gravity (m/s^2)

g_r: Steering gear ratio

I_z: Moment of inertia of entire vehicle about the yaw axis (kgm^2)

M_v: Total vehicle mass (kg)

W_f: Vehicle weight on the front axle

W_r: Vehicle weight on the rear axle

K_u: Desired understeer coefficient (deg/g)

T_w: Vehicle track width (m)

v_y: Lateral velocity of vehicle's center of gravity (m/s)

a_y: Lateral acceleration of vehicle's center of gravity (m/s^2)

a_{ym}:   Measured lateral acceleration at the vehicle's center of gravity (m/s^2)

v_{yd}:   Desired lateral velocity of vehicle's center of gravity (m/s)

$\dot{v}_{yd}$:   Desired lateral velocity rate of vehicle's center of gravity (m/s^2)

$V_{y\,gain}$:   Lateral velocity gain (m/s/rad)

$V_{y\,dss}$:   Steady sate desired lateral velocity of vehicle's center of gravity (m/s)

v_x: Longitudinal velocity of vehicle's center of gravity (m/s)

$x_i, i = 1, 2$:   State variables of the second-order filter

z_v: Negative of system zero for desired side slip velocity

$z_{\dot{\psi}}$:   Negative of system zero for desired yaw rate

δ: Steering angle of the front wheels (rad)

$\dot{\psi}$:   Yaw rate of vehicle (rad/s)

Φ: Bank angle (rad)

$\dot{\psi}_d$:   Desired yaw rate of vehicle (rad/s)

$\dot{\psi}_{dss}$:   Desired steady state yaw rate of vehicle (rad/s)

Ω_{gain}:   Yaw velocity gain (rad/s/rad)

ζ: Damping ratio of desired vehicle performance

ω_n: Natural frequency of desired vehicle performance (rad/s).

References

[1] D. Hoffman and M. Rizzo, "Chevrolet C5 corvette vehicle dynamic control system," SAE Paper 980233, 1998.

[2] K. Jost, "Cadillac stability enhancement," *Automotive Engineering*, 1996.

[3] H. Nakazato, K. Iwata, and Y. Yoshiyoka, "A new system for independently controlling braking force between inner and outer rear wheels," SAE Paper 890835, 1989.

[4] H. Inagaki, K. Akuzawa, and M. Sato, "Yaw rate feedback braking force distribution control with control by wire braking system," in *Proceedings of the International Symposium on Advanced Vehicle Control (AVEC '92)*, pp. 435–440, 1992.

[5] M. Salman, Z. Zang, and N. Boustany, "Coordinated control of four wheel braking and rear steering," in *Proceedings of the American Control Conference*, pp. 6–10, June 1992.

[6] T. Pilutti, G. Ulsoy, and D. Hrovat, "Vehicle steering intervention through differential braking," in *Proceedings of the American Control Conference*, June 1995.

[7] Y. -Shibahata, M. Abe, K. Shimada, and Y. Furukawa, "Improvement on limit performance of vehicle motion by chassis control," in *Proceedings of the 13th Symposium on the Dynamics of Vehicles on Roads and Tracks (IAVSD '95)*, 1995.

[8] C. A. Sawyer, "Controlling vehicle stability," *Automotive Industries*, 1995.

[9] Y. A. Ghoneim, W. C. Lin, D. M. Sidlosky, H. H. Chen, Y. K. Chin, and M. J. Tedrake, "Integrated chassis control system to enhance vehicle stability," *International Journal of Vehicle Design*, vol. 23, no. 1, pp. 124–144, 2000.

[10] H. Leffler, R. Auffhammer, R. Heyken, and H. Roth, "New driving stability control system with reduced technical effort for compact and medium class passenger cars," SAE Paper 980234, 1998.

[11] J. Ackermann and S. Turk, "A common controller for a family of plant models," in *Proceedings of the of the 21st IEEE Conference on Decision and Control (CDC '82)*, Orlando, Fla, USA, 1982.

[12] T. Pilutti, G. Ulsoy, and D. Hrovat, "Vehicle steering intervention through differential braking," in *Proceedings of the American Control Conference. Part 1 (of 6)*, pp. 1667–1671, June 1995.

[13] G. Thomas, "Fundamentals of vehicle dynamics," Society of Automotive Engineers, Inc. Second Printing, 1992.

[14] D. karnopp, *Vehicle Stability (Marcel Dekker)*, 2004.

3

Two-Lane Traffic Flow Simulation Model via Cellular Automaton

Kamini Rawat, Vinod Kumar Katiyar, and Pratibha Gupta

Department of Mathematics, IIT Roorkee, Roorkee 247667, India

Correspondence should be addressed to Kamini Rawat, rawatkamini@gmail.com

Academic Editor: Nandana Rajatheva

Road traffic microsimulations based on the individual motion of all the involved vehicles are now recognized as an important tool to describe, understand, and manage road traffic. Cellular automata (CA) are very efficient way to implement vehicle motion. CA is a methodology that uses a discrete space to represent the state of each element of a domain, and this state can be changed according to a transition rule. The well-known cellular automaton Nasch model with modified cell size and variable acceleration rate is extended to two-lane cellular automaton model for traffic flow. A set of state rules is applied to provide lane-changing maneuvers. S-t-s rule given in the BJH model which describes the behavior of jammed vehicle is implemented in the present model and effect of variability in traffic flow on lane-changing behavior is studied. Flow rate between the single-lane road and two-lane road where vehicles change the lane in order to avoid the collision is also compared under the influence of s-t-s rule and braking rule. Using results of numerical simulations, we analyzed the fundamental diagram of traffic flow and show that s-t-s probability has more effect than braking probability on lane-changing maneuver.

1. Introduction

Cellular automata (CA) is a mathematical machine that arises from very basic mathematical principles. Though they are remarkably simple at the start, CA has variety of applications. It has been used extensively for modeling of single-lane traffic. Some modifications are required to extend these models to two-lane traffic as these generally fail to explain the lane-changing behavior. First traffic flow model using the concept of one-dimensional CA was given by Nagel-Schrekenberg popularly known as Nasch model [1]. The simple mean-field theory approach assumes that the two neighboring sites are completely uncorrelated. A simple model for two-lane traffic was also investigated, but the update rules were not defined in the same manner as in Nasch model. It was found that the fundamental diagram for each lane is asymmetric but the maximum is shifted towards large values of vehicular density ρ ($\rho_{max} > 1/2$). The two-lane cellular automaton model based upon the single-lane CA introduced by Rickert et al. was examined [2]. It was concluded that for both symmetric and asymmetric version, maximum flow q_{max} is higher than twice the maximum flow of single-lane model. An introduction of small number of slow vehicles can initiate formation of clusters at smaller densities [3]. For asymmetric lane-changing rules, slow vehicles influence the system performance less than in symmetric case. In two-lane traffic flow model, if the car density is set within a range, the self-organization of the slow and fast lanes was observed in spite of symmetry between two lanes [4]. Several branches and hysteresis in flow-density graph are observed. Results relative to a simple CA model without periodic boundary condition for a highway with variable number of on-ramps were presented [5]. A 2D extended version of the 1D Fukui-Ishibashi model, elaborated by Wang et al. [6], was presented for single-lane traffic to take into account the exchange of vehicles between the first and the second lane [6]. In general lane-changing rule can be symmetric or asymmetric with respect to the lanes or to the vehicles. While symmetric rules treat both lanes equally, asymmetric rule sets especially have to be applied for the simulation of German highways, where lane changes are dominated by right lane preferences and a right lane overtaking ban [7]. A new CA model by introducing the Honk effect into the basic symmetric two-lane CA model was proposed in [8]. The set of lane-changing rules (STCA) suggested by Chowdhury et al. [9] was revised to take the Honk effect into account (H-STCA). Lane-changing frequency for fast and slow vehicles for both models, STCA and H-STCA, was compared, and it was found that introduction of the Honk enhances the performance of the mixed vehicle traffic in the intermediate density range. A simple lattice-based exclusion model which can be considered as a crude representation of traffic on a two-lane motorway was introduced [10]. The model was two-lane generalization of the asymmetric simple exclusion process which is known to reproduce some of the features of the single-lane traffic such as shocks and jams. Effect of an aggressive lane-changing behavior on a two-lane road in presence of slow vehicles and fast vehicles has been further studied [11]. Simulation results show that aggressive lane-changing behavior of fast vehicles can depress the plug formed by slow vehicles and improve traffic flow in mixed traffic in intermediate density region. A highway traffic flow model with blockage induced by an accident vehicle was introduced in which both symmetric and asymmetric lane-changing rules were adopted [12]. It is concluded by numerical simulation that accident vehicle not only causes a local jam behind it, but also causes vehicles to cluster in the bypass lane. Further it is found that vehicles will change lane more frequently when the traffic is inhomogeneous with an accident car. In presence of a signalized intersection, existence of a certain combination of density ρ and cycle time which optimizes the traffic efficiency in a two-lane model due to overtaking is investigated [13].

In the present study the cell size is reduced and variable acceleration rate (rather than 1) is taken into account [14]. Here we consider two-lane traffic and adopt the symmetric lane-changing rules, which take into account the incentive and safety criteria. A slow-to-start (s-t-s) rule used in the widely known Benjamin-Johnson-Hui (BJH) CA model for single lane traffic simulation [15] is implemented to two-lane traffic simulation. BJH model is also discussed, and a slightly modified deceleration rule in order to simulate braking behaviors of vehicles more correctly is taken into account [2]. The slow-to-start rule in BJH model is applicable only to stationary vehicles, that is, vehicles that are completely blocked by the leading vehicle in the previous time step. This rule is not applicable to those vehicles which are stopped due to randomization in the previous time step. We choose BJH model in the present study for the reason that it has extremely simple transition rules which are easy to implement on 2-lane road. We investigate the effect of s-t-s and braking rules over lane-changing maneuver among vehicles in two-lane road and a detailed comparison of effect of braking probability and s-t-s probability over two-lane traffic flow is carried out using simulation. Simulation results show that variability in traffic flow has significant effect on lane-changing behavior of vehicle.

2. Cellular Automaton Model

The Nagel-Schrekenberg model is a probabilistic CA model for the description of single-lane highway traffic. Mathematically it is expressed by four rules given as follows.

Rule 1. Acceleration: $v_i^{(t+\delta t/3)} = \min\{v_i^{(t)} + 1, V_{\max}\}$.

Rule 2. Deceleration: $v_i^{(t+2\delta t/3)} = \min\{v_i^{(t+\delta t/3)}, x_{i+1}^t - x_i^t - 1\}$.

Rule 3. Randomization: $v_i^{(t+\delta t)} = \max\{v_i^{(t+2\delta t/3)} - 1, 0\}$.

Rule 4. Movement: $x_i^{(t+\delta t)} = x_i^t + v_i^{(t+\delta t)}$.

Where x_i^t and v_i^t are the position and speed of ith vehicle at a time t and $x_{i+1}^t - x_i^t - 1$, the number of empty cells in front of ith vehicle at time t, is called distance headway. A time step of $\delta t = 1$ sec, and the typical reaction time of driver with a maximum speed $V_{\max} = 5$ cells/time step, that is, 135 km/h is taken in this model. Cell size is taken as 7.5 meters and maximum speed $V_{\max} = 5$ cells/time step which indicates the minimum headway of two moving vehicles, and time interval is one second. This indicates that changing speed will only be 7.5 m/sec, 15 m/sec, and 22.5 m/sec, and so on. To overcome this problem different cell sizes are modeled by recording observation from real word, and a reduced cell size of 0.5 meters and variable acceleration rate that depends upon the speed of particular vehicle, are taken into account. Under this fine discretization we can describe the vehicle moving process more properly. A light vehicle occupies 12 cells with $V_{\max} = 60$ cells which correspond to 108 km/h whereas heavy vehicle occupies 20 cells with $V_{\max} = 40$ cells which correspond to 72 km/h. Rule 1 of Nasch model is modified as follows.

Rule 1. Acceleration: $v_i^{(t+\delta t/3)} = \min\{v_i^{(t)} + a, V_{\max}\}$.

Where acceleration a is determined as follows:

$$a = \begin{cases} 4, & \text{if } v_n \le 12, \\ 3, & \text{if } 12 < v_n \le 22, \\ 2, & \text{if } v_n > 22. \end{cases} \tag{1}$$

2.1. Two-Lane Cellular Automata Traffic Model. Most of the major roads are two-lane one-way roads. We consider a two-lane model with periodic boundary conditions, where additional rules defining the exchange of vehicles between the lanes are applied. For our model we adapt the Nasch model to provide vehicle movements. Any vehicle may perform lane-changing maneuver based on three criteria: incentive criterion, improvement criteria, and safety criteria [11].

Incentive criteria:

$$d_{i,} < \min(v_i + a, V_{\max}). \tag{2}$$

Improvement criteria:

$$d_{i,\text{other}} > d_i. \tag{3}$$

Safety criteria:

$$d_{i,\text{back}} > V_{\max}, \tag{4}$$

where $d_{i,\text{other}}, d_{i,\text{back}}$ denote the number of empty cells between the ith vehicle and its two neighbor vehicles in the other lane at time t, respectively.

3. Slow-to-Start Rule

We now implement a further rule to two-lane model, which is referred to as slow-to-start rule [16]. This s-t-s rule is given by Benjamin-Johnson-Hui and popularly known as BJH CA model. S-t-s rule is applicable only to static vehicles. Mathematically this rule can be defined as

$$\text{If } v_i^{(t)} = 0, \qquad x_{i+1}^{t-1} - x_i^{t-1} - 1 = 0,$$
$$v_i^{(t+\delta t/3)} = v_i^{(t+2\delta t/3)} = v_i^{(t+\delta t)} = 0, \tag{5}$$

where v_i^t is the speed of ith vehicle at time t and δt is the time interval. The above-mentioned s-t-s rule is applied only to the stopped vehicles having 0 headway in the previous time step with s-t-s probability q. It implies that s-t-s rule has no effect on the vehicles stopped due to randomization in the previous time step.

4. New Stochastic CA Model for Two Lanes

CA model developed so far with modified cell size and variable acceleration rate and with implementation of slow-to-start rule for two-lane traffic is given as

Rule 1.

$$v_i^{(t)} = 0, \qquad x_{i+1}^{t-1} - x_i^{t-1} - 12 = 0, \tag{6}$$

$$v_i^{(t+\delta t/3)} = v_i^{(t+2\delta t/3)} = v_i^{(t+\delta t)} = 0 \quad \text{with s-t-s probability } q. \tag{7}$$

Rule 2.

$$v_i^{(t+\delta t/3)} = \min\left\{v_i^{(t)} + a, V_{\max}\right\}. \tag{8}$$

Rule 3.

$$v_i^{(t+2\delta t/3)} = \min\left\{v_i^{(t+\delta t/3)}, x_{i+1}^t - x_i^t - 12\right\}. \tag{9}$$

Rule 4.

$$v_i^{(t+\delta t)} = \max\left\{v_i^{(t+2\delta t/3)} - 1, 0\right\} \quad \text{with braking probability } p. \tag{10}$$

Rule 5.

$$x_i^{(t+\delta t)} = x_i^t + v_i^{(t+\delta t)}. \tag{11}$$

Together with lane-changing rules:

$$x_{i+1}^t - x_i^t - 12 < \min\left(v_i^{t+\delta t/3}, V_{\max}\right) \tag{12}$$

$d_{i,\text{other}} > d_i$ and $d_{i,\text{back}} > V_{\max}$ with lane changing probability s.

5. Numerical Simulation

The numerical simulation is carried out with randomly generated initial configurations on a closed track containing 10,000 cells which represents a simulated road section of 5 km. The periodic boundary condition is that N vehicles were randomly distributed on both lanes. For each initial configuration of vehicles, results are obtained by averaging over 3600 time steps. For each density ρ, results are averaged over 10 different initial configurations.

The computational formulas used in numerical simulation are given as follows:

$$\bar{\rho_i} = \frac{1}{T} \sum_{t=1}^{t=T} n_i(t), \tag{13}$$

$$\bar{q_i} = \frac{1}{T} \sum_{t=1}^{t=T} m_i(t), \tag{14}$$

where (13) represents the density of the vehicles on the ith site over a time period T. $n_i(t) = 0$ if ith site is empty and $n_i(t) = 1$ if ith site is occupied by a vehicle at time t. Equation (14) represents flow of vehicles on ith site; $m_i(t) = 1$, if at time $t - 1$, there was a vehicle behind or at ith site, and at time t, it is found after ith site (i.e., a vehicle is detected passing by ith site). Density and flow are measured and averaged out over a time period of T.

6. Lane-Changing Behavior

The behavior of lane-changing criteria can be explained if two criteria are fulfilled to initiate lane change. First, the situation on the other lane must be more convenient and second the safety rules must be followed. A probability $(1 - s)$ is prescribed for vehicle to stay in the current lane (left lane). There are two other parameters: braking probability and slow-to-start probability. We investigate the effect of two parameters p and q on lane-changing behavior of vehicles. It can be shown by simulation that initially vehicles change lanes frequently and the lane changing rate drops rapidly as time evolves. Figures 1(a) and 1(b) show independent effects of s-t-s probability and braking probability when acting alone, respectively, over the lane-changing behavior of a periodic two-lane system, whereas Figure 1(c) shows the combined effect of two probabilities. With smaller values of p and q, vehicles rarely change lanes. With increase in the value of parameters p and q, the lane-changing tendency among the vehicle increases. This is due to the cluster formation at different locations on both lanes. With higher values of p and q, there will always be cluster formation and between the two clusters there is sufficient space on the right lane for vehicles to change the lane, and lane change becomes more likely. Therefore maximum lane changes occur even at higher values of traffic density ($\rho > \rho_c$). Here we choose the lane-changing probability $s = 0.8$, and it remains constant throughout the simulation. From Figures 1(a) and 1(b), it can be observed that an introduction of nonzero q has a stronger influence than nonzero p on lane-changing rate of vehicles in two-lane road. As described in BJH model, an introduction of nonzero s-t-s probability makes the queues less fragmented and the interqueue regions widen, as a result vehicles find enough gaps in the target lane to change the

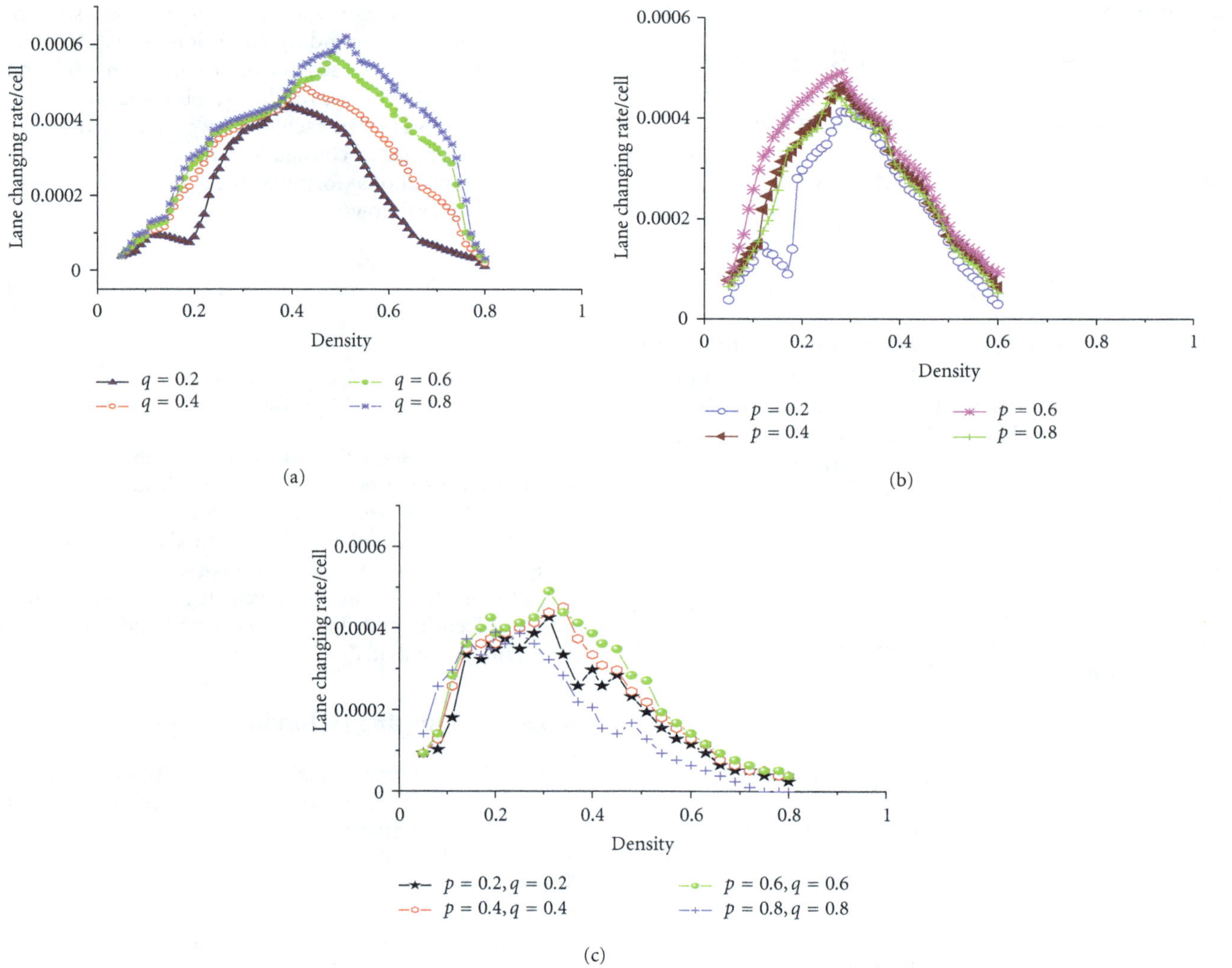

FIGURE 1: (a) Relationship between lane changing rate and density at braking probability $p = 0.0$. (b) Relationship between lane changing rate and density at s-t-s probability $q = 0.0$. (c) Relationship between lane changing rate and density at non zero values of braking probability p and s-t-s probability q.

lane. With higher values of parameters p and q, both criteria are fulfilled as the fluctuations of the distance between consecutive vehicles become larger. Figure 1(c) shows the lane-changing behavior of vehicles when both factors act together. With higher values of both parameters, safety criteria are not fulfilled and lane-changing rate again reduces drastically. Figure 2 describes the lane changing behavior of the Nasch model when simulated with implementation of s-t-s rule. Parameter q has more effect than parameter p on lane-changing behavior among vehicles. When both parameters are working together, the lane-changing tendency among vehicles increases. It can be seen that the results simulated from the Nasch model and modified cell model are close to each other except the magnitude of lane change rate/cell because of the reduced cell size.

7. Single Lane versus Two Lanes

A detailed comparison of the single-lane model with the corresponding two-lane model with effects of parameters p and

q is shown in Figure 3. The graph shows rise in maximum flow q_{max}, when simulating two-lane traffic as compared to one-lane system. This is due to the fact that some vehicles move into the other lane to avoid traffic jam and continuously contribute to flow. Now it is very interesting to observe how flow of the system is affected by the parameters p and q when acting alone. Since parameter p affects all the vehicles with equal probability, whereas parameter q affects only static vehicle that is, vehicles blocked by leading vehicles in the previous time step, the parameter p is more responsible than parameter q in reducing the throughput. However when values of p and q are high, this reduction can be significant. In the low-density region ($\rho < \rho_c$), the average velocity of the traffic system is close to maximum velocity V_{max}. Therefore there is not much difference in throughput of the system with nonzero p as shown in Figure 3. But with zero value of parameter p flow rises even in low-density region because parameter q comes into action only in high density region ($\rho > \rho_c$) as it affects only jammed vehicles.

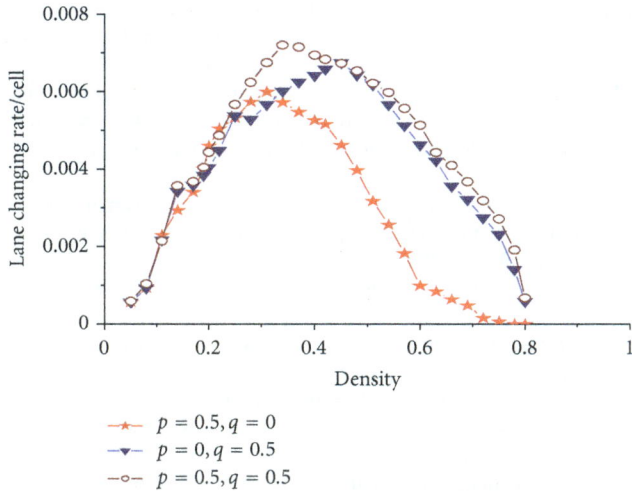

FIGURE 2: Relationship between lane-changing rate and density obtained from the Nasch model at various values of p and q.

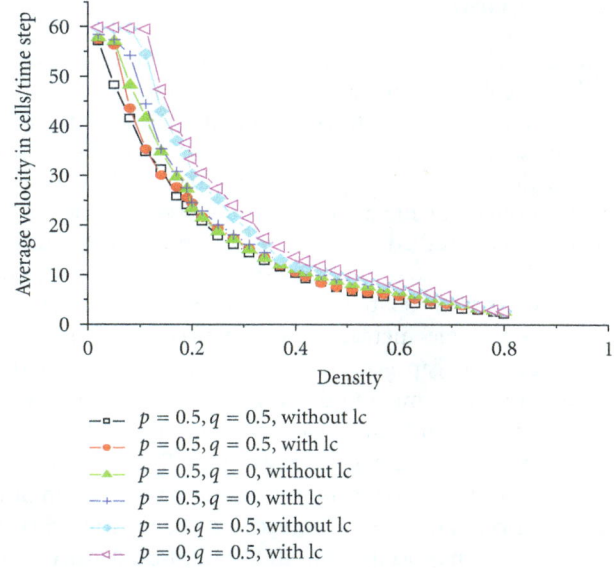

FIGURE 3: Relationship between density and flow with lane change and without lane change at various values of parameters p and q.

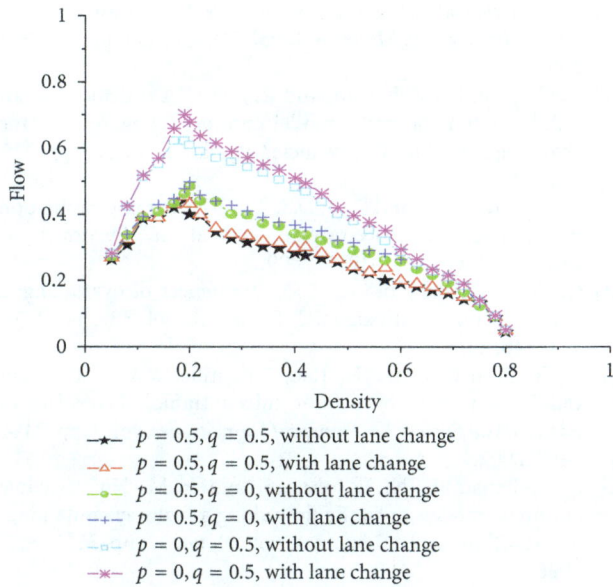

FIGURE 4: Relationship between density and average velocity with lane changing and without lane changing at various values of braking probability p and s-t-s probability q.

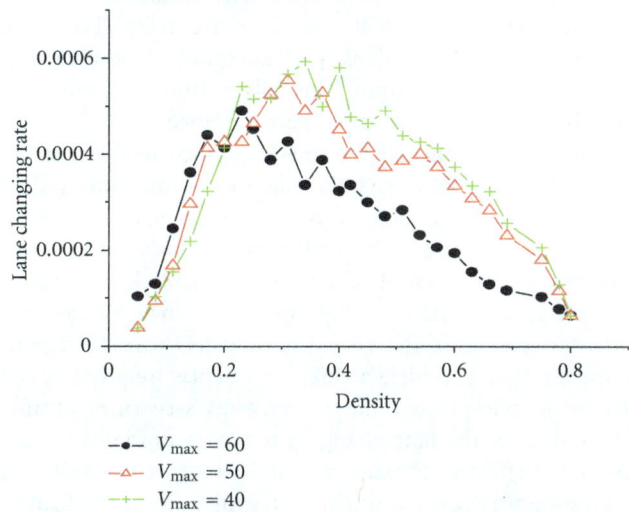

FIGURE 5: Relationship between density and lane-changing rate at various values of maximum velocity V_{max} with nonzero values of parameter p and q.

Figure 4 describes the effect of parameters p and q over average velocity of the two-lane system with maximum speed $V_{max} = 60$ cells/δt. In high-density region, average speed becomes the decreasing function of density. One can understand that average speed decreases linearly with the increase in the density of the system. But difference in speed variance is noticeable with implementation of nonzero parameters p and q. During the laminar flow phase, speed variance is negligible but starts to increase with onset of maximum flow. In the low-density region, average speed is near to maximum speed when parameter q acts alone and maximum speed variance is observed in maximum flow region. Thus with zero value of parameter p, the density of maximum flow is in the region of density of maximum

speed variance. But this result is not true for nonzero value of parameter p. In this case speed variance is observed even in low-density region. This is because parameter p comes into action long before the parameter q. Figure 5 shows the effect of maximum speed limit over lane-changing behavior of vehicles. In low-density region, increase in maximum speed limit V_{max} results in only limited amount of increase in lane-changing rate. But as density surpasses critical density ($\rho_c \sim 0.19$), the lane-changing rate decreases with V_{max}. This is due to the fact that in high-density region safety criteria are not fulfilled with higher values of V_{max}, that is, the condition for looking back in target lane for safety point of view is not fulfilled.

8. Conclusion

We have extended the BJH model to two-lane model with a reduced cell size and a variable acceleration rate. The reduced-cell-size CA model is more appropriate to describe the finer variability in traffic flow rather than the Nasch model with cell size $\delta x = 7.5\,\text{m}$. We investigated the effects of braking probability and s-t-s probability over lane-changing maneuver. We studied via simulation how vehicles fulfill both the incentive and safety criteria with higher values of braking and s-t-s probabilities. Combined effect of braking rule and s-t-s rule increases the effectiveness of the lane changes because gap acceptance is increased between the vehicles. We also compared the flow and average velocity of the two-lane system with single lane under the influence of braking probability and s-t-s probability. S-t-s probability has more effect than braking probability on lane-changing rate. It is also interestingly observed that the maximum lane changing frequency occurs long after the critical density ρ_c of maximum throughput. Lane-changing frequency increases with introduction of nonzero parameters p and q. This means that lane changing does little to increase throughput. Since more frequent lane changing means an increase in the likelihood of traffic accidents, the traffic should be operated at lower densities. In fact, the desired density should be smaller than both the density of maximum lane changing and maximum throughput. This will ensure traffic with few lane changes and with a small speed variance.

We also investigated the speed variance near the maximum flow and observed that density of maximum flow is in the region of density of maximum speed variance in case of nonzero values of braking probability. Thus presence of any amount of variability is sufficient to result in high speed variance. High speed variance means that different vehicles in the system have widely varying speeds. It means that a vehicle would experience frequent speed change per trip through the system. Since s-t-s rule in traffic flow enhances the lane-changing tendency among vehicles, combined effect of braking rule and s-t-s rule is significant in high-density region. Actually s-t-s rule reflects the feature of real driving and is distinct from general disorder rule. Therefore combined study of disorder rule with s-t-s rule is necessary for a safety point of view. Present model reveals all the features of two-lane traffic flow.

Acknowledgment

This work is partially supported by a grant-in-aid from the Council of Scientific and Industrial Research, India.

References

[1] K. Nagel and M. Schrekenberg, "A cellular automaton model for freeway traffic," *Journal de Physique I*, vol. 2, no. 12, pp. 2221–2229, 1992.

[2] M. Rickert, K. Nagel, M. Schreckenberg, and A. Latour, "Two lane traffic simulations using cellular automata," *Physica A*, vol. 231, no. 4, pp. 534–550, 1996.

[3] W. Knospe, L. Santen, A. Schadschneider, and M. Schrekenberg, "Disorder effects in cellular automata for two lane traffic," *Physica A*, vol. 265, no. 3-4, pp. 614–633, 1998.

[4] A. Awazu, "Dynamics of two equivalent lanes traffic flow model: self- organization of the slow lane and fast lane," *Journal of Physical Society of Japan*, vol. 64, no. 4, pp. 1071–1074, 1998.

[5] E. G. Campri and G. Levi, "A cellular automata model for highway traffic," *The Europian Phyisical Journal B*, vol. 17, no. 1, pp. 159–166, 2000.

[6] L. Wang, B. H. Wang, and B. Hu, "Cellular automaton traffic flow model between the Fukui-Ishibashi and Nagel-Schreckenberg models," *Physical Review E*, vol. 63, no. 5, Article ID 056117, 5 pages, 2001.

[7] W. Knospe, L. Santen, A. Schadschneider, and M. Schreckenberg, "A realistic two-lane traffic model for highway traffic," *Journal of Physics A*, vol. 35, no. 15, pp. 3369–3388, 2002.

[8] B. Jia, R. Jiang, Q. S. Wu, and M. B. Hu, "Honk effect in the two-lane cellular automaton model for traffic flow," *Physica A*, vol. 348, pp. 544–552, 2005.

[9] D. Chowdhury, L. Santen, and A. Schadschneider, "Statistical physics of vehicular traffic and some related systems," *Physics Report*, vol. 329, no. 4-6, pp. 199–329, 2000.

[10] R. J. Harris and R. B. Stinchcombe, "Ideal and disordered two-lane traffic models," *Physica A*, vol. 354, no. 1–4, pp. 582–596, 2005.

[11] X. G. Li, B. Jia, Z. Y. Gao, and R. Jiang, "A realistic two-lane cellular automata traffic model considering aggressive lane-changing behavior of fast vehicle," *Physica A*, vol. 367, pp. 479–486, 2006.

[12] H. B. Zhu, L. Lei, and S. Q. Dai, "Two-lane traffic simulations with a blockage induced by an accident car," *Physica A*, vol. 388, no. 14, pp. 2903–2910, 2009.

[13] C. Chen, J. Chen, and X. Guo, "Influences of overtaking on two-lane traffic with signals," *Physica A*, vol. 389, no. 1, pp. 141–148, 2010.

[14] C. Mallikarjuna and K. Rao, "Identification of a suitable cellular automata model for mixed traffic," *Journal of the Eastern Asia Society for Transportation Studies*, vol. 7, pp. 2454–2468, 2007.

[15] S. C. Benjamin, N. F. Johnson, and P. M. Hui, "Cellular automata models of traffic flow along a highway containing a junction," *Journal of Physics A*, vol. 29, no. 12, pp. 3119–3127, 1996.

[16] A. Clarridge and K. Salomaa, "Analysis of a cellular automaton model for car traffic with a slow-to-stop rule," *Theoretical Computer Science*, vol. 411, no. 38-39, pp. 3507–3515, 2010.

4

A New Movement Recognition Technique for Flight Mode Detection

Youssef Tawk,[1] **Aleksandar Jovanovic,**[1] **Phillip Tomé,**[1] **Jérôme Leclère,**[1]
Cyril Botteron,[1] **Pierre-André Farine,**[1] **Ruud Riem-Vis,**[2] **and Bertrand Spaeth**[2]

[1] *Electronics and Signal Processing Laboratory, Institute of Microengineering (IMT), École Polytechnique Fédérale de Lausanne, Breguet 2, 2000 Neuchâtel, Switzerland*
[2] *Jiiva, Stadtbachstrasse 40, 3012 Bern, Switzerland*

Correspondence should be addressed to Youssef Tawk; youssef.tawk@epfl.ch

Academic Editor: Aboelmagd Noureldin

Nowadays, in the aeronautical environments, the use of mobile communication and other wireless technologies is restricted. More specifically, the Federal Communications Commission (FCC) and the Federal Aviation Administration (FAA) prohibit the use of cellular phones and other wireless devices on airborne aircraft because of potential interference with wireless networks on the ground, and with the aircraft's navigation and communication systems. Within this context, we propose in this paper a movement recognition algorithm that will switch off a module including a GSM (Global System for Mobile Communications) device or any other mobile cellular technology as soon as it senses movement and thereby will prevent any forbidden transmissions that could occur in a moving airplane. The algorithm is based solely on measurements of a low-cost accelerometer and is easy to implement with a high degree of reliability.

1. Introduction

GSM localization or mobile phone tracking is a technology used to locate the position of a mobile phone. Localization uses the concept of multilateration of radio signals, where the phone must communicate wirelessly with at least three of the nearby radio base stations (RBSs). Knowing the position of the RBS's, and using a triangulation method, an approximation of the geographical location of the mobile phone can be calculated. This technology is based generally on four different techniques: network based, handset based, hybrid and subscriber identity module (SIM) based [1]. The SIM-based technique is of interest in this paper, where by using the SIM in mobile communication handsets it is possible to obtain raw radio measurements that include the serving cell ID, round trip time and signal strength. Different applications already use this service for localization, for example, resource tracking with dynamic distribution such as taxis, rental equipment, or fleet scheduling.

Within this context, Swisscom AutoID Services (SIS) in collaboration with La Poste Suisse aims to pioneer a service

that will allow clients to be able to track and trace their packages in near real-time mode anywhere in the world. The service will consist of including a tracker with the package that operates on the GSM network for location and communication [2]. Therefore, a key issue arises when the package is transported via airplane. Indeed, it is well known that aircraft remains one of the few places where the use of mobile communication signals is prohibited [3, 4]. In fact, the aircraft is not a good Faraday cage and cannot prevent transmissions to reach terrestrial cellular networks. Moreover, there is a possibility of interference with some avionics instruments in the aircraft. As a result, international regulation bodies have forbidden the use of mobile phones onboard an aircraft and hence any module used for tracking a package should be turned off as soon as it is located inside an airplane. In order to respect these regulations, the main idea we develop in this paper is to consider the use of low cost MEMS accelerometers, such as those found in today's smartphones, to allow the tracker to detect the movement and to turn off the GSM module automatically without any human intervention. For the envisioned application, this means that once a shipment equipped with a tracker is in the moving aircraft, the device will detect the aircraft's movements; it will automatically switch off the GSM module until the aircraft has come to a halt after landing and the device detects a static condition, allowing it to turn on the GSM module again.

In the recent literature, several papers already proposed the use of smartphone accelerometers for movement recognition. For example in [5], the authors investigated the suitability of the built-in smartphones accelerometer to provide good accuracy for common human activity recognition. In [6], the authors introduced a new method to implement a motion recognition process using a mobile phone fitted with an accelerometer to effectively recognize different human activities with a high-level accuracy. Human movement detection using an accelerometer was also the subject for several other papers [7–11]. The accelerometer was also used for several types of movement detection, for example, sign language detection [12], handwritten digit recognition [13], fast fall detection [14], and wireless motion sensing [15]. In these cases, the detection is based on different techniques and the degree of algorithm complexity varies depending on the application, alongside with the quality of the accelerometer. However, none of the above papers considered the use of smartphones' accelerometers for flight mode detection. Therefore, this paper fills a gap in the current literature by proposing a new movement recognition algorithm specifically tailored for flight mode detection, that is, taking in consideration the reliability requirements that must be satisfied for civil aviation applications. In particular, the proposed algorithm is specially conceived for detecting any movement that a package can be subject to inside an airplane, with a sufficiently high reliability to respect all the specifications and requirements from international airspace regulations. The analytical and statistical studies of the proposed algorithm are based on raw acceleration measurements acquired from static data and from dynamic data onboard different flights. These measurements are used to define the parameters required by the algorithm to detect either static or dynamic condition. Furthermore, the reliability is analyzed in terms of probability of misdetection and false alarm. In addition, implementation complexity of the algorithm is evaluated in terms of response time and resources, and finally power consumption considerations are presented.

It is important to note that the main contribution of the paper is not in devising the methods used in the algo-rithm, but instead results from the combination of these methods in one algorithm that is tailored to provide a highly reliable movement detection technique using low cost MEMS accelerometers. Moreover, the proposed algorithm is not intended to classify the source of movement (e.g., airplane, train, car, truck, etc.), as this would decrease the reliability of correctly detecting movement while in an airplane. Indeed, it is sufficient for the considered package tracking application to only get information about the position of the package when it is not moving and prohibit transmissions otherwise. Finally, the paper is not mainly involved from the theoretical point of view, but it has a clear engineering application while emphasizing the use of real acceleration measurements. For

TABLE 1: Main characteristics of LIS302DL accelerometer [16].

Noise ($1\,\sigma$)	2.23 mg
Measurement range (g)	±2 or ±8
Output data rate (Hz)	100 or 400
Temperature range	−40°C to +85°C
Power consumption	<1 mW

more details about the final product where the proposed algorithm is part of, the interested reader is referred to [2].

2. Measurement Data Collection

In order to analyze the accelerometer behavior under different conditions a database was created including acceleration measurements from different experiments consisted of real flights, of different duration and in different airplanes. An accelerometer is a device that measures the acceleration associated with the phenomenon of weight experienced by a test mass that resides in the frame of reference of the accelerometer device. The accelerometer chosen for conducting the analyses in this paper is the LIS302DL accelerometer, the same sensor as the one integrated inside an iPhone [16]. This accelerometer is an ST ultracompact low-power three-axis linear accelerometer. Its main characteristics are shown in Table 1.

The raw measurements were collected by putting an iPhone, which is set to flight-mode, inside a handbag and taking it onboard different flights. The X, Y, and Z measurements from the built-in accelerometer are saved using an application on the phone with a sampling frequency of 100 Hz. The recorded measurements are stored later using a data logging software in a database with a resolution of 16 bits. The measurements were taken before, during, and after every flight. In addition, with each series of flight data, there is a metadata description containing the flight duration time, the take-off time, landing time, as well as description about the turbulences, and other conditions occurring during the flight and all other data necessary to detect transitions and unusual events. The metadata were collected through the observations of the person holding the accelerometer onboard the flight. Hence, it has a couple of seconds of time accuracy, which is not critical because the proposed algorithm is independent of this information that is only used to better understand the behavior of the accelerometer. An example of a raw measurement collected is displayed in Figure 1, which shows the accelerometer three-axis output before, after, and during a flight from Geneva to London onboard an Airbus A319. The flight time is defined as the time from the push-back manoeuvre before takeoff until the time the plane engines are turned off after landing, which is the time when the GSM module should not transmit any signals. It can be noted that outside the flight time the accelerometer axes orientation is often changing and the measurements are so noisy. This is due to the fact that the used iPhone was inside a handbag that is subject to all types of movements that occur during normal activities of a travel.

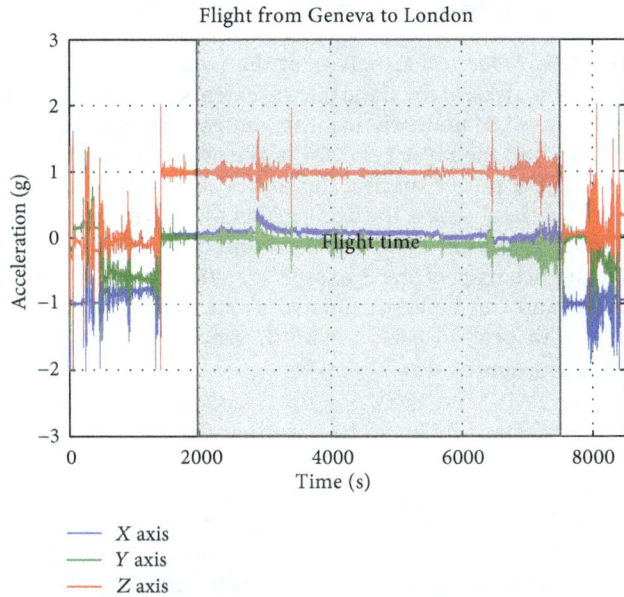

FIGURE 1: Acceleration measurements.

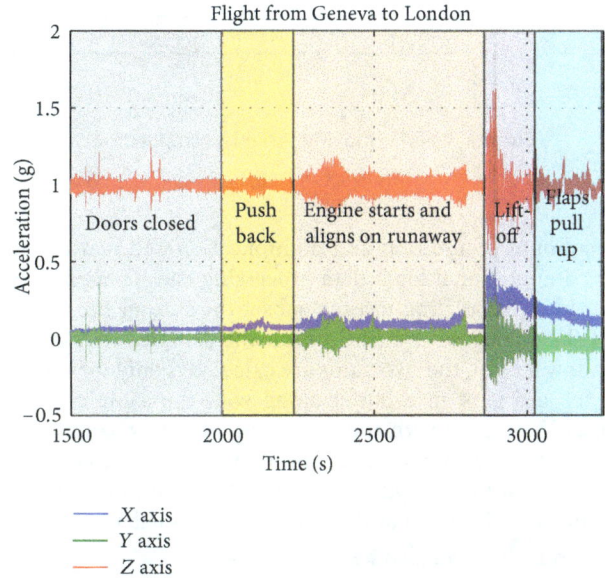

FIGURE 2: Takeoff metadata identification where the three traces represent the 3-axis acceleration during takeoff.

Static data was also collected when the accelerometer was not moving for a long period of time and stored in the database.

The first step in order to interpret the data collected was to identify the metadata provided with each set within the sequence of acceleration measurements. For example, for the flight above (to be considered throughout the paper), the sudden sharp jumps during flight time are due to strong turbulences. Also, the different parts of the takeoff manoeuvre are shown in Figure 2 where each part is identified based on the metadata provided. This helped to better understand the behavior of the accelerometer under different events. For instance, it can be seen that once the "aircraft push back maneuver" starts and the engines are turned on, the accelerometer outputs start to fluctuate and we can see immediately the change in the static condition of the sensor. Also the acceleration of the plane in the runway and the liftoff are identified easily as the amplitude of the acceleration on the 3 axes changes suddenly.

This short overview on the raw acceleration measurements shows that as soon as the sensor senses a movement, the amplitude and the variance of the accelerometer outputs increase. Bearing this in mind and computing a parameter that is dependent on these two values can provide a criterion to find the status of the sensor. Consequently, we chose the moving variance (MV) of the incoming measurements on the three axes as this parameter. But before analyzing the collected data with respect to this parameter, a short overview of the MV is given in the next section.

3. Moving Variance

A moving variance is used to analyze a set of data points by creating a series of variances of different subsets of the full data set [17]. Given a series of numbers and a fixed subset

size, the MV can be obtained by first taking the variance of the first subset. The fixed subset size is then shifted forward, creating a new subset of numbers, and consequently a new variance is computed. This process is repeated over the entire data series. The plot line connecting all the (fixed) variances is the moving variance. Thus, an MV is not a single number, but it is a set of values, each of which is the variance of the corresponding subset of a larger set of data points. An MV may also use unequal weights for each data value in the subset to emphasize particular values in the subset.

In this paper, two types of MV computations are used: (1) the normal moving variance (NMV) and (2) the exponential moving variance (EMV) also known as the exponentially weighted moving variance. The difference between the two is that the NMV calculates the variance of the fixed subsets without any weighting, in other words all the data in a specific subset have the same importance. As for the EMV, it applies weighting factors which decrease exponentially [17]. The weighting for each older data point decreases exponentially, never reaching zero. Alternatively, the EMV gives more weighting to the new data in the computation of the subset variance.

The formula for calculating the NMV is given by [17]:

$$\sigma_{\text{NMV}}^2(n) = \frac{1}{N} \sum_{k=n-N+1}^{n} [x(k) - \overline{x}(n)]^2$$

$$= \left[\frac{1}{N} \sum_{k=n-N+1}^{n} x^2(k)\right] - \overline{x}(n) = \overline{x^2}(n) - \overline{x}(n),$$

(1)

where $\overline{x}(n)$ corresponds to the average of the subset, $\overline{x^2}(n)$ is the sum of the squares over the number of samples in the subset N, and n is the subset's index number. To reduce

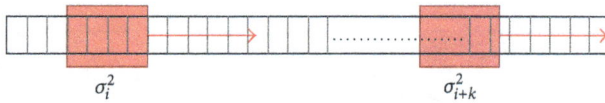

FIGURE 3: NMV moving variance computation.

the amount of addition and multiplication and hence lower the computational load, data processing can be applied. So, instead of calculating in each subset the sum of the squares and the square of the sum for all the samples over a fixed window length, the two sums are calculated only for the first subset and kept in a buffer along with the samples in the subset. Moreover with the upcoming of a new sample (i.e., moving to a new subset), the sample and its square are added to the two sums computed before and the old sample and its square are subtracted and a new variance is calculated. Thus $\overline{x}(n)$ and $\overline{x^2}(n)$ can also be expressed as

$$\overline{x}(n) = \frac{x(n) - x(n - N + 1)}{N} + \overline{x}(n - 1), \qquad (2)$$

$$\overline{x^2}(n) = \frac{x^2(n) - x^2(n - N + 1)}{N} + \overline{x^2}(n - 1). \qquad (3)$$

At the next iteration, the old sample is dropped from the buffer and the new sample replaces it. This operation is repeated at every new measurement (i.e., overlapping windows), and a new moving variance is computed as shown in Figure 3.

The EMV uses the same overall concept, but it differs in two main points: (1) it does not require the storage of all the samples of a subset, but only the last result and (2) it gives more weighting to new measurements. The EMV computation can be summarized as follows: there are two variables that need to be saved in the memory, the average and the variance of the samples within a subset. With the arrival of a new sample (i.e., moving to a new subset) the old average and variance are weighted and a new average and variance are computed taking in consideration the new sample. This way the samples do not need to be saved in the memory in a subset and consequently all what is needed for memory are two variables, the previous average and the variance. The formulas to calculate the EMV are given by [17]

$$\overline{x}_{\text{EMV}}(n) = \frac{(N - \alpha)\overline{x}_{\text{EMV}}(n - 1) + \alpha x(n)}{N}, \qquad (4)$$

$$\sigma^2_{\text{EMV}}(n) = \frac{(N - \alpha)\sigma^2(n - 1) + \alpha(x(n) - \overline{x}_{\text{EMV}}(n))^2}{N}, \qquad (5)$$

where α is the weighting factor, whose value depends on many factors as it will be seen in the following sections. In the following, both NMV and EMV of the raw acceleration measurements are computed, and performance comparison of these two methods are provided.

4. Acceleration Measurement Analysis

4.1. Static Mode. Our analysis of the recorded data started with the analysis of the static measurements of the accelerometer. Around 35 hours static measurements were collected in order to find the noise behavior of the accelerometer. Figure 4 shows the power spectral density of each axis where it is clear that the Y-axis is the axis sensing the gravity presenting a higher amplitude component around 0 Hz. It can be seen that the behavior of the 3 axes in a static condition is very similar and can be approximated as an independent white noise with peak-to-peak amplitude around 0.01 mg, and a variance around $5 \times 10^{-6}\,\text{g}^2$.

The NMV and EMV for the 3 accelerometer axes are shown in Figure 5. It can be seen that the NMV responses on the 3 axes are similar to the noise variance of the accelerometer and approximately equal to $5 \times 10^{-6}\,\text{g}^2$. As for the EMV, it can be noted that for the weighting factors of 1 and 5, the static behavior is very similar to the case of NMV and hence it is consistent with the noise level of the accelerometer. For a weighting factor of 10, we can see that the EMV has a slightly noisier response. This is due to the fact that the EMV gives more weighting to the new samples and in this case for a high weighting factor, the past information of the variance is less weighted than the new information that is added, and therefore, there is a loss in the smoothing effect on the variance computation.

Based on the results shown above, it can be seen that the output of the NMV and EMV in static mode can approximately provide the noise level of the sensor, and this value can be taken as a reference later to decide whether the accelerometer is in a static or dynamic mode.

4.2. Dynamic Mode. As in the static case, the PSD of the acceleration measurements was first computed in order to check if a low quality sensor as the one used for the tests herein could detect small movements and more specifically the vibrations of an airplane's motor. The power spectral densities of the measurement norm of the three accelerator axes during flights onboard different airplanes are shown in Figure 6. It is interesting to notice the existence of frequency peak components between 10 Hz and 30 Hz during all flights' stages. These peaks can be identified as the vibration frequency of the airplane and typically they characterize the vibration frequency of the motors. Analyses of different series with different types of planes show that most of the aircraft vibration frequencies and dynamic frequencies are within the band ranging from 0 Hz up to ~30 Hz. Therefore, the sampling frequency that has been used for the analysis (100 Hz) could be lowered down to 60 Hz if necessary. Thus, decreasing the sampling frequency in between 60 and 70 Hz seems feasible, and within this range all the vibration and dynamic frequencies are preserved.

Regarding the NMV and EMV of the dynamic flight mode, Figure 7 shows the NMV computation for the 3 accelerometer axes of the flight mentioned in Section 2. The existence of peak components with high amplitudes that represent sudden movements of the sensor can be noticed. A

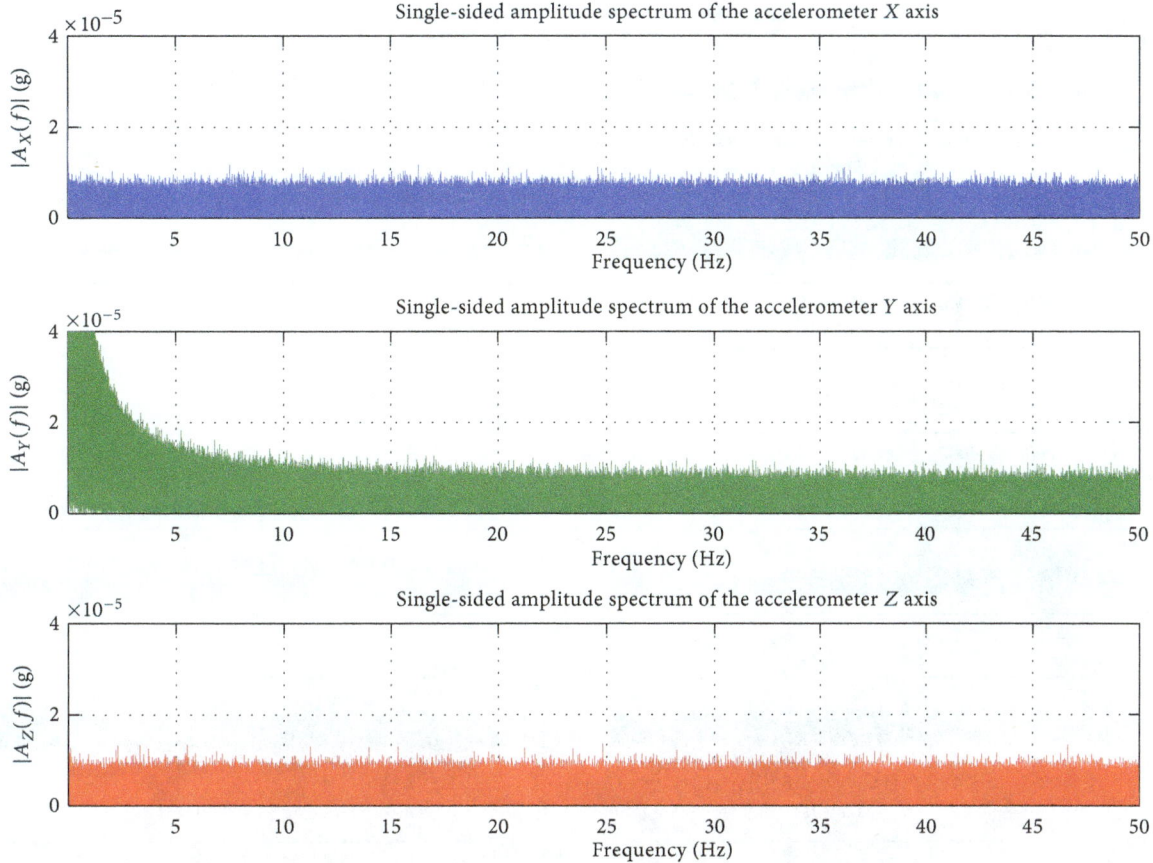

FIGURE 4: Power spectral density of the raw acceleration measurements in a static case.

closer look at the takeoff and landing phases during the flight is shown in Figure 8.

During the takeoff, it can be seen that as soon as the engines are on, the NMV outputs on the 3 axes increase, and also it can be noted the existence of high peaks during the plane acceleration and liftoff. During the landing, it can be seen that the NMV outputs have high amplitude peak components and, as soon as the engines are off, the outputs on the 3 axes decrease momentarily. Results of the EMV are similar to those of NMV and hence are not shown here.

Overall, these graphs show that both the NMV and EMV are able to detect movement of the sensor under different conditions. As a result, it can be concluded that by using one of these two methods, the state of the sensor can be found. Thus, the proposed algorithm for movement detection is based on these analyses and described in the next section.

5. Proposed Algorithm

The proposed algorithm for the detection of static/dynamic condition of the sensor is based on the computation of the moving variance of the raw accelerometer measurements as described before. The overall architecture is illustrated in Figure 9 and was implemented in Matlab in a first step. The algorithm can work in real time directly on measured data coming from the accelerometer, but as the analyses conducted during this work were limited to offline data processing, the measurements were taken from the database and postprocessed.

The algorithm starts by reading measurements during a specific window size (W_s) of time and computes the moving variance (MV) of the measurements for each accelerometer axis separately:

$$\text{MV}_i(n) = f\left(a_i(n-N), a_i(n-N+1), \ldots, a_i(n)\right), \quad (6)$$

where i represents the X, Y, or Z axis, N is the window size, f stands for a normal (NMV) or exponential (EMV) moving variance computed in (1) or (3), n is the subset's index number where the MV is computed, and a is the acceleration measurement during the specified period of time. Afterwards, two conditions are performed to set the status of the sensor. The first one is to check periodically every W_s second if the MV of each axis is smaller than a predefined threshold, as shown by

$$\text{cond}_\sigma^n = \left(\text{MV}_x(n) < T_\sigma^x\right) \text{ AND } \left(\text{MV}_y(n) < T_\sigma^y\right)$$
$$\text{AND } \left(\text{MV}_z(n), < T_\sigma^z\right), \quad (7)$$

where $\sigma_{m,\text{th}}^i$ is the MV threshold on the i axis. If this is not fulfilled, the condition is set to "nonstatic" and the next window measurements are read. Otherwise, the condition is

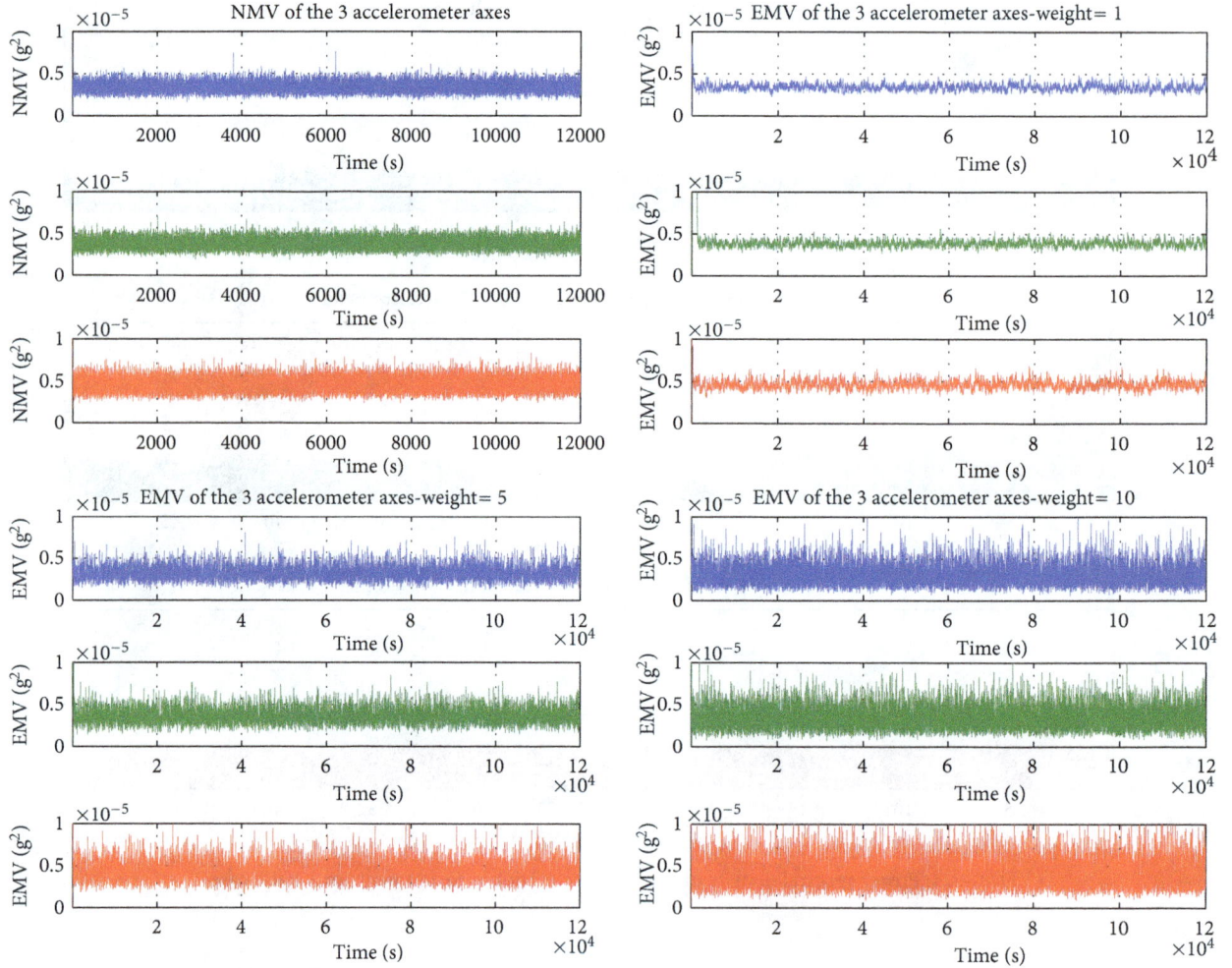

FIGURE 5: NMV and EMV of the 3 accelerometer axes in a static condition for a window size of 1 s or $N = 100$ samples.

set to a "temporary static condition" and another threshold related to time is being checked. This second threshold is performed in order to obtain a more secure and stable response. In fact, once a "temporary static condition" is declared, the algorithm goes backwards for a specific duration of time and examines if the status of the sensor within this duration was also "temporary static condition"

$$\text{cond}_{\text{time}}^n = \left(\text{cond}_\sigma^n\right) \text{ AND } \left(\text{cond}_\sigma^{n-1}\right) \text{ AND} \ldots \left(\text{cond}_\sigma^{n-T_{\text{th}}}\right), \quad (8)$$

where T_{th} represents the threshold in time or the duration of time in which the algorithm will look backward for a temporary static condition. If this is not fulfilled, a "nonstatic" condition is declared and the next window measurements are read. If the condition is fulfilled, the algorithm outputs a "static condition" and consequently at this time the module is allowed to transmit a wireless mobile communication signal if needed. T_{th} typically ranges from 1 up to several seconds. For example, if the threshold in time is set to 3 seconds and the MV condition is checked every 1 second, then in case of a "temporary static condition," the algorithm checks the state of the previous three conditions. If all the three of them were

of "temporary static condition," then the sensor is considered to be in a static state. Otherwise, the sensor is considered to be in dynamic state.

5.1. *Probability Analysis and MV Threshold Determination.* Our movement recognition algorithm can be seen as a binary detection problem, that is, deciding between two hypotheses whether the sensor is in a static or a dynamic mode. As a result, a statistical test, more specifically a binary hypothesis-testing problem [18], is set up. The two hypotheses are defined as H_0 where the sensor is in a dynamic mode and H_1 where the sensor is in a static mode. A binary test of H_0 versus H_1 takes the following form:

$$\phi(n) = \begin{cases} 0 \sim H_0, & \text{cond}_{\text{time}}^n \text{ is false}, \\ 1 \sim H_1, & \text{cond}_{\text{time}}^n \text{ is true}. \end{cases} \quad (9)$$

This equation can be read as the test function $\phi(n)$ equals 1, that is hypothesis H_0 is rejected and H_1 is accepted, if $\text{cond}_{\text{time}}^n$ is true (i.e., the MV on each of the accelerometer axes is smaller than the corresponding MV threshold during the duration of time specified by T_{th}). Otherwise (i.e., if at

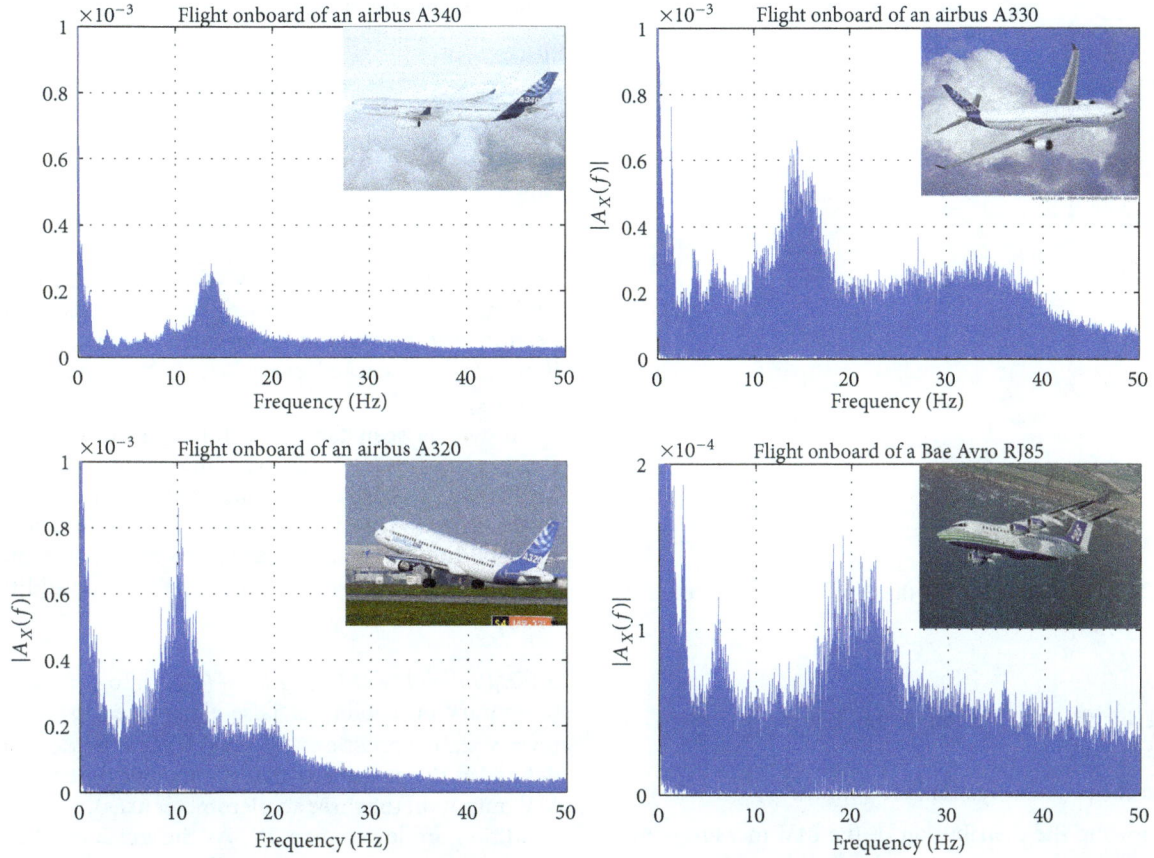

FIGURE 6: Single-sided amplitude spectrum of the accelerometer measurements norm during flights onboard different airplanes types.

least the MV of one of the accelerometer axes is higher than the MV threshold), $\text{cond}_{\text{time}}^n$ is false and the test function equals zero, that is hypothesis H_1 is rejected and H_0 is accepted. Figure 10 shows the binary hypothesis test and its corresponding probabilities.

If the sensor is static, then the probability of detection P_D is defined as the probability that $\text{cond}_{\text{time}}^n$ is true and H_1 is accepted (i.e., detecting final static condition), and the probability of misdetection P_M is defined as the probability that $\text{cond}_{\text{time}}^n$ is false and H_0 is accepted (i.e., detecting dynamic condition). They can be expressed as

$$P_D = P_{H_1}\left[\phi(n) = 1\right],$$
$$P_M = 1 - P_D = P_{H_1}\left[\phi(n) = 0\right]. \tag{10}$$

If the sensor is dynamic, then the probability of false alarm P_{FA} is defined as the probability that $\text{cond}_{\text{time}}^n$ is true and H_1 is accepted (i.e., detecting final static condition). It can be expressed as

$$P_{FA} = P_{H_0}\left[\phi(n) = 1\right]. \tag{11}$$

Note that our focus is on selecting a threshold that will lead to a sufficiently low P_{FA} that is compliant with the civil aviation regulations. For our particular application, the maximization of P_D (and thus minimization of P_M) is not so critical, as

long as static condition detection can still happen regularly to allow periodic location determination.

In the next two subsections, we seek to determine the probability density function for the NMV and EMV outputs in static and in dynamic mode, respectively, in order to apply the well-known classical detection theory (see, e.g., [18]) to theoretically determine a suitable threshold.

5.1.1. Probability of Misdetection (P_M). As shown in Section 4.1, the behavior of the 3 accelerometer axes while it is static is similar, with the variance on each axis approximately equal to $\sigma^2 = 5 \times 10^{-6}\,g^2$. The NMV and EMV outputs in this case are also very similar. Therefore, it is possible to define the same MV threshold for the three axes, that is,

$$T_\sigma^x = T_\sigma^y = T_\sigma^z = T_\sigma. \tag{12}$$

This threshold should be chosen as a function of two parameters: the noise behavior of the accelerometer and the probability of misdetection P_M, which is related to the distribution of the accelerometer measurements while it is static. In order to find the MV threshold, the sensor should be set in a static mode, and the subsequent approach can be followed:

(1) measure the noise variance on each axis of the accelerometer,

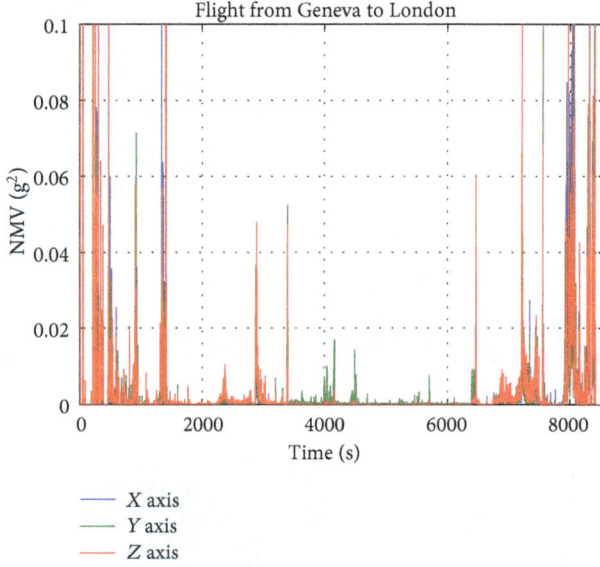

FIGURE 7: NMV computation for the 3 accelerometer axes of the flight from Geneva to London.

(2) derive the distribution of the MV measurements on each axis,

(3) set the probability of misdetection P_M,

(4) knowing the distribution of the MV measurements, then the theoretical formula of PM can be derived, and an MV threshold can be computed in function of the noise variance.

Once an MV threshold is computed, the correspondent P_{FA} should be derived to check if it is below the maximum value defined by the civil aviation regulations. If not, then P_M is increased and a new MV threshold is computed.

In the case of the accelerometer chosen in this paper [16], the output of the NMV or EMV algorithm applied to nonoverlapping datasets of the static accelerometer raw measurements can be approximated as independent and identically distributed (i.i.d.) with a Gaussian distribution, as can be seen in Figure 11 that depicts the probability density function (pdf) and the cumulative density function (cdf) for the Y axis. The measurements for the two other axes and from other static data sets have very similar probability distribution as the one plotted here and are thus not shown.

The probability of misdetection P_M in this case can be written as [18]

$$
\begin{aligned}
P_M &= P_{H_1}\left[MV \geq T_\sigma\right] \\
&= 1 - \theta\left(T_\sigma\right) \\
&= \frac{1}{2}\left[1 - \mathrm{erf}\left(\frac{T_\sigma - \mu}{\sqrt{2\sigma_{MV}^2}}\right)\right],
\end{aligned}
\tag{13}
$$

where μ, σ_{MV}^2, and $\theta(x)$ are the mean, the variance, and the cdf of the MV distribution, and erf is the error function.

TABLE 2: Probability of misdetection versus different MV thresholds.

T_σ	P_M	Time_{Error}
$2\sigma^2$	6×10^{-3}	166 s
$2.5\sigma^2$	8×10^{-6}	35 hours
$3\sigma^2$	2×10^{-11}	1585 years
$4\sigma^2$	6×10^{-19}	5×10^{10} years

Table 2 shows the probability of misdetection for different MV thresholds, where Time_{Error} corresponds to the period of time for the algorithm to missdetect a dynamic condition if the MV is checked every 1 s, that is Time_{Error} is equal to $1/P_M$.

It can be seen that for a threshold of $2.5\sigma^2$ or above, P_M is already very small, yielding sufficiently large values for Time_{Error} (considering our application). We now look at the threshold requirements to yield a sufficiently small probability of false alarm when we are in dynamic mode, starting with the above thresholds as well as an additional one of $5\sigma^2$.

5.1.2. Probability of False Alarm (P_{FA}). To compute P_{FA}, the temporary probability of false alarm (i.e., to detect a temporary static condition) P_{FA}^t should first be derived. P_{FA}^t is defined as the probability that at a specific time the computed MV outputs of the three accelerometer axes during a dynamic condition are lower than T_σ. As the noise on the 3 axes is uncorrelated (i.e., consecutive MV output measurements and measurements from different axes are assumed independent since we assume that they are obtained from nonoverlapping datasets of the static accelerometer raw measurements), then P_{FA}^t and P_{FA} can be expressed as

$$
\begin{aligned}
P_{FA}^t &= P_{H_0}\left[MV \leq T_\sigma\right] = P_{FA}^x P_{FA}^y P_{FA}^z, \\
P_{FA} &= \left(P_{FA}^t\right)^{T_{th}/W_s},
\end{aligned}
\tag{14}
$$

where P_{FA}^i is the probability that the computed MV on the "i" axis is lower than T_σ during a dynamic condition. In order to compute these probabilities, the distribution of the NMV and EMV outputs in dynamic conditions should be found. However, by analyzing all the collected measurements, we could neither determine a theoretical nor an empirical model for the dynamic mode (such a model would need to depend on many parameters such as the type of aircraft, the weather, the packaging of the sensor, etc.). Therefore, we could not provide an analytical expression for the MV threshold as a function of the desired probability of false alarm. Thus, we decided to proceed empirically as follows:

(i) for each data set, only the part of the series where the sensor is subject to a dynamic movement is considered and used to compute the EMV and NMV;

(ii) then the corresponding pdf and cdf of the EMV and NMV outputs are computed;

(iii) finally, the probability that the NMV or EMV outputs of each axis are lower than the threshold is computed.

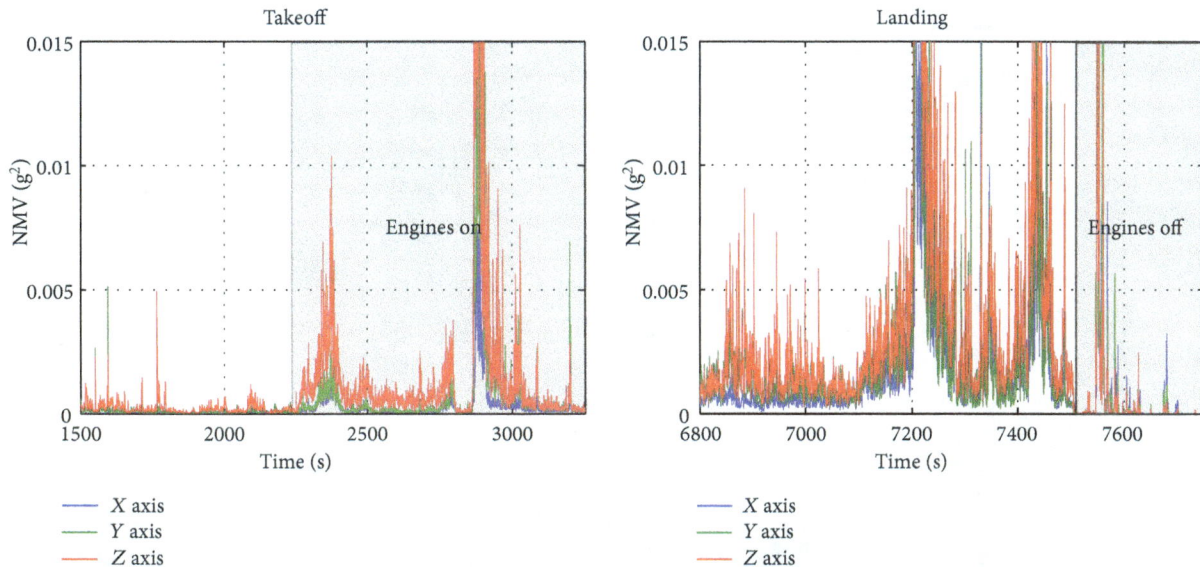

FIGURE 8: NMV computation for the 3 accelerometer axes during landing and takeoff of a flight from Geneva to London.

Following this approach, P_{FA} was empirically computed for different MV thresholds and thresholds in time.

5.2. Threshold in Time. As it was described above, the threshold in time T_{th} stands as a second condition after testing the MV on the three axes. In fact, the threshold in time ensures that the sensor was static for the period of time that is specified, and once this period is exceeded, a final static condition is declared that can be used as an indication for the allowance of transmission of mobile communication signals. The main reason behind this second condition is to decrease P_{FA} in order to meet the specific requirements for the civil aviation regulations, which is the most critical point especially during flight time. Similar to the MV threshold, we analyzed empirically different threshold in times on different measurements from different flights, in order to find a minimum threshold in time that ensures a reliable performance of the algorithm under different conditions for the chosen accelerometer.

5.3. Impact of Window Size (W_s). The window size (W_s) is defined as the duration of time where the algorithm is computing the MV and the period at which $cond_\sigma^n$ is checked. If the accelerometer measurement rate is 100 Hz and W_s is equal to 1 s, it means that the MV over 100 samples is computed and compared to the MV threshold every 1 s. When decreasing the window size, the MV bandwidth is increased and the noise also increases. Consequently, the threshold should be increased. Table 3 shows the MV average for different window sizes considering static measurements. For 1, 2, or 3 s window size, the moving variance is almost the same; meaning that no significant additional information will be gained when increasing the window size more than 1 s for static data. For 0.5 s the response starts to be slightly noisier and in this case the threshold should be slightly increased. Overall, a small change in window size does not have a great

TABLE 3: MV average for static data for different window sizes.

Window size (W_s)	0.5 s	1 s	2 s	3 s
MV $\cong$	0.6	0.5	0.51	0.51

impact on the moving variance of the static condition and hence the MV threshold is not so affected.

Another aspect that should be taken in consideration is the effect of different window sizes on the MV during dynamic conditions. Figure 12 shows the acceleration measurement on one axis and its corresponding NMV computation. In the graphs on the left, it can be seen that when the window size is increased, the moving variance becomes less sensitive to dynamic change. Also, a smaller window size results in a noisier response and in this case it would be harder to detect a static condition. However, one exception is noticed regarding this point. In fact, it can be seen from the graphs on the right of Figure 12 that there is an opposite behavior when increasing the window size. Normally, for dynamic movements resulting in an acceleration measurement having sudden jumps with a very short duration (a), the NMV output for a 1 s window size (c) is higher than for a 3 s window size. This is expected, because for a window size of 1 s the variance is computed using less information and any short duration movement within this 1 s can have more effect than in the case of a 3 s window size where more information are taken in consideration. In a different case, when a movement occurs resulting in an acceleration having a jump that lasts more than 3 s (b), it can be seen that a window size of 3 s results in a higher variance (d). In general, it has been noticed that this type of movement rarely occurs and even in this case a 1 s window size can still detect it.

Overall, it is observed that varying the window size has a limited impact on the MV threshold and a general value for different window sizes can be used. Regarding sensitivity to

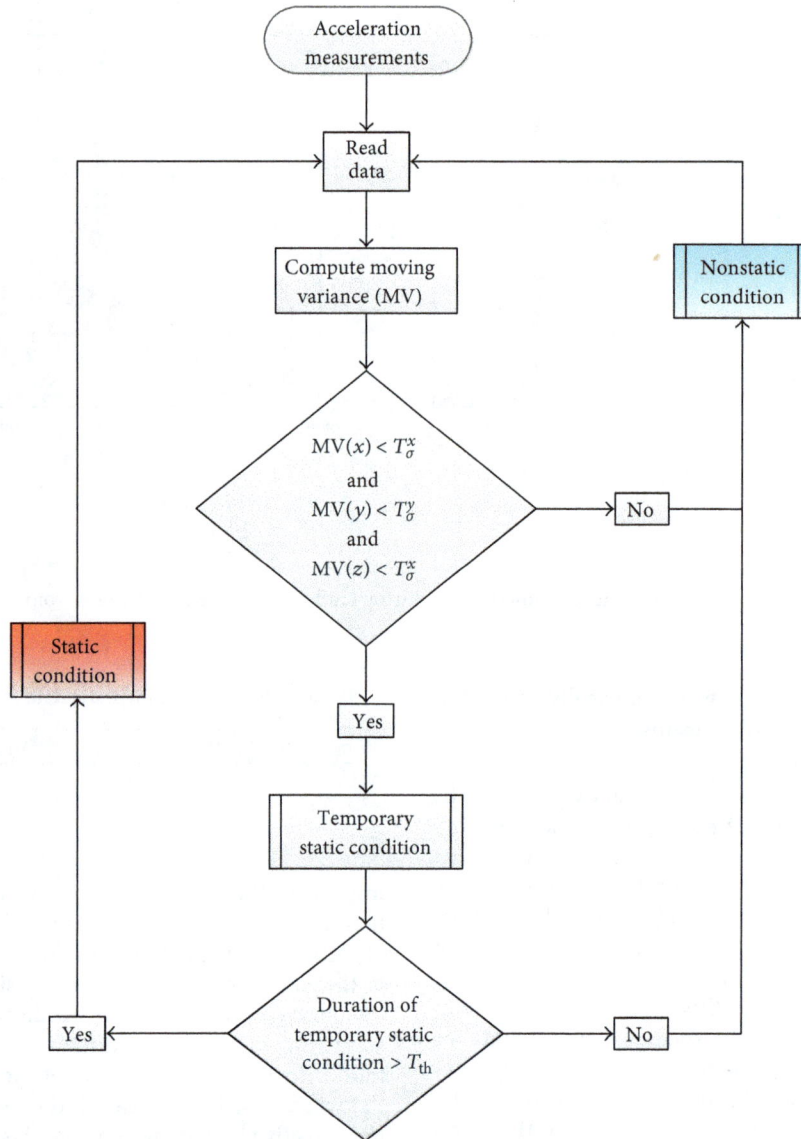

FIGURE 9: Proposed algorithm for detecting the status condition of the sensor.

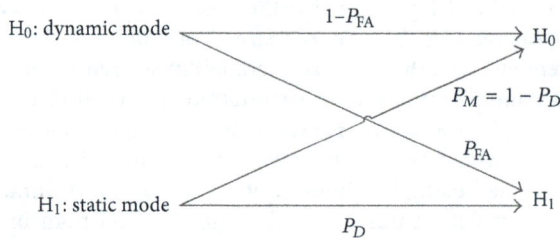

FIGURE 10: Binary hypotheses for the movement recognition algorithm and its corresponding probabilities.

dynamic movement, it is shown that increasing the window size results in a variance that is less sensitive to dynamic detection and decreasing it results in a variance less sensitive to static detection. As a tradeoff between these two results, we have found that a window size between 1 and 2 s works well.

6. Simulation Results

For testing and optimization, the algorithm is implemented in Matlab where simulation results are obtained using measurements from all the flights datasets. Hereafter, only results from two selected flights are shown for illustration and comparison purposes.

(i) Flight 1, which is the flight from Geneva to London already described in Section 2, is chosen as a representative of a typical flight.

(ii) Flight 2, from Helsinki to Zurich onboard a Bae Avro RJ85, is chosen because it was the most challenging flight for detecting dynamic conditions during flight time.

Figures 13 and 14 show the algorithm detection output applied on measurements from both flights using NMV

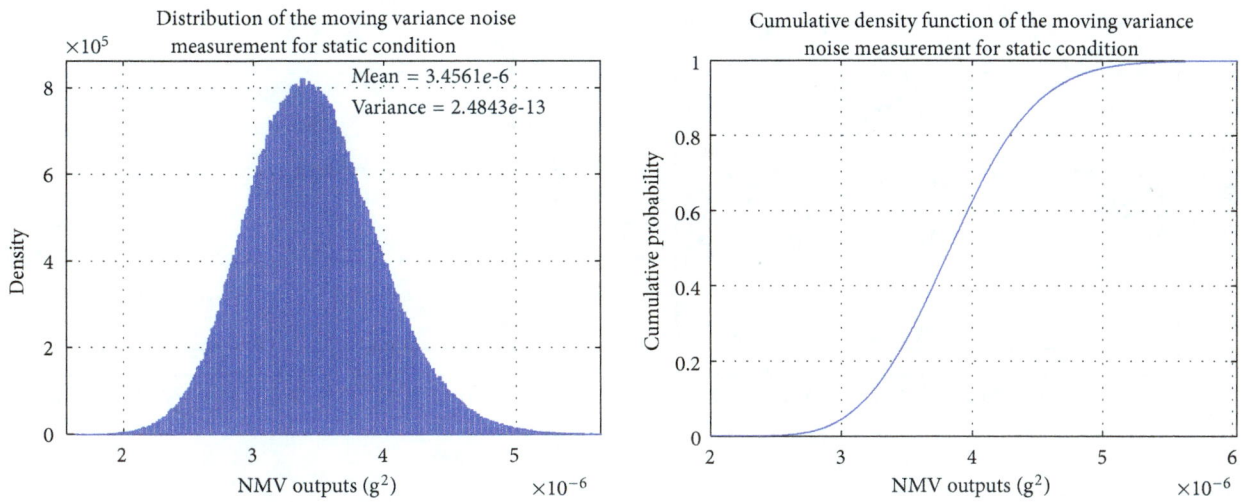

FIGURE 11: pdf and cdf of NMV (or EMV) output in static condition.

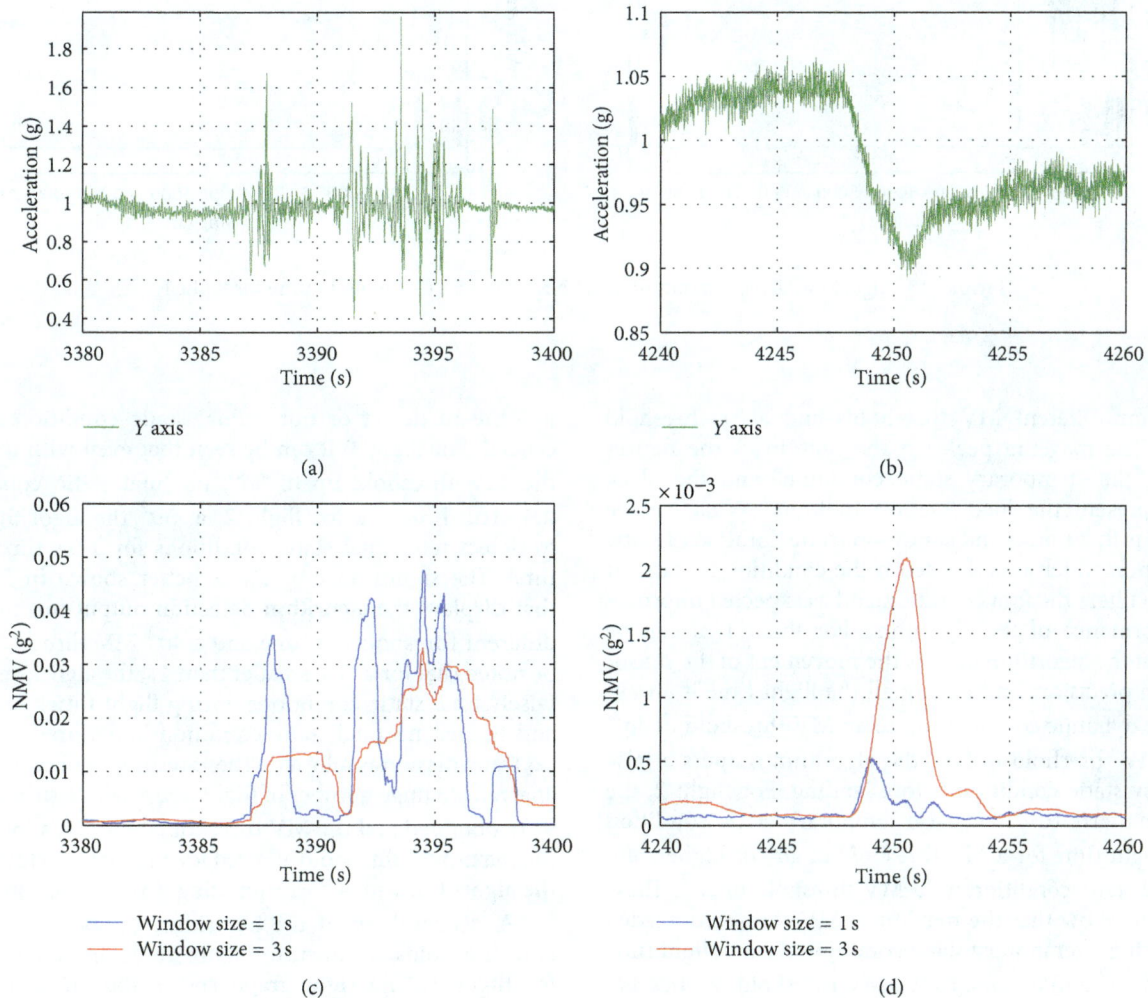

FIGURE 12: NMV behavior under different dynamic conditions.

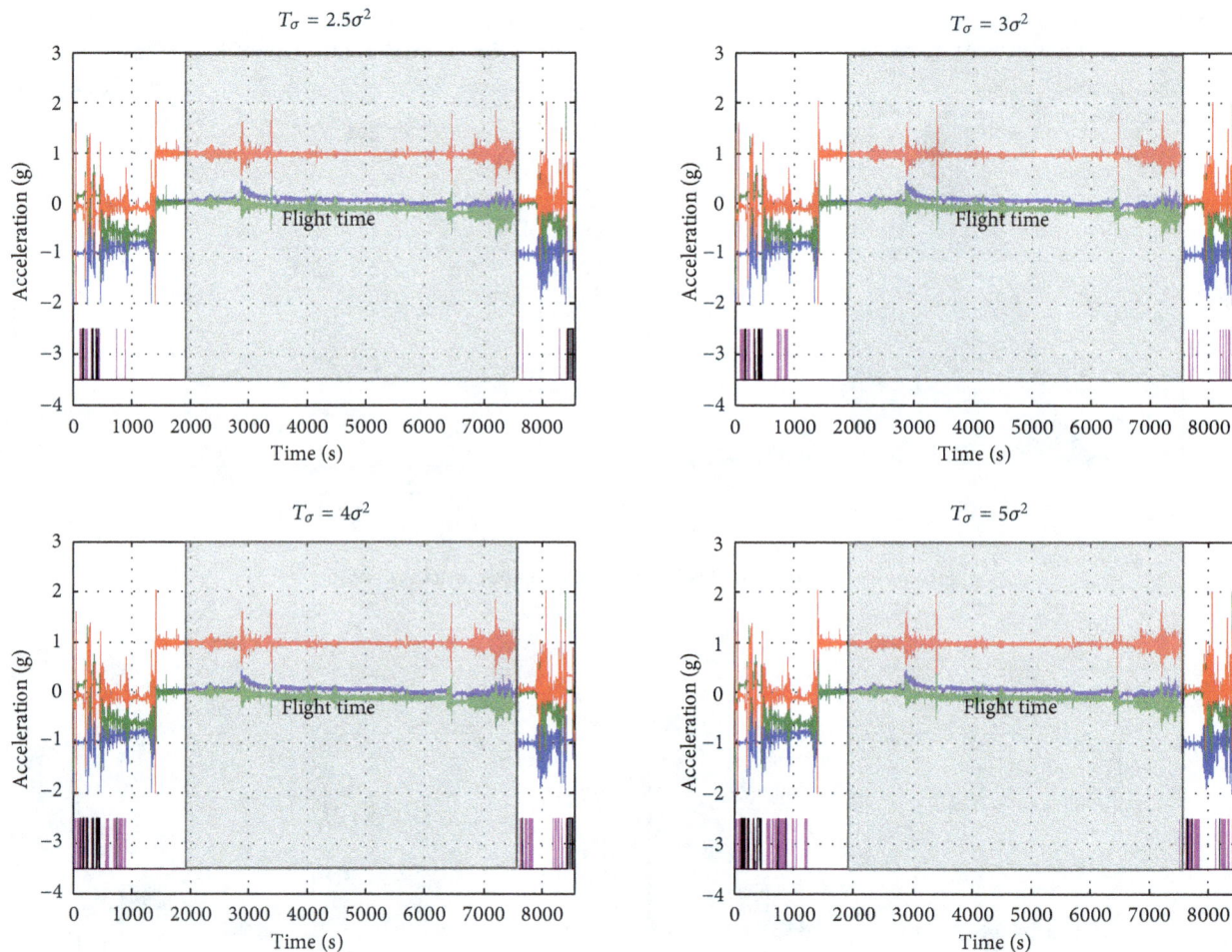

FIGURE 13: Algorithm detection output using NMV and 5 s threshold in time for flight 1.

method for different MV thresholds and a 5 s threshold in time. The magenta peaks in the bottom of the figures represent the "temporary static condition" and the black peaks represent the final "static condition." What can be observed is that the second condition (time duration of static period upon which it is decided if the condition is static or dynamic) filters the first condition and as expected improves the performance of the algorithm. For flight 1, it can be noted that the algorithm detects the movement of the sensor with high precision, and during all the flight time it always outputs a dynamic condition up to an MV threshold of $4\sigma^2$. For an MV threshold of $5\sigma^2$, the algorithm outputs a false temporary static condition before landing. For flight 2, the algorithm starts to detect false temporary static condition during flight time for a MV threshold of $4\sigma^2$ or higher, and false final static condition for a MV threshold of $5\sigma^2$. These two graphs show that the maximum MV threshold needed for detecting a temporary static condition during flight time is around $3\sigma^2$ and as soon as the MV threshold reaches $4\sigma^2$ the false temporary static conditions start to appear. At this MV threshold level or higher, the importance of the threshold

in time to detect or not a final static condition becomes crucial. For flight 1, it can be seen that even with increasing the MV threshold up to $5\sigma^2$, no final static condition is detected; however, for flight 2, at $5\sigma^2$ the algorithm starts to detect false final static conditions for a 5 s threshold in time. The significance of T_{th} is better shown in Figure 15 that displays the algorithm detection output for flight 2 for different thresholds in time and a $4\sigma^2$ MV threshold. It can be noted that for a T_{th} smaller than 4 s, the algorithm detects falsely final static conditions during flight time. And for 4 s and higher, no final static condition is declared and hence no transmission is allowed. These results were confirmed by analyzing a huge number of flight hours where similar results were obtained and the MV threshold of $4\sigma^2$ was found to be the maximum threshold allowed for a reliable performance of the algorithm with a corresponding threshold in time of 4 s.

A deeper look of the algorithm behavior using these two thresholds is shown in Figure 16 during boarding time for flight 1. The right graph shows the NMV output of the 3 accelerometer axes and the two dimensions check (i.e., amplitude and time). Once both conditions are satisfied,

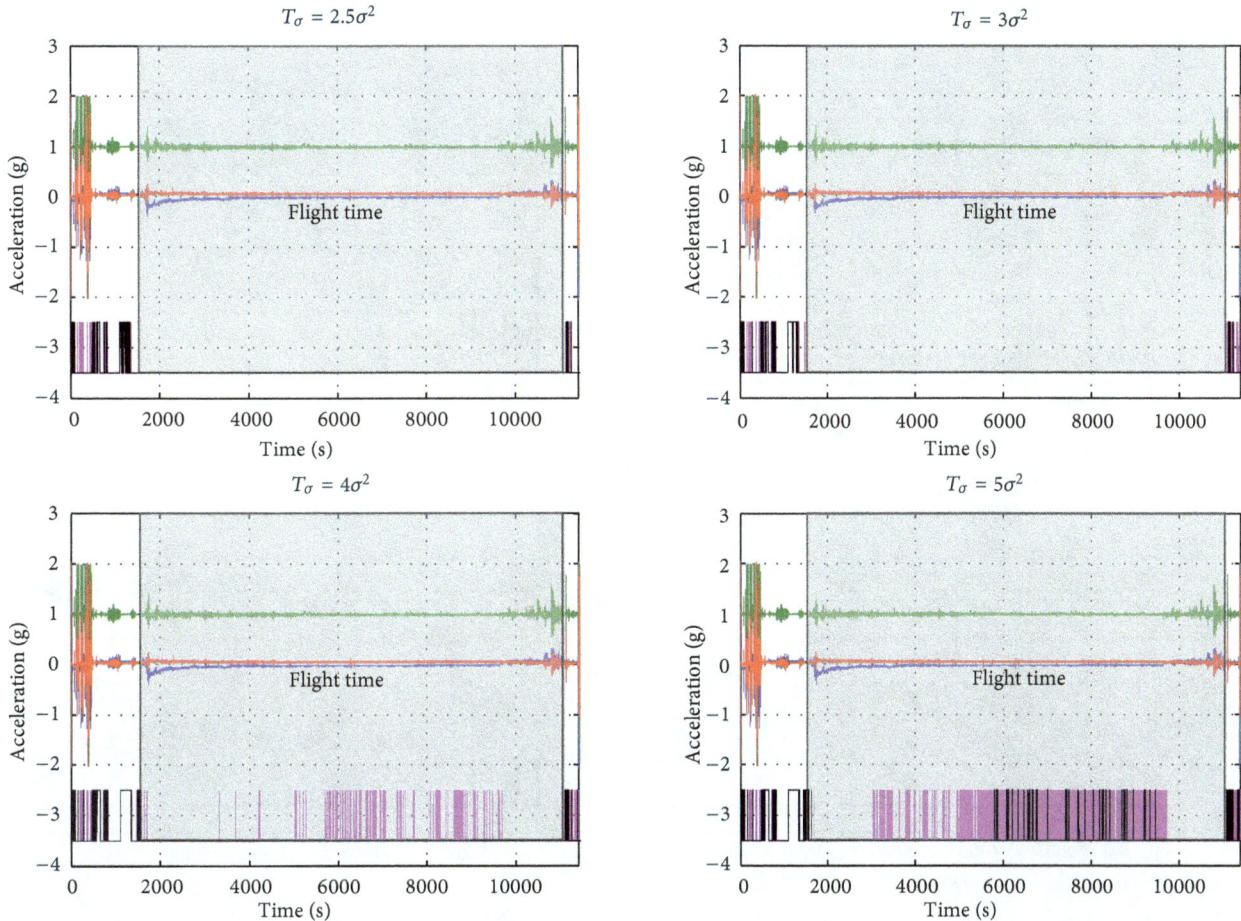

FIGURE 14: Algorithm detection output using NMV and 5 s threshold in time for flight 2.

the algorithm outputs a static condition where it can be seen as the black peak in the left graph.

The output of the algorithm using EMV method is similar when using NMV for small weighting; however, when increasing the weighting factor the EMV starts to be more sensitive to static condition detection. In fact when increasing the weight, and consequently increasing the importance of new measurements, the response of the algorithm starts to be more fluctuating. This can be explained by looking at Figure 17, where it is shown the NMV and EMV outputs for a measurement resulting from a dynamic movement. It can be seen that for EMV, with increasing the weighting, the amplitude variation increases; this is because the new sample has more weight, and if it represents a sample coming from a dynamic movement it will increase the output variations. As for the NMV where there is no weighting, the variance is more stable as it results from an average of data having similar weight. Therefore, the effect of increasing the weight too much in the EMV method has a disadvantage. For example, if the flight is very calm, then the MV output could potentially be close to the MV threshold, and if the weight is increased, then the EMV output will be noisier and there is a possibility due to the variation that it crosses the threshold and thus the

algorithm outputs a static detection during the flight. This type of scenario should be avoided, and this is why it is not recommended to increase the weight above 10.

Regarding the probability of false alarm, and following the approach in Section 5.1.2, Tables 4 and 5 show the temporary probability of false alarm and P_{FA} during flight time for flights 1 and 2. It can be seen that the performance of the algorithm varies from one flight to another, and in both cases P_{FA} is considerably low. Also the table shows that if a lower P_{FA} is required, it is sufficient to either decrease the MV threshold or increase the threshold in time. Probability analyses from other flights were similar to the assessment discussed in this paper and the algorithm showed a high fidelity in detecting the state of the accelerometer.

To conclude, this section shows that the proposed algorithm has a very precise detection of static/dynamic conditions of the sensor using EMV or NMV for a MV threshold less than or equal to $4\sigma^2$. It has been also shown that the detection performance is very good especially during flight time where P_{FA} is very low. For EMV it is not recommended to increase the weighting more than 10, as the detection output starts to be biased by the weighting factor on the new measurement and can lead to false detection.

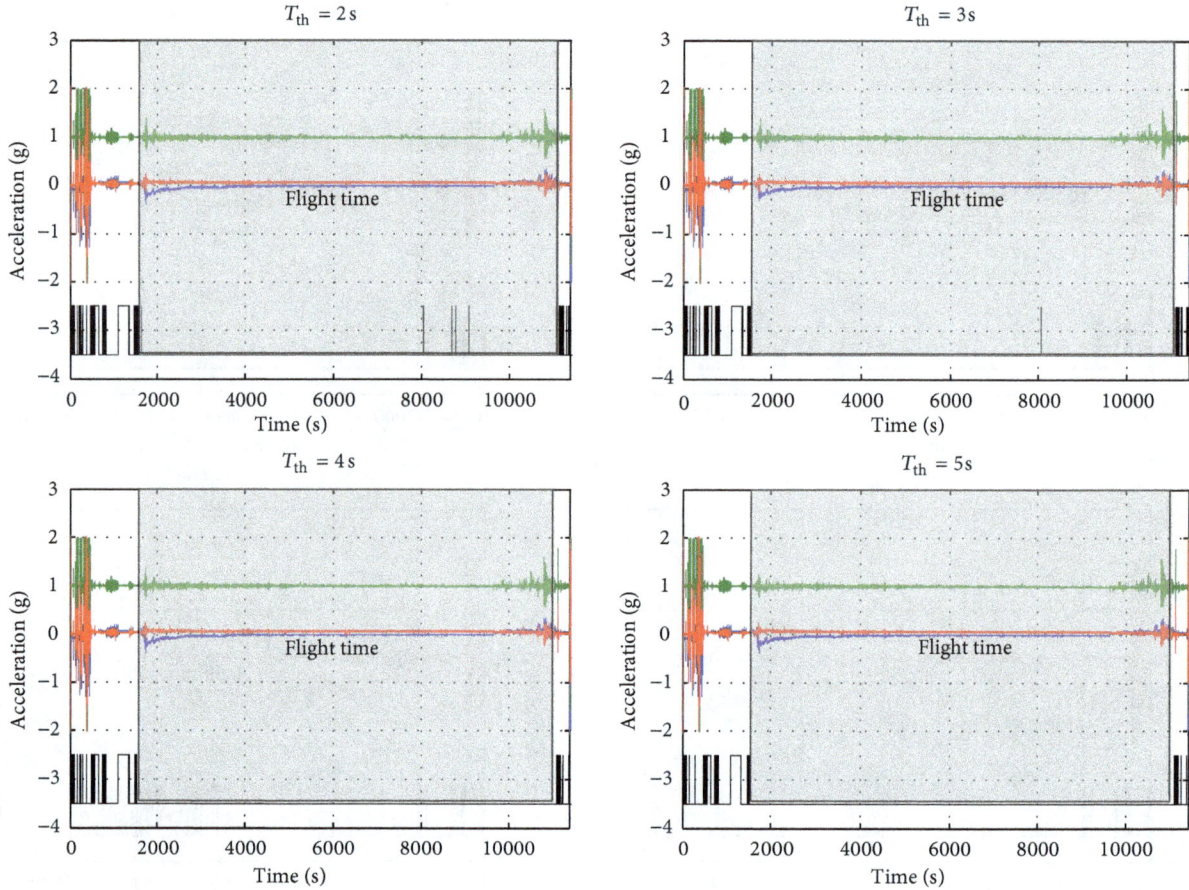

FIGURE 15: Algorithm detection output using NMV and $4\sigma^2$ MV threshold for flight 2.

TABLE 4: P_{FA} during flight time for flight 1 using NMV method.

Flight			1			
T_σ	$3\sigma^2$		$4\sigma^2$		$5\sigma^2$	
T_{th} (s)	3	4	3	4	3	4
P_{FA}^x	5×10^{-2}	5×10^{-2}	8×10^{-2}	8×10^{-2}	1×10^{-1}	1×10^{-1}
P_{FA}^y	7×10^{-4}	7×10^{-4}	2×10^{-3}	2×10^{-3}	4×10^{-3}	4×10^{-3}
P_{FA}^z	9×10^{-7}	9×10^{-7}	2×10^{-6}	2×10^{-6}	2×10^{-4}	2×10^{-4}
P_{FA}^t	2×10^{-10}	2×10^{-10}	3×10^{-10}	3×10^{-10}	8×10^{-8}	8×10^{-8}
P_{FA}	6×10^{-33}	1×10^{-42}	3×10^{-29}	1×10^{-38}	5×10^{-22}	4×10^{-29}

7. Implementation Analysis

The implementation of the algorithm is assessed using three criteria: the response time, resources, and practical considerations.

7.1. Response Time. An important point to study is the response time of the algorithm, which we define as the time required for the algorithm to output a static condition at turn-on assuming that the sensor is static. Indeed, minimizing this response time will also extend the battery life.

For the NMV algorithm and assuming a window size of 1 s, it will first take 1 s for the data window to be full

of measurements before the algorithm can output the first temporary condition. Assuming a threshold in time of 4 s, the final static condition will be generated after 4 s.

For the EMV algorithm, the behavior is different and depends on the initialization state of the accelerometer before startup. For example, Figure 18 shows the EMV outputs for different weightings compared to the NMV output (i.e., no weighting) for one axis of an accelerometer in static mode when the algorithm is initialized with an acceleration variance of $10\sigma^2$ (this value is chosen to guarantee that the algorithm will not provide a temporary static condition at startup). The cyan dotted line represents a MV threshold of $4\sigma^2$.

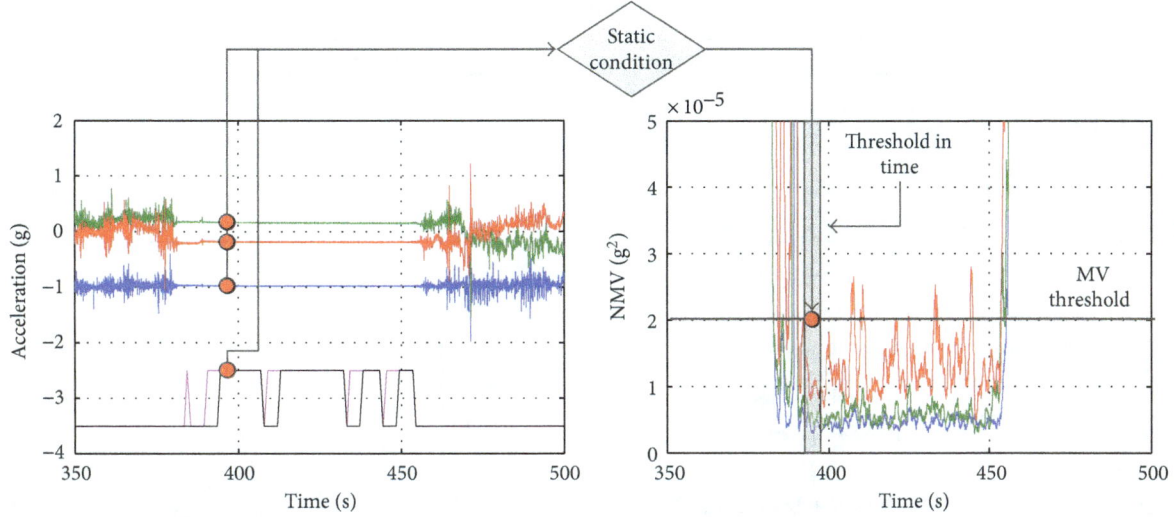

FIGURE 16: Algorithm behavior during boarding. Left side is the algorithm output in comparison to the acceleration on the 3 axes. Right side is the corresponding NMV outputs.

TABLE 5: P_{FA} during flight time for flight 2 using NMV method.

Flight				2		
T_σ	$3\sigma^2$		$4\sigma^2$		$5\sigma^2$	
T_{th} (s)	3	4	3	4	3	4
P_{FA}^x	3×10^{-1}	3×10^{-1}	6×10^{-1}	6×10^{-1}	7×10^{-1}	7×10^{-1}
P_{FA}^y	1×10^{-2}	1×10^{-2}	1×10^{-1}	1×10^{-1}	4×10^{-1}	4×10^{-1}
P_{FA}^z	7×10^{-3}	7×10^{-3}	7×10^{-2}	7×10^{-2}	2×10^{-1}	2×10^{-1}
P_{FA}^t	2×10^{-5}	2×10^{-5}	4×10^{-3}	4×10^{-3}	6×10^{-2}	6×10^{-2}
P_{FA}	9×10^{-15}	2×10^{-19}	7×10^{-8}	3×10^{-10}	2×10^{-4}	1×10^{-5}

It can be noted that with increasing the weighting, the response time decreases as expected and consequently the output of the EMV algorithm reaches faster the MV threshold than the NMV. It is also important to note that increasing the variance of the initial dynamic condition will increase the response time. Therefore this factor should be chosen carefully to guarantee a shorter response time while minimizing the probability of false alarm.

7.2. Implementation Resources. The NMV and EMV algorithms have shown similar results in terms of performance. In this section, we explore the hardware implementation of these two algorithms in order to select the most appropriate one.

7.2.1. NMV Implementation. The basic element of the implementation of the NMV method is a moving averager (MA) shown in Figure 19 and its z-domain transfer function is

$$H(z) = \frac{1}{N}\left(1 + z^{-1} + z^{-2} + \cdots + z^{-N+1}\right). \quad (15)$$

This element is then used to compute the NMV of a sequence as defined by (1) and depicted in Figure 20.

7.2.2. EMV Implementation. The same procedure is used for the EMV method. The basic element of the implementation

is an exponential averager (EA) which is defined by (4) and depicted in Figure 21. This element is then used to compute the EMV of a sequence as defined by (5) and depicted in Figure 22.

7.2.3. Optimization and Comparison. It can be seen that the implementation of these two algorithms requires only four different operations: (1) delay; (2) addition/subtraction; (3) multiplication by a constant; (4) squaring. When an adder has one input which is the delayed version of its output, as it is the case in the MA and EA, this is equivalent to an integrator. Table 6 summarizes the resources required by these operations in terms of logic cell available in CPLDs and FPGAs, when the input signal of the operation has a resolution of R bits.

The implementation of the multiplication by a constant is highly dependent on the value of the constant. If the constant is a power of two, it corresponds to a simple shift and does not require resource at all. If this rule is too strict, it is also possible to implement the multiplication using two shifts and a subtraction when the constant is in the form $2^L - 2^M$ [17]. Table 7 summarizes the total resources required by the algorithm, when the resolution of the input signal is R bits, the number of samples in a MA or EA is N, and considering that the multiplications are all implemented by shifts.

TABLE 6: Resources required by the basis operations.

Operation	Logic utilization	Register utilization	Number of logic cell
Delay	0	R	R
Addition/subtraction	R	0	R
Integrator	R	R	
Multiplication by a constant	0 or R	0	0 or R
Squaring	$R^2 + 4R$	0	$R^2 + 4R$

TABLE 7: Resources required by the algorithms.

Operation	Number of logic cell
Moving averager	$R(N+2) + \log_2(N) + 1$
Exponential averager	$2R + 1$
Moving variance network	$2R^2 + R(3N + 16) + 2\log_2(N) + 1$
Exponential moving variance network	$R^2 + 13R + 12$

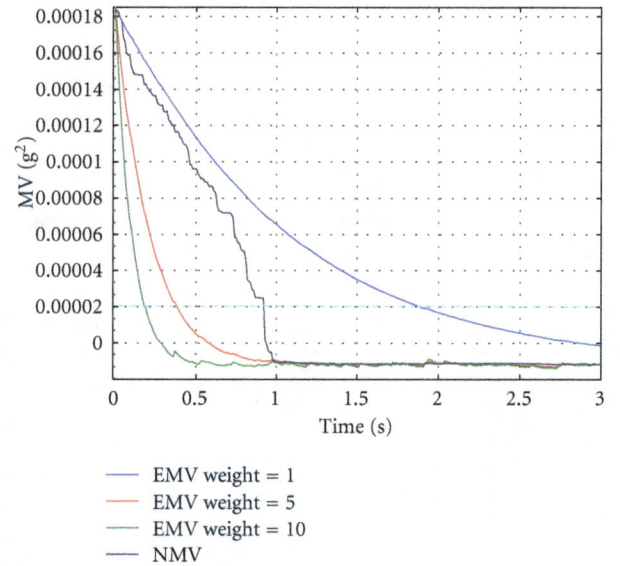

FIGURE 18: Comparison of time responses of NMV and EMV with different weights.

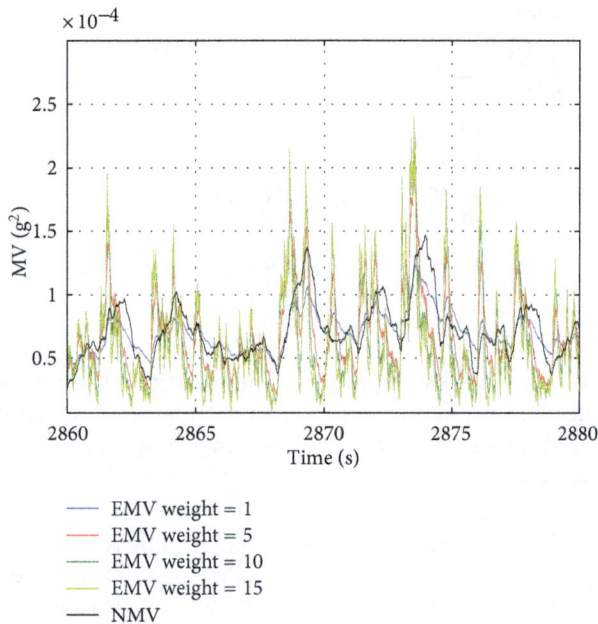

FIGURE 17: NMV and EMV output during dynamic conditions.

It can be seen that the EMV implementation requires less resources than the NMV. This is due to two points: (1) it does not need to store N samples; (2) it requires only one squaring against two for the NMV implementation. For $N = 64$, Table 8 shows a comparison for implementation of both methods in terms of logical elements, where a block includes all the operations needed to process measurements from one axis of the accelerometer.

7.3. Power Consumption Considerations. Package tracking applications are not necessarily very demanding in terms of positioning availability. In fact, it is typically sufficient to obtain a position fix every few hours in order to track a package from its starting point to its destination. Therefore, the proposed algorithm does not need to work all the time. Instead, a timer can periodically wake up the tracking module algorithm for limited time duration. If during this time a final static condition is detected, then the module can make a transmission to enable its localization. Otherwise, the module goes back to sleep until the next timer occurs. By doing so, the power consumption of the algorithm can be reduced drastically. In addition, even when the system is awake, it has been shown that the algorithm only requires few logical elements and that the sampling frequency needed is very low (<70 Hz), which will result in a power consumption of less than 4 mW considering, for example, an Altera MAX V CPLD [19]. Also, taking into consideration the fact that MEMS-based accelerometers are not power hungry (e.g., <1 mW for the LIS302DL accelerometer [16]), we can infer that the total power consumption of the detection module will be less than 5 mW while active.

Regarding the power consumption of the transmission module, it will consist of the power to establish the connection, and the power to maintain the transmission. As it was already shown in [20], module communication radios have low connection maintenance energy, but high energy per bit transmission cost and low bandwidth. Therefore, the

TABLE 8: Comparison for implementation resources of different methods for $N = 64$.

R	NMV						EMV	
	Total of LEs without memory		Total of LEs with memory		Total of bits with memory		Total of LEs	
	1 block	3 blocks	1 block	3 blocks	1 block	3 blocks	1 block	3 blocks
8	1807	5421	271	813	1536	4608	180	540
10	2295	6885	375	1125	1920	5760	242	726
12	2799	8397	495	1485	2304	6912	312	936
14	3319	9957	631	1893	2688	8064	390	1170
16	3855	11565	783	2349	3072	9216	476	1428
18	4407	13221	951	2853	3456	10368	570	1710

TABLE 9: Main parameter characterization of the proposed algorithm using NMV or EMV methods.

Algorithm	NMV	EMV
T_σ	$2.5\sigma^2$–$4\sigma^2$	$2.5\sigma^2$–$4\sigma^2$
T_{th}	$4\,\text{s} \leq T_{th} \leq 5\,\text{s}$	$4\,\text{s} \leq T_{th} \leq 5\,\text{s}$
α	\	$1/N < \alpha \leq 5/N^*$
W_s	$1\,\text{s} \leq W_s \leq 2\,\text{s}^{**}$	$1\,\text{s} \leq W_s \leq 2\,\text{s}^{**}$
P_M	$\geq 6 \times 10^{-19}$	$\geq 6 \times 10^{-19}$
P_{FA}^t	$\leq 4 \times 10^{-3}$	$\leq 4 \times 10^{-3}$
P_{FA}	$\leq 3 \times 10^{-10}$	$\leq 3 \times 10^{-10}$
Sampling frequency	60 Hz–70 Hz	60 Hz–70 Hz
Implementation resources	$813 \leq \text{LEs} \leq 3405$	$540 \leq \text{LEs} \leq 2016$

$^*N$ is the number of samples per window size, and the result of dividing the weighting factor by N should be a multiple of 2 for optimal implementation.
**The number of samples in a window size should be a multiple of 2.

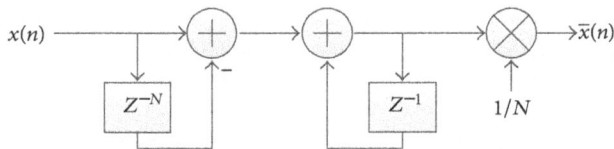

FIGURE 19: Implementation of an N-point moving averager [18].

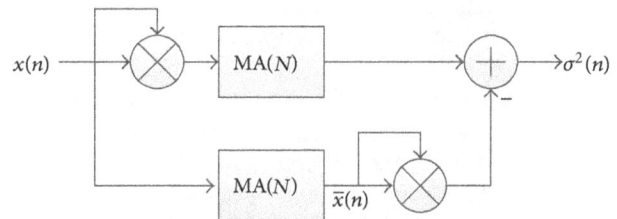

FIGURE 20: Implementation of a recursive N-point moving variance network [18].

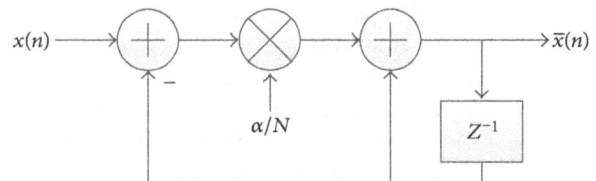

FIGURE 21: Implementation of an exponential averager [18].

transmission power will be the dominant factor. Since no data is sent during the sleep mode of the tracking module and the size of the data packets sent during the active mode will only consist of the protocol data units (PDUs), that are small in size (16 bytes), the required power consumption should be typically less than 10 mW, according to [20]. Therefore, if we assume the usage of a standard Li-Ion battery such as the one used in an iPhone holding an energy of 5.3 Wh, then we can conclude that the module can support thousands of transmissions, which roughly means many days of battery lifetime.

8. Summary

Based on the analysis shown in this paper, the parameters characterizations of the proposed algorithm using either NMV or EMV methods are summarized in Table 9.

9. Conclusions

In this paper, we proposed a movement recognition algorithm using a low quality MEMS accelerometer sensor integrated in a module that is meant to be used onboard an airplane for flight mode detection. The algorithm successfully detects the state of the sensor and prevents the module to transmit mobile communication signals during the flight. The proposed algorithm is based on computing the normal moving variance (NMV) or exponential moving variance (EMV) during a specific window of time of the acceleration measurements and comparing the result to a threshold. In addition, consecutive results from different time windows are

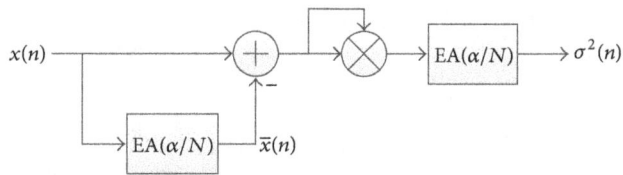

FIGURE 22: Implementation of an exponential averaging network [18].

further compared to a threshold in time to declare a final static condition. The main characteristics of the algorithm are summarized as follows. Generality, as it is independent of the accelerometer quality or type, where only the noise behavior of the sensor is needed to define all the algorithm parameters. Reliability, as it provides a very low probability of false alarm to meet the requirements needed by the airspace regulations. Flexibility, where it is sufficient to vary the MV threshold and the threshold in time to adapt it for different application. Simplicity, as it requires low resources, and thus, it can be implemented on a small FPGA or a CPLD. Consequently, the detection module including the accelerometer and the hardware where the algorithm is implemented will have very low power consumption, typically less than 5 mW.

References

[1] S. Wang, J. Min, and B. K. Yi, "Location based services for mobiles: technologies and standards," in *Proceedings of IEEE International Conference on Communication (ICC '08)*, Beijing, China, May 2008.

[2] "Jiiva Tracing Solutions," August 2012, http://www.jiiva.com/.

[3] "Wireless Devices on Airplane," Federal Communications Commission FCC, May 2011, http://www.fcc.gov/guides/wireless-devices-airplanes.

[4] C. S. Miguelez, "GSM Operation Onboard Aircraft," European Telecommunications Standards Institute (ETSI), Cedex, France, January 2007.

[5] S. L. Lau and K. David, "Movement recognition using the accelerometer in smartphones," in *Future Network and Mobile Summit*, pp. 1–9, June 2010.

[6] D. Fuentes, L. Gonzalez-Abril, C. Angulo, and J. A. Ortega, "Online motion recognition using an accelerometer in a mobile device," *Expert Systems with Applications*, vol. 39, no. 3, pp. 2461–2465, 2012.

[7] T. Brezmes, J. Gorricho, and J. Cotrina, "Activity recognition from accelerometer data on a mobile phone," in *Proceedings of the 10th International Work-Conference on Artificial Neural Networks: Part II: Distributed Computing, Artificial Intelligence, Bioinformatics, Soft Computing, and Ambient Assisted Living*, vol. 5518 of *Lecture Notes in Computer Science*, pp. 796–799, Springer, Salamanca, Spain, June 2009.

[8] T. Zhang, J. Wang, P. Liu, and J. Hou, "Fall detection by embedding an accelerometer in cellphone and using KFD algorithm," *International Journal of Computer Science and Network Security*, vol. 6, no. 10, pp. 277–284, 2006.

[9] D. M. Karantonis, M. R. Narayanan, M. Mathie, N. H. Lovell, and B. G. Celler, "Implementation of a real-time human movement classifier using a triaxial accelerometer for ambulatory monitoring," *IEEE Transactions on Information Technology in Biomedicine*, vol. 10, no. 1, pp. 156–167, 2006.

[10] T. R. Burchfield and S. Venkatesan, "Accelerometer-based human abnormal movement detection in wireless sensor networks," in *Proceedings of the 5th International Conference on Mobile Systems, Applications and Services (HealthNet '07)*, pp. 67–69, San Juan, Puerto Rico, June 2007.

[11] J. Shin, D. Shin, D. Shin, S. Her, S. Kim, and M. Lee, "Human movement detection algorithm using 3-axis accelerometer sensor based on low-power management scheme for mobile health care system," in *Advances in Grid and Pervasive Computing*, vol. 6104 of *Lecture Notes in Computer Science*, pp. 81–90, Springer, 2010.

[12] Y. Li, X. Chen, X. Zhang, K. Wang, and Z. J. Wang, "A sign-component-based framework for Chinese sign language recognition using accelerometer and sEMG data," *IEEE Transactions on Biomedical Engineering*, vol. 59, no. 10, pp. 2695–2704, 2012.

[13] J. S. Wang, Y. L. Hsu, and J. N. Liu, "An inertial-measurement-unit-based pen with a trajectory reconstruction algorithm and its applications," *IEEE Transactions on Industrial Electronics*, vol. 57, no. 10, pp. 3508–3521, 2010.

[14] Q. Li, J. A. Stankovic, M. A. Hanson, A. T. Barth, J. Lach, and G. Zhou, "Accurate, fast fall detection using gyroscopes and accelerometer-derived posture information," in *Proceedings of the 6th International Workshop on Wearable and Implantable Body Sensor Networks (BSN '09)*, pp. 138–143, June 2009.

[15] D. T. W. Fong, J. C. Y. Wong, A. H. F. Lam, R. H. W. Lam, and W. J. Li, "A wireless motion sensing system using ADXL MEMS accelerometers for sports science applications," in *Proceedings of the 5th World Congress on Intelligent Control and Automation, Conference Proceedings (WCICA '04)*, pp. 5635–5640, June 2004.

[16] STMicroelectronics, "Analog, Sensors and MEMS, Accelerometers, LIS302DL," August 2012, http://www.st.com/internet/analog/product/152913.jsp.

[17] R. G. Lyons, *Understanding Digital Signal Processing*, Prentice Hall, New York, NY, USA, 3rd edition, 2010.

[18] L. L. Scharf, *Statistical Signal Processing Detection, Estimation, and Time Series Analysis*, Addison Wesley, Reading, Mass, USA, 1991.

[19] Altera, "PowerPlay Early Power Estimators (EPE) and Power Analyzer," August 2012, http://www.altera.com/support/devices/estimator/pow-powerplay.jsp.

[20] A. Rahmati and L. Zhong, "Context-for-wireless: context-sensitive energy-efficient wireless data transfer," in *Proceedings of the 5th International Conference on Mobile Systems, Applications and Services (MobiSys '07)*, pp. 165–178, San Juan, Puerto Rico, June 2007.

5

Modeling and Analysis of Connected Traffic Intersections Based on Modified Binary Petri Nets

Omar Yaqub[1,2] and Lingxi Li[1,2]

[1] Department of Electrical and Computer Engineering, Indiana University-Purdue University Indianapolis (IUPUI),
   Indianapolis, IN 46202, USA
[2] Transportation Active Safety Institute (TASI), Indiana University-Purdue University Indianapolis (IUPUI),
   Indianapolis, IN 46202, USA

Correspondence should be addressed to Lingxi Li; ll7@iupui.edu

Academic Editor: Aboelmagd Noureldin

We propose an approach for the modeling and analysis of two connected traffic intersections based on Petri nets (PNs). We first use a PN to model an isolated four-way signalized intersection; then we extend it to model two successive signalized intersections. We find that this model has unbounded places, which in turn results in some confliction problems. Hence, we introduce the concept of modified binary petri nets (MBPNs) to overcome the limitation and resolve the confliction problem when we design our model and its controller. This MBPN model is a powerful tool and can be useful for the modeling and analysis of many other traffic applications.

1. Introduction

As a powerful tool that consists of a combined graphical and mathematical representation, Petri nets have been used for the modeling, control, and analysis in different applications including sensor networks [1], power systems [2, 3], manufacturing systems [4–7], and many other practical systems. In traffic management problems, Petri nets have been used to model the traffic network in different ways for a variety of purposes. It can be concluded that when vehicle flow has been studied, hybrid Petri nets (HPNs) are a suitable modeling tool because they consist of both continuous and discrete nodes that work together to reflect the dynamics of the overall traffic system. Continuous nodes are suitable for modeling continuous events such as vehicle flow while discrete nodes are used to represent discrete events such as phase change in traffic signal and enabling/disabling vehicle movement because of occurrences of emergent events such as accidents and the blocking of the road. In [8], a general HPN model for transportation system was developed. Traffic flow was described by continuous nodes and the events that affect the traffic dynamics were modeled through discrete nodes. In [9], a simple HPN was used to model the intersection of two one-way streets, while in [10] a continuous Petri net was

used to model a nonsignalized intersection, and then were added discrete nodes that are essential to represent a four-way intersection with two-phase traffic light through an HPN. In [11], the authors developed an HPN model to improve the performance of special and emergency vehicles.

In the aforementioned works on traffic network modeling and management, HPN models were adopted because they are more accurate to reflect the dynamics of the entire traffic network for certain applications. On the other hand, however, in some traffic network applications such as control and monitoring, only events are critical to be studied and analyzed. For these problems, discrete Petri nets as well as timed/colored Petri nets are the suitable modeling tools. In [12], the authors proposed a discrete Petri net model which describes the phase change of traffic light signal for an intersection. In [13], a discrete Petri net model for a small transportation system was developed to estimate the optimal travel route between the starting and destination points. In [14, 15], timed colored Petri nets were used to represent urban traffic light control systems and then a real-world supervisor of the modeled urban traffic light system was implemented. Deterministic and stochastic Petri nets were used in [16] to model railroad level crossing traffic control systems to identify and avoid the critical scenarios.

Compared with HPN models, discrete Petri nets are much simpler in terms of modeling and analysis. Therefore, for applications that do not need to capture the detailed vehicle flow dynamics, discrete Petri nets are a good tool to use. In this paper, we introduce a discrete Petri net model for two connected traffic intersections. This model is based mainly on higher-level event occurrences without considering details of lower-level continuous dynamics. For instance, it can tell us that the event of vehicles crossing an intersection during a specific traffic light phase from a specific entrance to a known destination has occurred, but without considering information such as how many vehicles have crossed the intersection or how much time this process has taken. The advantage is that this model can give us an abstract view of the event correlations about the entire complex traffic network. Therefore, it can be used to analyze the effect of the occurrence of a specific event at a specific node on other parts of the network. In addition, we propose a modified binary Petri net (MBPN) model that is suitable for the analysis and control of these connected intersections.

The contributions of this paper are two-fold: (1) we develop a comprehensive discrete Petri net model for two connected intersections by considering both signal phase change and vehicle flow directions; and (2) we propose a modified binary Petri net model to resolve the potential confliction problem for the two-intersection PN.

This paper is organized as follows. In Section 2, the basics of discrete Petri nets are briefly reviewed. A discrete Petri net model for a single signalized intersection is described in detail in Section 3. In Section 4, two connected intersections are modeled by extending the single-intersection model and are analyzed in detail. In Section 5, we propose the modified binary Petri net for the purpose of monitoring and control. Conclusions and directions for future work are presented in Section 6.

2. Petri Nets Basics

In this section, the basic principles of (discrete) Petri net are briefly reviewed. The reader can find more details about Petri nets and their extensions in [17–19].

A Petri net is a bipartite directed graph comprising of two types of nodes, called *places* (drawn as circles) and *transitions* (drawn as bars), which are connected through arrows that are called *arcs*. A Petri net is called a marked Petri net if a nonnegative integer number of tokens (drawn as black dots) is assigned to each place.

Thus, a marked Petri net is defined as

$$\text{PN} = \left(P, T, B^-, B^+, m_0\right), \qquad (1)$$

where $P = \{p_1, p_2, \ldots, p_n\}$ is a finite set of places consisting of n elements; $T = \{t_1, t_2, \ldots, t_m\}$ is a finite set of transitions consisting of m elements; B^- is the input incident matrix, which captures the arc weights directed from places to transitions. For instance, $B^-(p_i, t_j)$ is the weight of the arc directed from the place p_i to transition t_j. If there is no such arc, then the element of $B^-(p_i, t_j)$ is set to be zero; B^+ is the output incident matrix, which captures the arc weights directed from

transitions to places. For instance, $B^+(p_i, t_j)$ is the weight of the arc directed from the transition t_j to place p_i. If there is no such arc, then the element of $B^+(p_i, t_j)$ is set to be zero; m_0 is the initial marking (denotes the number of tokens in each place initially) of the Petri net.

It is not difficult to see that the sizes of the input and output incident matrices are $n \times m$.

A transition is said to be enabled if each of its input places contains a number of tokens that is greater than or equal to the arc weight that connects that specific place to the transition. In other words, transition t_j is enabled at a specific marking m_k if and only if

$$m_k\left(p_i\right) \ge B^-\left(p_i, t_j\right), \qquad (2)$$

for all input places p_i to transition t_j. The inequality is taken elementwise. An enabled transition may fire. The firing of a transition is the mechanism of removing a specific number of tokens from each input place (which is equal to the weight of the arc directing from this specific input place to the transition) and depositing a specific number of tokens to each output place (which is equal to the weight of the arc directing the transition to this specific output place).

Let S be a transition firing sequence; then we define s as the firing vector of sequence S in such a way that each element s_j in s captures the number of times that transition t_j fires in S. For instance, if $S = t_2 t_3 t_1 t_1$ and the Petri net has four transitions, then $s = [2 \ 1 \ 1 \ 0]^T$. For a given Petri net marked with a specific marking m_k, we say that the firing sequence S is enabled if and only if a specific marking $m_{k'}$ exists in such a way that when the firing sequence S is applied to the given Petri net marked with m_k it will produce the marking $m_{k'}$. Mathematically, transition firings can be captured by state equation given as follows:

$$m_{k'} = m_k + B \cdot s, \qquad (3)$$

where B is the incident matrix of the Petri net and it is defined as $B = B^+ - B^-$ (B^- is the input incident matrix and B^+ is the output incident matrix of the Petri net as defined earlier).

Example 1. Consider the Petri net shown in Figure 1. The net consists of four places: p_1, p_2, p_3, and p_4 and four transitions: t_1, t_2, t_3, and t_4. Since p_1 is the only place that has a token, the initial marking is written as $m_0 = [1 \ 0 \ 0 \ 0]^T$. The input and output incident matrices (B^-, B^+) are given by

$$B^- = \begin{bmatrix} 1 & 0 & 0 & 0 \\ 0 & 1 & 0 & 0 \\ 0 & 0 & 1 & 0 \\ 0 & 0 & 0 & 1 \end{bmatrix}, \quad B^+ = \begin{bmatrix} 0 & 0 & 0 & 1 \\ 1 & 0 & 0 & 0 \\ 1 & 0 & 0 & 0 \\ 0 & 1 & 1 & 0 \end{bmatrix}. \qquad (4)$$

So the incident matrix B can be calculated as

$$B = B^+ - B^- = \begin{bmatrix} -1 & 0 & 0 & 1 \\ 1 & -1 & 0 & 0 \\ 1 & 0 & -1 & 0 \\ 0 & 1 & 1 & -1 \end{bmatrix}. \qquad (5)$$

According to (2), it is not difficult to see that the only enabled transition is t_1 at the initial marking. When t_1 fires, one token

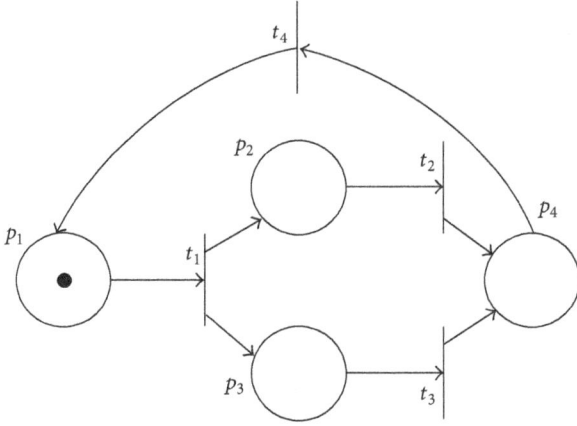

FIGURE 1: A simple Petri net model.

is removed from place p_1 and one token is added to p_2 and p_3, respectively, to result in the new marking $m = [0 \ 1 \ 1 \ 0]^T$. Mathematically, the new marking can also be obtained by using (3), where m_k is the initial marking m_0, $m_{k'}$ is the new marking, and the firing vector s is given by $[1 \ 0 \ 0 \ 0]^T$ since we are looking for the result of firing t_1 only once. Therefore, we have

$$m_{k'} = m_k + B \cdot s$$

$$= \begin{bmatrix} 1 \\ 0 \\ 0 \\ 0 \end{bmatrix} + \begin{bmatrix} -1 & 0 & 0 & 1 \\ 1 & -1 & 0 & 0 \\ 1 & 0 & -1 & 0 \\ 0 & 1 & 1 & -1 \end{bmatrix} \begin{bmatrix} 1 \\ 0 \\ 0 \\ 0 \end{bmatrix} = \begin{bmatrix} 0 \\ 1 \\ 1 \\ 0 \end{bmatrix}. \quad (6)$$

3. Petri Net Model for a Single Intersection

3.1. Structure of a Single Intersection. In Figure 2, a four-bidirectional-road signalized intersection is shown with its accessible roads. Direction symbols {n, s, e, w} are used to define the four bidirectional roads. They represent the four directions north, south, east, and west, respectively. The set $\{R_w^{in}, R_e^{in}, R_n^{in}, R_s^{in}, R_w^{out}, R_e^{out}, R_n^{out}, R_s^{out}\}$ represents the roads connecting to the intersection; the subset $\{R_w^{in}, R_e^{in}, R_n^{in}, R_s^{in}\}$ represents the incoming roads while the subset $\{R_w^{out}, R_e^{out}, R_n^{out}, R_s^{out}\}$ represents the outgoing roads.

The Petri net model, which will be introduced in the following subsection, is designed based on the assumption that this intersection's traffic light signals have four phases as a case of study as shown in Table 1. However, note that the design concept can also be applied to other phase planes. In general, the number of phases for a four-bidirectional-signalized intersection is no more than eight, as shown in [12]. In particular, two- and six-phase plans [12, 20] are also applicable for such an intersection. Our PN model here can also be modified to account for two-phase and six-phase signalized traffic intersection. A phase is defined as an event of giving particular permissions to vehicles coming from specific directions to cross the physical area of the intersection towards specific destinations.

TABLE 1: Four phases of the signalized intersection considered in this paper.

Phase	Enabled movements
1	
2	
3	
4	

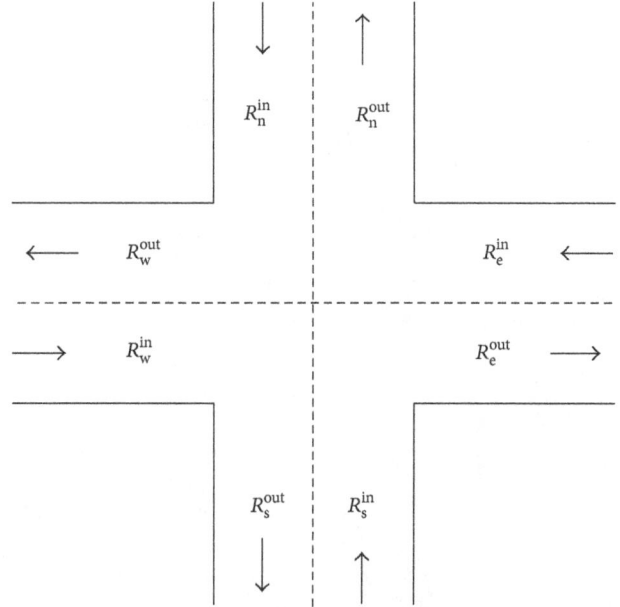

FIGURE 2: The structure of a single intersection.

As shown in Table 1, during Phase 1, vehicles flowing from east to south and from west to north are permitted to cross the intersection; during Phase 2, vehicles flowing from east to west (and north) and from west to east (and south) are permitted to cross the intersection; during Phase 3, vehicles flowing from north to east and from south to west are permitted to cross the intersection; and during Phase 4, vehicles flowing from north to south (and west) and from south to north (and east) are permitted to cross the intersection. It is not difficult to check that these signal phases do not conflict with each other and they comprise of a complete signal cycle.

3.2. Petri Net Model of the Single Intersection. The discrete Petri net model for the four-bidirectional road (shown in Figure 2) is depicted in Figure 3. The overall PN model is divided into 13 sub-PNs (modules), where each one is bounded by a dashed rectangular box. Eight of these modules represent the movement of vehicles while entering and crossing the intersection from left (N_l^{in}, S_l^{in}, E_l^{in}, and W_l^{in}) or from forward/right (N_f^{in}, S_f^{in}, E_f^{in}, and W_f^{in}). On the other hand, the four sub-PNs (N^{out}, S^{out}, E^{out}, and W^{out}) represent movement of vehicles while leaving the intersection toward

FIGURE 3: Petri net model for the single signalized intersection shown in Figure 2.

the four directions. Finally, the sub-PN (T) models the phase change of the traffic light signal.

The sub-PN (T) consists mainly of four places (p_{p1}, p_{p2}, p_{p3}, and p_{p4}) and four transitions (t_{p1}^s, t_{p2}^s, t_{p3}^s, and t_{p4}^s) that represent the four different phases mentioned in Table 1. The other nodes in module T exist only to make sure that the model offers a safe operation (no directional conflicts between vehicles while crossing the intersection). Having a token at the place p_{pk} indicates that the kth phase is taking place while losing this token means the ending of this phase. Furthermore, firing of transition t_{pk}^s reflects the starting of the kth phase. It is easy to note that the condition $p_{p1} + p_{p2} + p_{p3} + p_{p4} = 1$ has to be satisfied all the time.

Events related to vehicle movement through the intersection are described by other 12 sub-PNs; eight of them are used to model vehicles entering and crossing the intersection as I_j^{in} where I takes the notations (N, S, E, and W) and j takes either l (left) or f (forward and right). The subnet I_j^{in} consists of five places and four transitions; having a token in place p_{ij}^{in} implies that vehicles are entering the intersection from the ith incoming direction with the intention to take the direction of j. Similarly, having a token in place p_{ij}^q indicates the event that there is a queue of vehicles waiting in the incoming direction i with the intention to cross the intersection directed j (left or forward/right). Furthermore, a token at the places p_{ij}^p means that the permission is given to those vehicles in the queue to cross the intersection while a token existing in p_{ij}^c represents the event that those vehicles are crossing the intersection.

Finally, the other four sub-PNs (N^{out}, S^{out}, E^{out}, and W^{out}) are quite simple. Each of them consists of a place p_i^{out} and a transition t_i^{out}. A token in the aforementioned places describes the event that vehicles are leaving the intersection at the outgoing direction i.

3.3. Dynamics of the Petri Net Model. The state evolution of the PN model shown in Figure 3 is performed based mainly on the state equation given in (3) and on the assumption that whenever a transition is enabled it fires immediately. The fact that permission is given to vehicles entering the intersection from a specific direction towards a specific destination during a specific phase results in the consideration that this permission has to be canceled at the end of this phase no matter some vehicles cross the intersection or not. Consequently, eight conflicts might appear in the model between each coupled transitions (t_{ij}^p, t_{ij}^s). The proposed algorithm specifies a priority execution for each conflict. If the initial marking of the Petri net model is the one given in Figure 3, the only enabled transition is t_{p1}^s. Therefore, it fires immediately through removing a token from each input place to this transition (p_{p4}, p_{p4}^2) and adding a token to each output place of this transition (p_{p1}, p_{p1}^1, p_{wl}^p, and p_{el}^p). It simply means that the first phase starts (indicated via the token in place p_{p1}^{in}) and permission is given to vehicles to cross the intersection.

To clarify the firing sequence, let us stick only to those vehicles that enter from west towards north. As shown in Figure 3, initially a token exists in place p_{wl}^q, so vehicles are

waiting in the west entrances to take left with the intersection. However, a conflict appears at this step between transitions (t_{wl}^p, t_{wl}^s) since both of them are enabled and only one token is marked in their common input place p_{wl}^p. At this point, the algorithm will give the priority to transition t_{el}^p, whose firing will result in removing tokens from the permission and queuing places (p_{wl}^p, p_{wl}^q) and adding a token to the place p_{wl}^c, which indicates that the event of vehicles crossing the intersection from west to north is taking place simultaneously with the occurrence of the event that vehicles are crossing the intersection from east to south.

In the following step, three transitions are fired together, named (t_{wl}^c, t_{el}^c, and t_{p2}^s). Firing the transitions t_{wl}^c, t_{el}^c implies that vehicles are leaving the intersection towards south and north (through removing tokens from p_{wl}^c, p_{el}^c) and adding a token to each one of p_s^{out}, p_n^{out} indicates the event that vehicles are leaving intersection towards south and north directions. Simultaneously, in the sub-PN, the transition t_{p2}^s fires will indicate the starting of the second phase.

The necessity of having the sink transition t_{wl}^s is to make sure that vehicles cross the intersection safely; if we assume that no vehicles are waiting in the queue to cross the intersection from west to north during the first phase, the permission will still be available through the existence of a token in p_{wl}^p, which will lead to unsafe operation that vehicles coming from the west cross the intersection even after the ending of the first phase. Therefore, the role of the sink transition t_{wl}^s is to take away the permission if it has not been used during the first phase.

We developed an algorithm to identify the enabled transitions at each time step and fire the enabled transitions immediately based on the state equation and the aforementioned priority rule. Given the initial marking, input and output incident matrices, the algorithm will firstly construct the column matrix A (which initially is equal to the initial marking); then it will calculate the subsequent markings by state equation given in (3) and compare them with the existing marking; if it is not equal to any of them, it will be saved as a new column in matrix A and will be used to find the subsequent markings. Otherwise, if the newly calculated marking is equal to any previous ones (any column in A), the algorithm will stop. Matrix A at this point consists of all possible markings in their order of occurrences. The block diagram of the algorithm (Algorithm 1) for capturing the state evolution of the PN is shown in Figure 4.

In Table 2, we capture the state (marking) evolution of the Petri net model shown in Figure 3 related to each step of transition firing. Note that this state evolution is partial since we only show the number of tokens in some important places. The first column in the table represents the initial marking of the Petri net. When a specific event occurs at a specific firing step, it will be represented through the value "1" in the table. Similarly, the change of value from "1" to "0" in a specific table cell implies the ending of that specific event. Since the model consists of 56 places, it is not possible to include all of them in the table. We focus only on the places

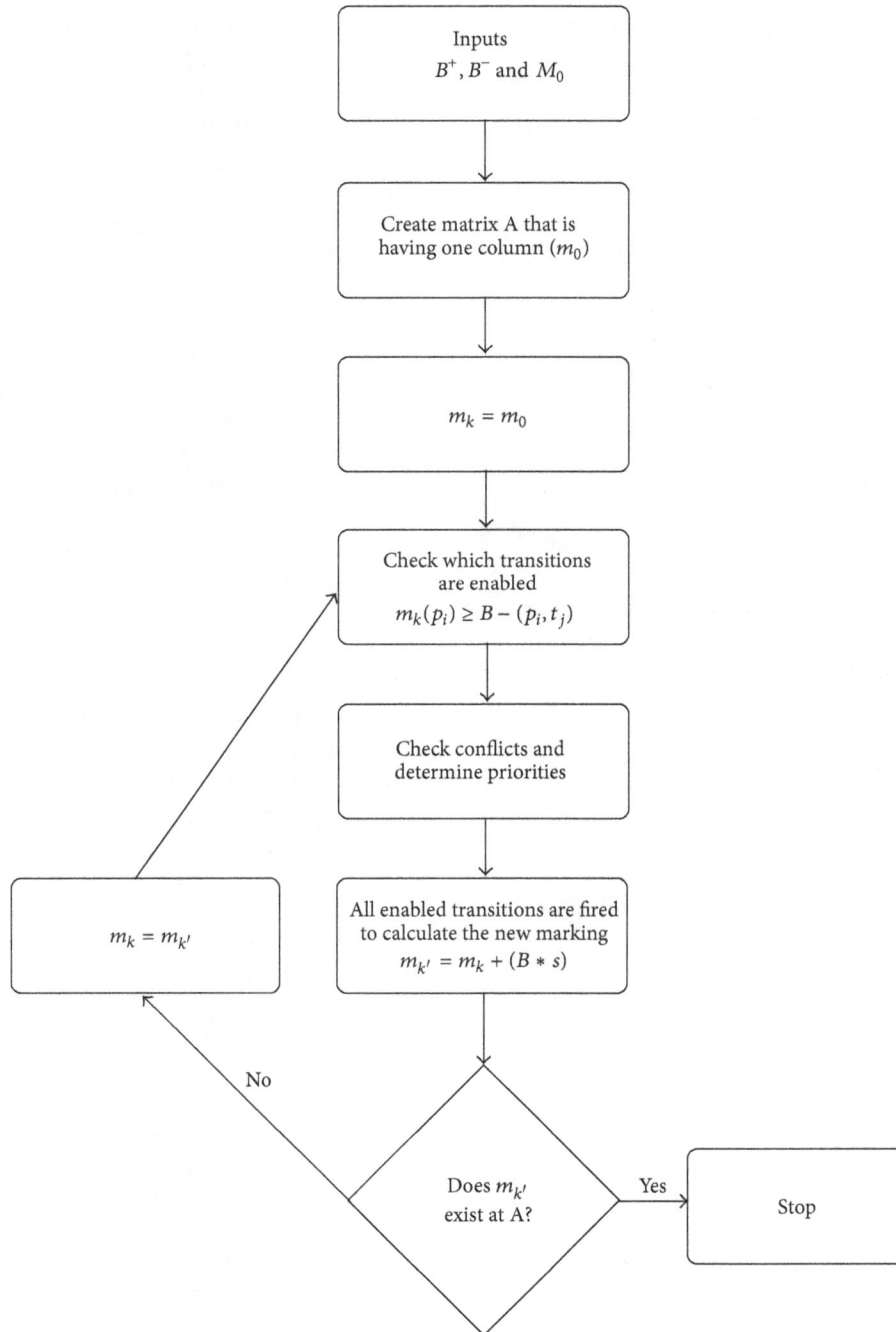

FIGURE 4: Block diagram of Algorithm 1 for capturing the state evolution of the Petri net shown in Figure 3.

with essential meanings. It is to be noted that at each firing step, one and only one of the four phase places (p_{p1}, p_{p2}, p_{p3}, and p_{p4}) will take the value "1" (which is expected as previously mentioned). It is also not difficult to notice that during the 10th step, the current marking goes back to the second generated marking in Table 2, which indicates that the proposed PN model is live and reversible.

If we use a different initial marking, such as a marking that represents the case where no vehicles enter the intersection from north, south, and east, the value "1" will never exist at the place p_w^{out} during any firing step, which means that the

event vehicles leave the intersection towards west will not take place. This is an expected logical result to avoid the conflicts of events.

4. Petri Net Model for Two Connected Intersections

In this section, we generalize the PN model obtained from the previous section to two connected intersections. The structure of two connected intersections is shown in Figure 5; if we call the intersection on the left side *Intersection-1* and

TABLE 2: Partial state evolution of the single-intersection Petri net model shown in Figure 3.

	m_0	1st	2nd	3rd	4th	5th	6th	7th	8th	9th	10th
p_{p1}	0	1	1	0	0	0	0	0	0	1	1
p_{p2}	0	0	0	1	1	0	0	0	0	0	0
p_{p3}	0	0	0	0	0	1	1	0	0	0	0
p_{p4}	1	0	0	0	0	0	0	1	1	0	0
p_{ij}^{in}	1	1	1	1	1	1	1	1	1	1	1
p_{wl}^c, p_{el}^c	0	0	1	0	0	0	0	0	0	0	1
p_{wf}^c, p_{ef}^c	0	0	0	0	1	0	0	0	0	0	0
p_{nl}^c, p_{sl}^c	0	0	0	0	0	0	1	0	0	0	0
p_{nf}^c, p_{sf}^c	0	0	0	0	0	0	0	0	1	0	0
p_w^{out}	0	0	0	0	0	1	0	1	0	1	0
p_e^{out}	0	0	0	0	0	1	0	1	0	1	0
p_n^{out}	0	0	0	1	0	1	0	0	0	1	0
p_s^{out}	0	0	0	1	0	1	0	0	0	1	0

the intersection on the right side *Intersection-2*, then we can say that the east output of *Intersection-1* is the west input to *Intersection-2* while the west output of *Intersection-2* is the east input to *Intersection-1*.

For this case, the Petri net model for traffic network shown in Figure 5 consists of two-single intersection PN models connected to each other. However, minor modifications are done for each model. For *Intersection-1* model, the arcs connecting t_{el}^{in} to p_{el}^{in} and t_{ef}^{in} to p_{ef}^{in} have been removed; while for *Intersection-2* model, arcs connecting t_{wl}^{in} to p_{wl}^{in} and t_{wf}^{in} to p_{wf}^{in} have been removed. To distinguish between the PN nodes belonging to each intersection, we added (2) as the subscript for each node symbol in *Intersection-1*, similarly adding (3) for *Intersection-2*. The two-single intersection models are connected to each other through their east and west inputs/outputs nodes, the PN model for the connection between two intersections is shown in Figure 6.

During the event that vehicles are leaving *Intersection-1*, a token existing at $p_{e(1)}^{out}$ will result in the firing of transition $t_{e(1)}^{out}$, which removes the token from $p_{e(1)}^{out}$ and adds a token to both places $p_{wf(2)}^{in}$ and $p_{wl(2)}^{in}$ that represents the occurrence of the event that vehicles are entering *Intersection-2* from the west. However, this step does not perform very smoothly when we execute the two-intersection model using Algorithm 1.

The case is that, during one-phase cycle (the four phases took place once successively), transitions $t_{e(1)}^{out}$ and $t_{w(2)}^{out}$ fire three times because places $p_{e(1)}^{out}$ and $p_{w(2)}^{out}$ are charged three times per cycle since there are vehicles leaving *Intersection-1* east directed and *Intersection-2* west directed during three phases in a cycle. On the other hand, the four places, $p_{ef(1)}^{in}$, $p_{el(1)}^{in}$, $p_{wf(2)}^{in}$, and $p_{wl(2)}^{in}$, gain three tokens each per cycle and release only one of these tokens since the transitions downstream fire only once per cycle. It is obvious that this firing mechanism will lead the Petri net to be unbounded. Physically, the model will not be considered as a good representation for event occurrences because at some point if no vehicle enters the network through one of the three entrance of *Intersection-1*: west, north, and south, there will

be still vehicles traveling from *Intersection-1* to *Intersection-2* which is not a real case. That is because places $p_{ef(1)}^{in}$, $p_{el(1)}^{in}$ will have tokens to feed the net of *Intersection-2*. The same case of unreal representation will exist for vehicles traveling from *Intersection-2* towards *Intersection-1*.

To tackle this modeling issue, we have tried several approaches. Firstly, a Petri net controller was designed based on [21] to keep the number of tokens in each of the four places not exceeding one. This controller worked for these four places but it transferred the unbounded problem to the upstream places $p_{e(1)}^{out}$ and $p_{w(2)}^{out}$. Adding another controller places to control these two nodes will not solve the problem because it causes conflicts which stop the execution of Algorithm 1 when it is used for the new model (with the controller). Designing a controller for the six places in one step will cause similar problem. Therefore, we propose the concept of the modified binary Petri net (MBPN), which will be presented in detail in the next section.

5. Modified Binary Petri Nets

5.1. Introduction to Modified Binary Petri Nets. As it can be seen from the previous discussions, we use PNs to determine whether an event has occurred or not based on the occurrence of other events. Modeling in such a way can be easily done through representing the occurrence of a specific event by associating a token in a specified place that describes this event. However, modeling in this traditional way will lead to problems in our models as discussed above.

Thus, we propose the concept of the modified binary Petri net (MBPN). The idea is to follow the traditional Petri net firing mechanisms, but with a restriction that no places can be marked with more than one token. Thus, the MBPN can be defined as follows.

(i) The weights of all arcs in the MBPN are ones, which indicates that the net is ordinary and all elements in the input and output incidents matrices can only be zeros or ones.

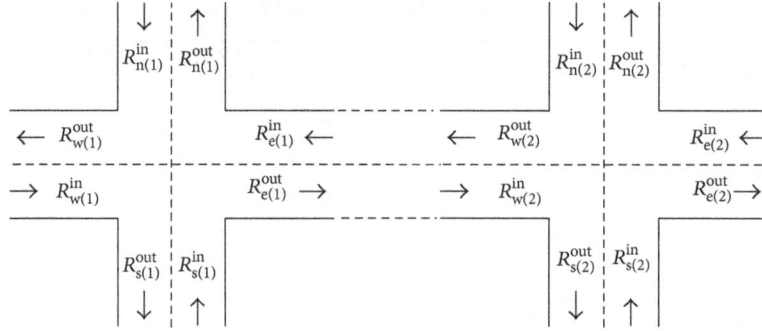

FIGURE 5: The structure of two connected intersections.

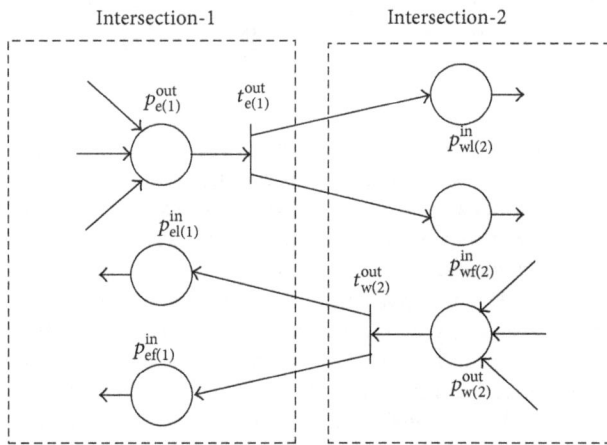

FIGURE 6: Petri net model for the connections between two intersections.

(ii) The initial marking of the MBPN can only be zero or one for each place.

(iii) State evolution from one marking to another is performed through the following two steps:

(1) we calculate marking $m_{k'}$ according to traditional state equation of Petri nets:

$$m_{k'} = m_k + B \cdot s; \tag{7}$$

(2) we update the marking in places according to the number of tokens as follows:

$$m_{k'}(p) = \begin{cases} 0 & \text{for places } p \text{ such that } m_{k'}(p) = 0, \\ 1 & \text{for places } p \text{ such that } m_{k'}(p) \geq 1. \end{cases} \tag{8}$$

Based on this definition, it is easy to see that for MBPN we have the following:

(i) elements of incident matrices take only three values −1, 0, or 1;

(ii) any modified binary Petri net is bounded since the number of states is less than or equal to 2^n, where n is the number of places.

To illustrate the differences between the traditional PN and the MBPN, we can use the example provided in Figure 1 for illustration. If Algorithm 1 is applied for traditional PN, the marking evolution will be given by

$$
\begin{bmatrix} 1 \\ 0 \\ 0 \\ 0 \end{bmatrix} \xrightarrow{t_1} \begin{bmatrix} 0 \\ 1 \\ 1 \\ 0 \end{bmatrix} \xrightarrow{t_2,t_3} \begin{bmatrix} 0 \\ 0 \\ 0 \\ 2 \end{bmatrix} \xrightarrow{t_4} \begin{bmatrix} 1 \\ 0 \\ 0 \\ 1 \end{bmatrix} \xrightarrow{t_1,t_4} \begin{bmatrix} 1 \\ 1 \\ 1 \\ 0 \end{bmatrix}
$$

$$
\xrightarrow{t_1,t_2,t_3} \begin{bmatrix} 0 \\ 1 \\ 1 \\ 2 \end{bmatrix} \xrightarrow{t_2,t_3,t_4} \begin{bmatrix} 1 \\ 0 \\ 0 \\ 3 \end{bmatrix} \xrightarrow{t_1,t_4} \begin{bmatrix} 1 \\ 1 \\ 1 \\ 2 \end{bmatrix}
$$

$$
\xrightarrow{t_1,t_2,t_3,t_4} \begin{bmatrix} 1 \\ 1 \\ 1 \\ 3 \end{bmatrix} \xrightarrow{t_1,t_2,t_3,t_4} \begin{bmatrix} 1 \\ 1 \\ 1 \\ 4 \end{bmatrix} \cdots \begin{bmatrix} 0 \\ 1 \\ 1 \\ r-1 \end{bmatrix}
$$

$$
\xrightarrow{t_1,t_2,t_3,t_4} \begin{bmatrix} 1 \\ 1 \\ 1 \\ r \end{bmatrix} \xrightarrow{t_1,t_2,t_3,t_4} \begin{bmatrix} 1 \\ 1 \\ 1 \\ r+1 \end{bmatrix} \cdots . \tag{9}
$$

As it is shown above, the PN is unbounded due to place p_4, in which the number of tokens will keep increasing after the marking $[1 \ \ 1 \ \ 1 \ \ 2]^T$.

Now we consider the MBPN. It is not difficult to see that the marking evolution is given by

$$
\begin{bmatrix} 1 \\ 0 \\ 0 \\ 0 \end{bmatrix} \xrightarrow{t_1} \begin{bmatrix} 0 \\ 1 \\ 1 \\ 0 \end{bmatrix} \xrightarrow{t_2,t_3} \begin{bmatrix} 0 \\ 0 \\ 0 \\ 1 \end{bmatrix} \xrightarrow{t_4} \begin{bmatrix} 1 \\ 0 \\ 0 \\ 0 \end{bmatrix} . \tag{10}
$$

Thus, MBPN has bounded markings and avoids the potential problems in creating extra tokens for our connected traffic intersection model. In general, for traffic applications that can be represented through binary-state nodes (0 or 1), MBPN is a suitable modeling tool and can guarantee the smooth representation for event occurrences in these applications.

5.2. Using MBPN Firing Mechanism for Studying the Two Connected Intersections. Now we investigate the PN model

TABLE 3: A partial set of marking evolution of MBPN model for two connected intersections.

	m_0	1st	2nd	3rd	4th	5th	6th	7th	8th	9th	10th	11th	12th	13th	14th	15th
$P_{p1(1)}, P_{p1(2)}$	0	1	1	0	0	0	0	0	0	1	1	0	0	0	0	0
$P_{p2(1)}, P_{p2(2)}$	0	0	0	1	1	0	0	0	0	0	0	1	1	0	0	0
$P_{p3(1)}, P_{p3(2)}$	0	0	0	0	0	1	1	0	0	0	0	0	0	1	1	0
$P_{p4(1)}, P_{p4(2)}$	1	0	0	0	0	0	0	1	1	0	0	0	0	0	0	1
$P_{e(1)}^{out}$	0	0	0	0	0	1	0	1	0	1	0	0	0	1	0	1
$P_{wl(2)}^{in}$	0	0	0	0	0	0	1	0	1	1	1	0	0	0	1	1
$P_{wf(2)}^{in}$	0	0	0	0	0	0	1	0	1	1	1	1	1	0	1	1
$P_{w(2)}^{out}$	0	0	0	0	0	1	0	1	0	1	0	0	0	1	0	1
$P_{el(1)}^{in}$	0	0	0	0	0	0	1	0	1	1	1	0	0	0	1	1
$P_{ef(1)}^{in}$	0	0	0	0	0	0	1	0	1	1	1	1	1	0	1	1

shown in Figure 6. We use Algorithm 1 with a small modification that takes into account (8) mentioned above by MBPN. The algorithm performed 15 steps before it stopped due to the condition of having repeated markings. Because of the size of the PN model ($P = 112$, $T = 88$), we do not show markings for all places in the net. Instead, in Table 3, we show the markings of the places which represent the connections between two intersections. It is easy to note that the MBPN framework resolves the issue of unbounded states and is a more realistic tool for the modeling and analysis of certain traffic applications other than traditional PNs.

5.3. Discussions on Potential Extensions to Traffic Networks. Starting from two connected intersections model, we can expand the approach proposed in this paper to study traffic network with much more complex dynamics. To see this implication, assuming that we have a complex traffic network consisting of n intersections, we can derive its input incident matrix B^- composed from input incident block matrices as:

$$B^- = \begin{bmatrix} B_1^- & 0 & \cdots & 0 \\ \vdots & B_2^- & 0 & \vdots \\ 0 & \vdots & \ddots & 0 \\ 0 & 0 & \cdots & B_n^- \end{bmatrix}, \quad (11)$$

where $B_1^-, B_2^-, \ldots, B_n^-$ are the input incident matrices for intersections $I_1, I_2, \ldots, I_n$, respectively. The sizes of these block matrices may be different because they depend on the intersection specifications such as the number of incoming and outgoing roads connected to them and the number of phases applied to them. Note that the off-diagonal entries are zeros because intersections are connected through transitions to places.

For the output incident matrix, the diagonal block matrices will be the output incident matrices $B_1^+, B_2^+, \ldots, B_n^+$ for the intersections $I_1, I_2, \ldots, I_n$. Off-diagonal block matrices represent the connection between two intersections. For instance, B_{1-2}^+ represents the connections from intersection I_1 to intersection I_2, B_{2-1}^+ represents the connection from I_2

to I_1, and so forth. The general form of the output incident matrix for the overall traffic network can be written as

$$B^+ = \begin{bmatrix} B_1^+ & B_{2-1}^+ & \cdots & B_{n-1}^+ \\ B_{1-2}^+ & B_2^+ & \cdots & \vdots \\ \vdots & \vdots & \ddots & \vdots \\ B_{1-n}^+ & B_{2-n}^+ & \cdots & B_n^+ \end{bmatrix}. \quad (12)$$

From the discussions above, we observe that we can easily expand the two-intersection model proposed in this paper to a large traffic network with much more intersections. In addition, the MBPN model proposed can guarantee the boundedness of places in the net; thus, it is suitable for capturing the event occurrences of the overall traffic network.

6. Conclusions and Discussions

In this paper, we developed a discrete Petri net model for a single traffic intersection and then extended it to represent a traffic network consisting of two connected intersections. This model was based on traditional Petri nets in order to analyze the effect of occurrence/nonoccurrence of specific events on other network nodes without considering the detailed dynamics. An algorithm was designed to capture transition firings based on the state equation and the priority assignment. We noticed that the traditional Petri net model for two connected intersections will lead to unbounded places in the model, which are not realistic for the monitoring and control. Thus, we introduced a modification to the traditional Petri nets to resolve this issue. We called this new type of Petri nets the *modified binary Petri nets*, which have a good representation for traffic network in terms of events occurrences and traffic dynamic correlations.

One direction for our future work is to enhance the model with timing information by using timed transitions. By incorporating with time information, the new model will be suitable for event simulations in large traffic networks. Another interesting research direction is to identify a wider range of applications (e.g., routing, scheduling, etc.) of modified binary Petri nets in traffic systems and networks.

References

[1] L. Li and D. S. Kim, "Least-cost path estimation in wireless ad hoc sensor networks using Petri nets," in *Proceedings of the 5th ACM International Conference on Ubiquitous Information Management and Communication (ICUIMC '11)*, Seoul, South Korea, February 2011.

[2] C. N. Hadjicostis and G. C. Verghese, "Power system monitoring using Petri net embeddings," *IEE Proceedings C*, vol. 147, no. 5, pp. 299–303, 2000.

[3] N. Lu, *Power system modeling using Petri nets [Ph.D. thesis]*, 2002, http://www.ima.umn.edu/~mali/thesis.pdf.

[4] F. S. Hsieh and J. B. Lin, "Context-aware workflow management for virtual enterprises based on coordination of agents," *Journal of Intelligent Manufacturing*, 2012.

[5] F. S. Hsieh and C. Y. Chiang, "Collaborative composition of processes in holonic manufacturing systems," *Computers in Industry*, vol. 62, no. 1, pp. 51–64, 2011.

[6] F. S. Hsieh, "Design of reconfiguration mechanism for holonic manufacturing systems based on formal models," *Engineering Applications of Artificial Intelligence*, vol. 23, no. 7, pp. 1187–1199, 2010.

[7] L. Li, C. N. Hadjicostis, and R. S. Sreenivas, "Designs of bisimilar Petri net controllers with fault tolerance capabilities," *IEEE Transactions on Systems, Man, and Cybernetics A*, vol. 38, no. 1, pp. 207–217, 2008.

[8] A. di Febbraro and S. Sacone, "Hybrid modelling of transportation systems by means of Petri nets," in *Proceedings of the IEEE International Conference on Systems, Man, and Cybernetics*, pp. 131–135, October 1998.

[9] A. di Febbraro, D. Giglio, and N. Sacco, "Modular representation of urban traffic systems based on hybrid petri nets," in *Proceedings of the IEEE International Conferences on Intelligent Transportation Systems*, pp. 866–871, August 2001.

[10] C. R. Vázquez, H. Y. Sutarto, R. Boel, and M. Silva, "Hybrid Petri net model of a traffic intersection in an urban network," in *Proceedings of the IEEE International Conference on Control Applications (CCA '10)*, pp. 658–664, Yokohama, Japan, September 2010.

[11] A. di Febbraro, D. Giglio, and N. Sacco, "Urban traffic control structure based on hybrid petri nets," *IEEE Transactions on Intelligent Transportation Systems*, vol. 5, no. 4, pp. 224–237, 2004.

[12] G. F. List and M. Cetin, "Modeling traffic signal control using Petri nets," *IEEE Transactions on Intelligent Transportation Systems*, vol. 5, no. 3, pp. 177–187, 2004.

[13] Y. Qu, L. Li, Y. Liu, Y. Chen, and Y. Dai, "Travel routes estimation in transportation systems modeled by Petri nets," in *Proceedings of the IEEE International Conference on Vehicular Electronics and Safety*, pp. 73–77, Qingdao, China, July 2010.

[14] Y. S. Huang and T. H. Chung, "Modeling and analysis of urban traffic lights control systems using timed CP-nets," *Journal of Information Science and Engineering*, vol. 24, no. 3, pp. 875–890, 2008.

[15] Y. S. Huang and P. J. Su, "Modelling and analysis of traffic light control systems," *IET Control Theory and Applications*, vol. 3, no. 3, pp. 340–350, 2009.

[16] Y. S. Huang, Y. S. Weng, and M. Zhou, "Critical scenarios and their identification in parallel railroad level crossing traffic control systems," *IEEE Transactions on Intelligent Transportation Systems*, vol. 11, no. 4, pp. 968–977, 2010.

[17] C. G. Cassandras and S. Lafortune, *Introduction to Discrete Event Systems*, Springer, New York, NY, USA, 2008.

[18] T. Murata, "Petri nets: properties, analysis and applications," *Proceedings of the IEEE*, vol. 77, no. 4, pp. 541–580, 1989.

[19] R. David and H. Alla, *Discrete, Continuous, and Hybrid Petri Nets*, Springer, New York, NY, USA, 2005.

[20] Y. S. Huang, "Design of traffic light control systems using statecharts," *The Computer Journal*, vol. 49, no. 6, pp. 634–649, 2006.

[21] J. Moody, K. Yamalidou, M. Lemmon, and P. J. Antsaklis, "Feedback control of petri nets based on place invariants," *Automatica*, vol. 32, no. 1, pp. 15–28, 1996.

6

A Driver Face Monitoring System for Fatigue and Distraction Detection

Mohamad-Hoseyn Sigari,[1] Mahmood Fathy,[2] and Mohsen Soryani[2]

[1] Control and Intelligent Processing Center of Excellence (CIPCE), School of Electrical and Computer Engineering, College of Engineering, University of Tehran, Tehran 14399, Iran
[2] Computer Engineering Department, Iran University of Science and Technology, Tehran 16846, Iran

Correspondence should be addressed to Mohamad-Hoseyn Sigari; hoseyn_sigari@ieee.org

Academic Editor: Chyi-Ren Dow

Driver face monitoring system is a real-time system that can detect driver fatigue and distraction using machine vision approaches. In this paper, a new approach is introduced for driver hypovigilance (fatigue and distraction) detection based on the symptoms related to face and eye regions. In this method, face template matching and horizontal projection of top-half segment of face image are used to extract hypovigilance symptoms from face and eye, respectively. Head rotation is a symptom to detect distraction that is extracted from face region. The extracted symptoms from eye region are (1) percentage of eye closure, (2) eyelid distance changes with respect to the normal eyelid distance, and (3) eye closure rate. The first and second symptoms related to eye region are used for fatigue detection; the last one is used for distraction detection. In the proposed system, a fuzzy expert system combines the symptoms to estimate level of driver hypo-vigilance. There are three main contributions in the introduced method: (1) simple and efficient head rotation detection based on face template matching, (2) adaptive symptom extraction from eye region without explicit eye detection, and (3) normalizing and personalizing the extracted symptoms using a short training phase. These three contributions lead to develop an adaptive driver eye/face monitoring. Experiments show that the proposed system is relatively efficient for estimating the driver fatigue and distraction.

1. Introduction

Improvement of public safety and the reduction of accidents are of the important goals of the Intelligent Transportation Systems (ITS). One of the most important factors in accidents, especially on rural roads, is the driver fatigue and monotony. Fatigue reduces driver perceptions and decision making capability to control the vehicle. Researches show that usually the driver is fatigued after 1 hour of driving. In the afternoon early hours, after eating lunch and at midnight, driver fatigue and drowsiness is much more than other times. In addition, drinking alcohol, drug addiction, and using hypnotic medicines can lead to loss of consciousness [1, 2].

In different countries, different statistics were reported about accidents that happened due to driver fatigue and distraction. Generally, the main reason of about 20% of the crashes and 30% of fatal crashes is the driver drowsiness

and lack of concentration. In single-vehicle crashes (accidents in which only one vehicle is damaged) or crashes involving heavy vehicles, up to 50% of accidents are related to driver hypovigilance [1, 3–5]. According to the current studies, it is expected that the amount of crashes will be reduced by 10%–20% using driver face monitoring systems [6].

The driver face monitoring system is a real-time system that investigates the driver physical and mental condition based on the processing of driver face images. The driver state can be estimated from the eye closure, eyelid distance, blinking, gaze direction, yawning, and head rotation. This system will alarm in the hypovigilance states including fatigue and distraction. The major parts of the driver face monitoring system are (1) imaging, (2) hardware platform, and (3) the intelligent software.

In the driver face monitoring systems, two main challenges can be considered: (1) "how to measure the fatigue?"

and (2) "how to measure the concentration?". These problems are the main challenges of a driver face monitoring system.

The first challenge is how to define fatigue exactly and how to measure it. Despite the progress of science in physiology and psychology, there is still no precise definition for fatigue. Certainly, due to the lack of precise definition of fatigue, there is not any measurable criterion or tool [3]. However, a precise definition for fatigue is not defined yet, but there is a relationship between fatigue and some symptoms including body temperature, electrical resistance of skin, eye movement, breathing rate, heart rate, and brain activity [2, 3, 7, 8]. One of the first and most important symptoms of fatigue appears in the eye. There is a very close relationship between Psychomotor Vigilance Task (PVT) and the percentage of eyelid closure over time (PERCLOS). PVT shows the response speed of a person to a visual stimulation. Therefore, almost in all driver face monitoring systems, eye closure detection is the first symptom used to measure fatigue.

The second challenge is measuring the driver attention to the road. The driver attention can be partly estimated from the driver head and gaze direction. The main problem is that if the head is forward and looking toward the road, the driver does not necessarily pay attention to the road. In other words, looking toward the road is not paying attention to it [3].

In this paper, a new driver face monitoring system is proposed which extracts the hypovigilance symptoms from driver face and eye adaptively. Then, the symptoms are analyzed by a fuzzy expert system to determine the driver state. The remainder of paper is organized as follow. In Section 2, some previous researches are reviewed. The proposed system is described with details in Section 3. In Section 4, the experimental results and discussions are presented. Section 5 is related to the conclusions.

2. Previous Works

The driver face monitoring systems can be divided into two general categories. In one category, driver fatigue and distraction is detected only by processing of eye region. There are many researches based on this approach. The main reason of this large amount of researches is that the main symptoms of fatigue and distraction appear in the driver eyes. Moreover, the processing of the eye region instead of the processing of the face region has less computational complexity. In the other category, the symptoms of fatigue and distraction are detected not only from eyes, but also from other regions of the face and head. In this approach, in addition to processing of eye region, other symptoms including yawning and head nodding are also extracted.

Driver face monitoring system includes some main parts: (1) face detection, (2) eye detection, (3) face tracking, (4) symptom extraction, and (5) driver state estimation. These main parts are reviewed in different systems in the current section.

In the most of driver face monitoring systems, the face detection is the first part of the image processing operations. Face detection methods can be divided into two general categories [9]: (1) feature-based and (2) learning-based methods.

In the feature-based methods, the assumption is that the face in the image can be detected based on applying heuristic rules on features. These methods are usually used for detecting one face in the image. Color-based face recognition is one of the fast and common methods. In these methods, the face is detected based on the color of skin and the shape of face. Color-based face detection may be applied on different color-space including RGB [10, 11], YCbCr [12], or HIS [13]. In noisy images or in the images with low illuminations, these algorithms have low accuracy.

Learning-based face detection uses statistical learning methods and training samples to learn the discriminative features. These methods benefit from statistical models and machine learning algorithms. Generally, learning-based methods have less error rates for face detection, but these methods usually have more computational complexity. Viola and Jones [14] presented an algorithm for object detection, which is very fast and robust. This algorithm was used in [15–17] for face detection.

Almost in all driver face monitoring systems, because of the importance of symptoms related to eye, the eye region is always processed for extracting the symptoms. Therefore, before the processing of eye region, eye detection is required. Eye detection methods can be divided into three general categories: (1) methods based on the imaging in the infrared spectrum, (2) feature-based methods, and (3) other methods.

One of the fast and relatively accurate methods for eye detection is the method based on the imaging in the infrared (IR) spectrum. In this method, physiological and optical properties of the eye in the IR spectrum are used. The eye pupil reflects IR beams, and it seems as a bright spot when the angle of IR source and imaging device are suitable. According to this interesting property, pupil and eye are detected. The systems proposed in [4, 18–20] used such method for eye detection.

Feature-based eye detection approach includes various methods. Image binarization [5, 21, 22] and projection [23, 24] are two feature-based eye detection methods which assume that the eye is darker than the face skin. Usually, more complicated processing is needed to detect the proper location of eyes, because these methods are simple and have high error rate.

There are few methods for eye detection based on other approaches which were used in driver face monitoring systems. In [10], a geometrical face model with some feature-based methods was used to detect eyes. In addition, some systems such as [15] used hybrid methods for eye detection. In [15], elliptical gray-level template matching and IR imaging system were used for eye detection in day and night, respectively.

Usually, the entire image is searched for detecting the face/eye. Searching the entire image increases the computational complexity of the system. Therefore, usually after early detection of the face/eyes, in the next frames, face/eye tracking is performed. In the most of driver face monitoring systems, Kalman filter [4, 19, 25] or extended versions of Kalman filter such as Unscented Kalman Filter (UKF) [23] were used. However, in some researches, search window [18] and particle filter (PF) [26] were used for tracking.

In the driver face monitoring systems, useful symptoms for fatigue and distraction detection can be divided into three general categories:

(i) symptoms related to the eye region;

(ii) symptoms related to the mouth region;

(iii) symptoms related to the head.

Eye is the most important area of the face where the symptoms of fatigue and distraction appear in it. Therefore, many of the driver face monitoring systems detect driver fatigue and distraction only based on the symptoms extracted from the eyes. The symptoms related to eye region include PERCLOS [3, 4, 10, 15], eyelid distance [25, 27], eye blink speed [4, 10], eye blink rate [4, 19], and gaze direction [4].

Yawning is one of the hypovigilance symptoms related to the mouth region. This symptom was extracted by detecting the open mouth in [11, 16]. These systems detect the mouth based on the color features of the lips in the image.

Some fatigue and distraction symptoms are related to head. These symptoms include head nodding [5, 19] and head orientation [4, 10, 19]. Head nodding can be used for fatigue detection, and head orientation can be used for both fatigue and distraction detection. Driver nodding and lack of driver attention to the road can be detected by estimating the angle of head direction.

After symptom extraction, the driver state has to be determined. The determination of the driver state is considered as a classification problem. The simplest method for detecting the driver fatigue or distraction is based on applying a threshold on extracted symptom [22].

Another method for determining the driver state is knowledge-based approaches. In a knowledge-based approach, decision making about the driver fatigue and distraction is based on the knowledge of an expert which the knowledge usually appears in the form of if-then rules. In [19, 25], fuzzy expert systems were used as knowledge-based approach for estimating the driver state.

More complicated approaches such as Bayesian network [4] and nave dynamic Bayesian network [26] were used for driver state determination. These approaches are usually more accurate than threshold-based and knowledge-based approaches; however, they are more complicated.

3. The Proposed System

The proposed system is a driver face monitoring system that can detect driver hypovigilance (both fatigue and distraction) by processing of eye and face regions. Flowchart of our system is shown in Figure 1. After image acquisition, face detection is the first stage of processing. Then, symptoms of hypovigilance are extracted from face image. However, an explicit eye detection stage is not used to determine the eye in the face, but some of important symptoms related to eye region (top-half segment of the face) are extracted. Additionally, a template matching method is used for detecting the head rotation. Finally, we used a fuzzy expert system to estimate driver hypo-vigilance.

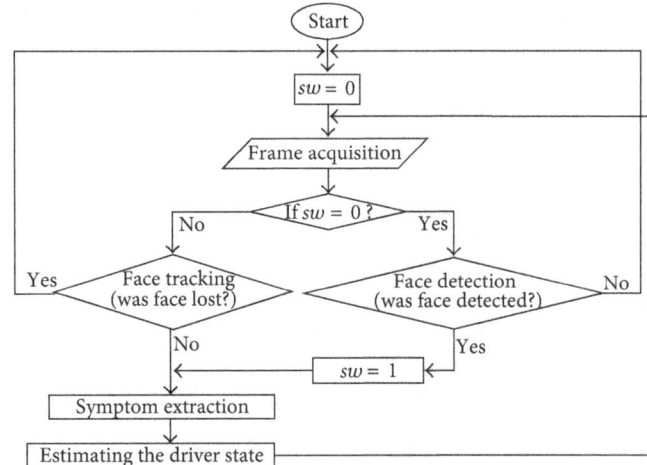

FIGURE 1: Flowchart of the proposed system.

Performing the face detection algorithm for all frames is computationally complex. Therefore, after face detection in the first frame, face tracking algorithms are used to track driver face in the next frames unless the face is lost. Therefore, we use an auxiliary variable denoted by sw for determination of face tracking status in Figure 1. If sw is 0, the face is lost, and face detection algorithm must be performed to localize the driver face. In contrast, if sw is 1, it shows that face is tracked successfully by face tracking method. For system initialization, sw is 0. It means that the system must perform face detection algorithm for first frame.

We used Haar-like features and adaptive boosting method proposed by Viola and Jones [14] for face detection. Face detection algorithm was trained by about 3000 faces and about 300000 nonfaces. For face tracking, full search method is used to find the driver face image in the new frame. The search region is around the center of face image in the last frame which the size of search region is changed according to the size of face image (1.5 times bigger than the size of face image). Then, correlation coefficient between the face image and the subwindows of search region is used as the matching criteria.

3.1. The Symptom Extraction. In the proposed system, two types of symptoms are extracted: (1) the symptoms related to eye region and (2) the symptom related to face region. The symptoms related to eye region are PERCLOS, eyelid distance changes with respect to the normal eyelid distance (ELDC), and eye closure rate (CLOSNO). The symptom related to face region is head rotation (ROT).

3.1.1. The Symptoms Related to Eye Region. The proposed system uses horizontal projection in top-half segment of face image to extract symptoms of driver hypovigilance. Our proposed method uses a spatiotemporal approach without explicit eye detection for feature extraction which is not very sensitive to illumination, skin color, and wearing glasses, because it is an adaptive method. This method is based on changing the horizontal projection of top-half segment of face image during time. Horizontal projection in image I is computed by

$$HP(j) = \sum_i I(i, j). \qquad (1)$$

Length of HP is equal to the height of I. In our proposed system, only horizontal projection of top-half segment of face image is used, so the length of horizontal projection will be equal to half height of driver face image. Before extracting the symptoms related to eye region, system needs to be trained. Because of different eyelid behavior in different individuals, estimating driver vigilance level based on absolute values is not suitable for robustness of driver face monitoring systems. Therefore, for developing a robust and adaptive system, normal values of the vigilance symptoms must be estimated by training phase. In our proposed method, "training" has a little different definition in comparison with general machine learning systems. In the proposed method, training means extracting normal value of vigilance symptoms of driver. Therefore, training phase is a short period of time that we assume that driver is fully aware and looking forward. In training phase, the normal values of PERCLOS, CLOSNO, and ELDC are calculated. Normal values of PERCLOS and CLOSNO are denoted by $PERCLOS_N$ and $CLOSNO_N$, respectively. Because the eye is not detected explicitly, the eyelid distance and normal eyelid distance are estimated implicitly. The eyelid distance is estimated by the horizontal projection of top-half segments of face; therefore, the average horizontal projection of top-half segments of face is computed during training phase to estimate the normal eyelid distance.

Training duration is about 1-2 minutes. In the first 100 frames of training sequence, we suppose that driver eyes are usually open. So, horizontal projection of open-eyes can be estimated by computing average of horizontal projections of first 100 frames. Horizontal projection of open eyes was named HP_{LO}, and it can be computed by (2). In (2), HP_i is the horizontal projection of frame i and N is 100. Consider

$$HP_{LO} = \frac{\sum_{i=1}^{N} HP_i}{N}. \qquad (2)$$

Eye closure can be detected by computing the correlation of horizontal projection of current frame (HP_i) and HP_{LO}. The correlation of HP_i and HP_{LO} is denoted by CHP_i. If CHP_i is larger than th_{CHP}, eye is open in frame i, otherwise, the eye is closed. Consider

$$CHP_i = Corr(HP_{LO}, HP_i),$$

$$\begin{cases} eye\ is\ closed & if\ CHP_i < th_{CHP} \\ eye\ is\ open & if\ else. \end{cases} \qquad (3)$$

After computing the HP_{LO} as horizontal projection of open eyes, a copy of HP_{LO} is named as HP_O. HP_O will be updated during acquisition of new frames using fuzzy running average method [28], while HP_{LO} is not updated. In fuzzy running average method, updating HP_O is dependent to the matching degree (correlation coefficient) of HP_O

and HP_i. Fuzzy running average is shown in (4). In (4), α represents the weighting factor and is calculated based on CHP_i as shown in (5). Consider

$$HP_O = \alpha \cdot HP_O + (1 - \alpha) HP_i, \qquad (4)$$

$$a = 1 - (1 - \alpha_{min}) \exp(-5^* CHP_i). \qquad (5)$$

In (5), α_{min} is a constant (0.8 in our system) and represents the minimum value of α. According to (5), α varies in range [0.8, 1]. A higher α updates HP_O slower. Therefore, HP_O is updated during driving based on the changes of HP_i.

Eye closure state is saved in a circular list ($L_{eye\_closure}$). If eye is open, the current element of $L_{eye\_closure}$ will be 1, else, the current element of $L_{eye\_closure}$ will be 0. When $L_{eye\_closure}$ is full, the oldest data is replaced by new data. Length of $L_{eye\_closure}$ (N_L) must be equal to the number of training frames (about 1500–3000). $L_{eye\_closure}$ is helpful for computing PERCLOS and CLOSNO, but ELDC is computed using correlation of current horizontal projection (HP_i) and HP_{LO}. HP_{LO} shows the eyelid distance of driver in normal state implicitly.

PERCLOS shows the percentage of eye closure during last frames computed by

$$PERCLOS = \frac{N_L - \sum L_{eye\_closure}}{N_L}. \qquad (6)$$

CLOSNO shows eye blink rate (frequency) in a given duration. If $DL_{eye\_closure}$ is the first derivation of $L_{eye\_closure}$, CLOSNO can be computed based on $DL_{eye\_closure}$. According to (7), $DL_{eye\_closure}$ indicates the start and stop frames of eye closure events by +1 and −1, respectively, and other elements of $DL_{eye\_closure}$ are zero. Therefore, CLOSNO is computed by (8). Consider

$$DL_{eye\_closure}(i) = \begin{cases} L_{eye\_closure}(1) - L_{eye_{closure}}(N_L), & i = 1, \\ L_{eye\_closure}(i) - L_{eye_{closure}}(i-1), & i \neq 1. \end{cases} \qquad (7)$$

$$CLOSNO = \frac{\sum |DL_{eye\_closure}|}{2}. \qquad (8)$$

ELDC is computed based on correlation between current horizontal projection of open eyes (HP_O) and horizontal projection of open eyes in training phase (HP_{LO}) according to

$$ELDC = 1 - Sigm(Corr(HP_O, HP_{LO}), \alpha_S, \beta_S). \qquad (9)$$

In (9), Sigm is the sigmoid function, and α_S and β_S are the parameters of sigmoid function. α_S and β_S show the slope and displacement of sigmoid function respectively. General form of sigmoid function is shown in

$$Sigm(x, \alpha_S, \beta_S) = \frac{1}{1 + \exp(\alpha_S \cdot (x - \beta_S))}. \qquad (10)$$

In the proposed system, $\alpha_S = 5$ and $\beta_S = 0.5$. Because the range of sigmoid function is [0, 1], ELDC is always in range [0, 1]. If ELDC is near to zero, distance of eyelids is normal, but if ELDC approaches to one, distance of eyelids approaches to zero (eye is closed).

3.1.2. The Symptom Related to Face Region. Head rotation is a symptom of distraction which is extracted from face region in the proposed system. The head rotation is estimated based on the changes of face image with respect to the frontal face template. In order to compute the frontal face template, we assume that the driver face is in frontal mode during the first 100 frames. The average face image during these frames is computed as frontal face template. Then, the absolute difference of face image in the current frame and the frontal face template is named D_{Face}. Therefore, the head rotation (ROT) is estimated by

$$ROT = 1 - \exp\left(-0.1 \times D_{\text{Face}}\right). \qquad (11)$$

ROT changes in range $[0, 1]$. When D_{Face} is near to zero, the ROT is near to zero too, and when D_{Face} is near to one, ROT is near to 1. Greater ROT value indicates more head rotation. The proposed method for head rotation estimation cannot determine the angle of rotation.

3.2. Fatigue and Distraction Detection. In the proposed system, driver fatigue and distraction detection is estimated using a fuzzy expert system (Figure 2). A fuzzy expert system is an expert system that uses fuzzy logic instead of Boolean logic. In other words, a fuzzy expert system is a collection of membership functions, inference engine, and rules that are used to reason about inputs and generate proper outputs. At first, a fuzzy expert system fuzzifies crisp inputs by predefined membership functions to generate fuzzy inputs. Then, fuzzy inputs are processed by an inference engine. In inference engine, the truth value for each rule of rule-base is computed using a fuzzy implication method (usually Mamdani or Larsen methods) and applied to the conclusion part of each rule. These results are assigned to each output variable for each rule as a fuzzy subset. Then, all of the fuzzy subsets assigned to each output variable are combined together to form a single fuzzy subset for each output variable. Finally, the fuzzy subset of each output variable is defuzzified to generate the crisp output.

The proposed fuzzy expert system processes four inputs and generates two outputs. The inputs are (1) PERCLOS, (2) ELDC, (3) CLOSNO, and (4) ROT, and outputs are (1) fatigue estimation and (2) distraction estimation. In order to build a fuzzy expert system, Mamdani fuzzy inference method (also called min-max method) is applied on a set of fuzzy rules. The fuzzy rules are shown in Tables 1 and 2. These rules are extracted by an expert. However, these rules are not very complicated, and they are clear to understand.

The fuzzy membership functions of the inputs are depicted in Figures 3–6. According to Figures 3 and 4, the membership function of PERCLOS and CLOSNO is defined based on the $PERCLOS_N$ and $CLOSNO_N$, respectively. Additionally, ELDC and ROT are two symptoms that were normalized during the computation, and they always vary in range $[0, 1]$ (Figures 5 and 6). Therefore, the defined membership functions for the inputs are fully adaptive and normalized. The membership functions for the outputs are singleton and are depicted in Figures 7 and 8. The number

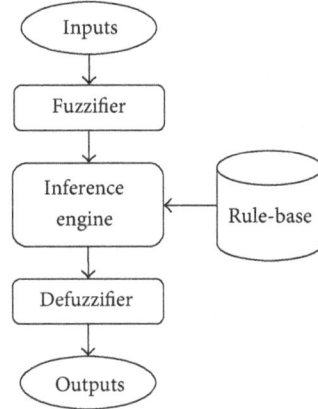

FIGURE 2: Block diagram of a fuzzy expert system.

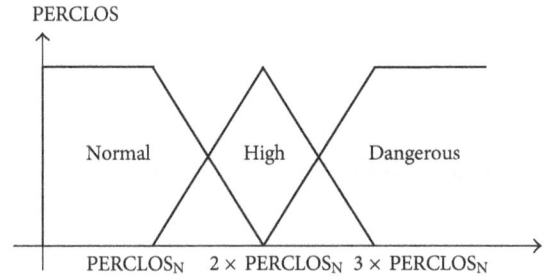

FIGURE 3: Fuzzy membership functions for PERCLOS.

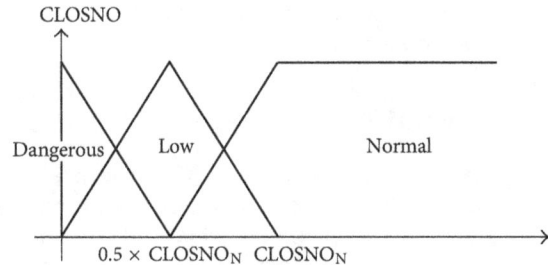

FIGURE 4: Fuzzy membership functions for CLOSNO.

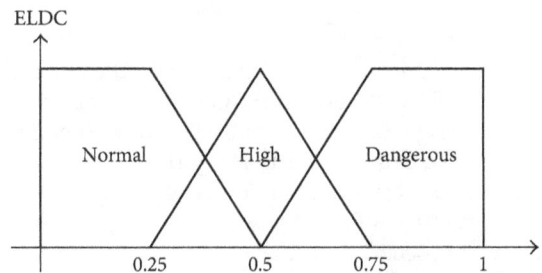

FIGURE 5: Fuzzy membership functions for ELDC.

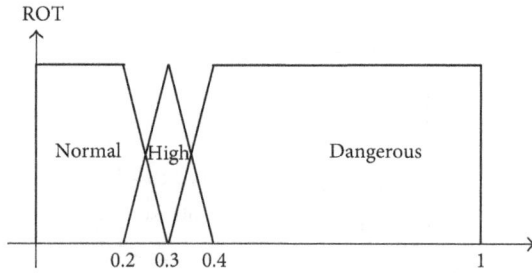

FIGURE 6: Fuzzy membership functions for ROT.

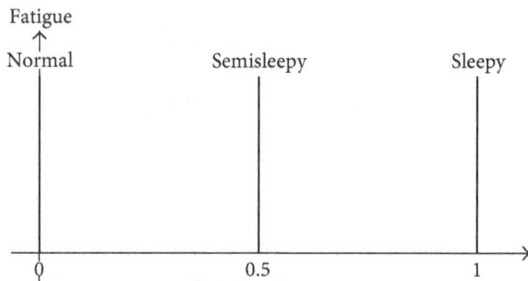

FIGURE 7: Fuzzy membership functions for fatigue estimation.

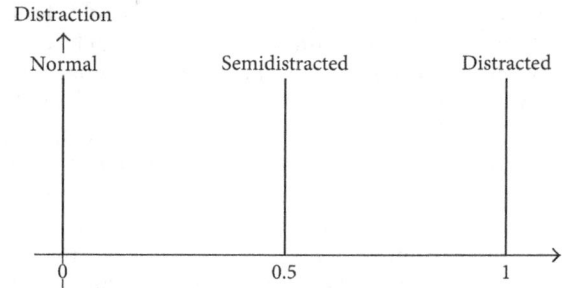

FIGURE 8: Fuzzy membership functions for distraction estimation.

of fuzzy subsets for each membership function is 3. A larger number of fuzzy subset leads to define more rules in rulebase, and this issue makes the system more complicated. In contrast, a smaller number of fuzzy subset leads to decrease the accuracy of driver state estimation.

The defuzzification method in the proposed method is Center Of Gravity (COG). This method is the most familiar and useful method for defuzzification.

4. Experimental Results

The proposed system was tested on 27 sequences which lasted about 76 minutes. The sequences were captured in both laboratory conditions (indoor) and real conditions (in vehicle) from 5 different individuals using a digital camera.

There is no tool for measuring the fatigue and distraction; therefore, objective evaluation is not possible for evaluating the proposed system directly. In this section, the proposed methods for extracting the symptoms are evaluated at first, and then an example sequence is investigated to evaluate the system subjectively.

4.1. Experiments on Symptom Extraction. The accuracy of computing PERCLOS and CLOSNO is directly dependent to the accuracy of eye closure detection algorithm. Therefore, we evaluate the eye closure detection algorithm in this section. Evaluation of eye closure detection is based on two criteria: false positive rate (FPR) and false negative rate (FNR). False positive error occurs when eye is open but the system detected it as closed eye. False negative error occurs when eye is closed but the system detected it as open eye. Table 3 shows

FPR and FNR of the proposed algorithm for eye closure detection in different states.

According to Table 3, the FNR of eye closure detection for drowsy state without glasses is greater than normal state without glasses. In drowsy state, the eyelid distance is reduced and blinking speed is slow. Then, horizontal projection of consecutive frames in drowsy state changes slowly. Therefore, many of eye closure events are not detected, and FNR in drowsy state is greater than normal state. But the FPR of eye closure detection in drowsy state is very low with respect to normal state.

According to Table 3, both FPR and FNR of eye closure detection for normal state with glasses are greater than normal state without glasses. In normal state with glasses, the reflection of glasses may appear in the image as a bright spot near the eye. Therefore, detection of changes of horizontal projection of top-half segment of face is difficult, and eye closure detection will have more error rate.

For investigating the accuracy of ELDC, we tested our method on 9-minute-long sequence. Figure 9 shows four sample frames of this sequence in which the driver is being drowsy after 7 minutes. Figure 10 shows the measured ELDC for this sequence. According to Figure 10, the ELDC can indicate the driver drowsiness correctly.

Accuracy of the proposed method for head rotation detection is investigated by applying a threshold on ROT. If ROT is more than 0.3, the head rotation is detected. According to this experiment, FPR and FNR of 9.2% and 12.1% were achieved for head rotation detection, respectively. In Figure 11, some sample frames of a 2-minute-long sequence are shown in which driver rotated his head to different directions. In this figure, (a) image shows the driver face without any rotation, and other images show head rotation of driver in different directions. The result of head rotation detection by the proposed method for the given video sequence is depicted in Figure 12.

4.2. Experiments on Driver State Estimation. Evaluation of driver state estimation is a difficult task because there is not any criterion for measurement of fatigue and distraction. Therefore, objective evaluation is not possible for driver state estimation.

In this section, the extracted symptoms from a sample sequence are plotted, and fatigue and distraction levels in

FIGURE 9: Four sample frames of a 9-minute-long sequence in which driver is drowsy. (a) Frame at $t = 1$ min, (b) frame at $t = 5$ min, (c) frame at $t = 7$ min, and (d) frame at $t = 9$ min.

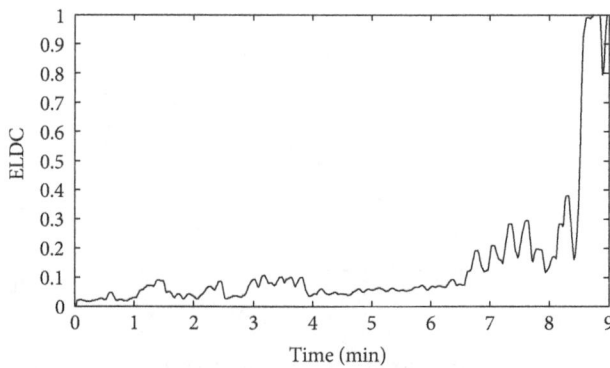

FIGURE 10: Measurement of ELDC for a sample sequence.

the sequence are estimated by the proposed system. At this experiment, ten-minute-long sequence is used. The first minute of the sequence is used for training. According to the training phase, $PERCLOS_N$ is 0.02 and $CLOSNO_N$ is 13 times per minute. The curvature of PERCLOS, CLOSNO, ELDC, and ROT related to this sequence are plotted in Figures 13, 14, 15, and 16.

The estimated levels of fatigue and distraction are shown in Figures 17 and 18. According to Figure 17, the driver has been semidistracted at about the 3rd minute. The estimated level of distraction seems true, because the CLOSNO was decreased with respect to the $CLOSNO_N$ during this time. In addition, the driver has been drowsy after 7 minutes. The drowsiness state was estimated based on two symptoms: (1) increasing the PERCLOS during the time from 7th to 8th

minute and (2) increasing the ELDC after 8 minutes. These symptoms are depicted in Figures 13 and 15.

4.3. The Processing Speed. The proposed method was implemented in MATLAB R2008a and was tested on a personal computer with Intel Core2 Dou 2.66 GHz and 2 GB RAM memory. The processing speed of the proposed method is more than 5 frames per second. Over 85% of computational complexity of the system is related to face tracking.

4.4. Comparison with Other Methods. In this section, we compare our system with other previous systems. Unfortunately, we cannot compare accuracy of different driver state estimation algorithms, because there is not any scientific and precise criterion to measure fatigue and distraction. Therefore, we only compare the accuracy of different system for symptom extraction.

For eye closure detection, the proposed algorithm is compared with other algorithms presented in [10, 19, 21]. The results of comparison are depicted in Table 4. This table shows that the performance of our proposed method is very good in comparison to other methods, while the experimental setup of our system is more realistic, and we used longer video sequences for our experiments.

For head rotation detection, the proposed method is compared with the algorithm presented in [19]. Unfortunately, the accuracy of other methods for head rotation detection was not reported. For example, accuracy of the methods presented in [4, 10] was not reported. In these papers, only the ability of system to measure head rotation in different direction and in a specific interval was reported.

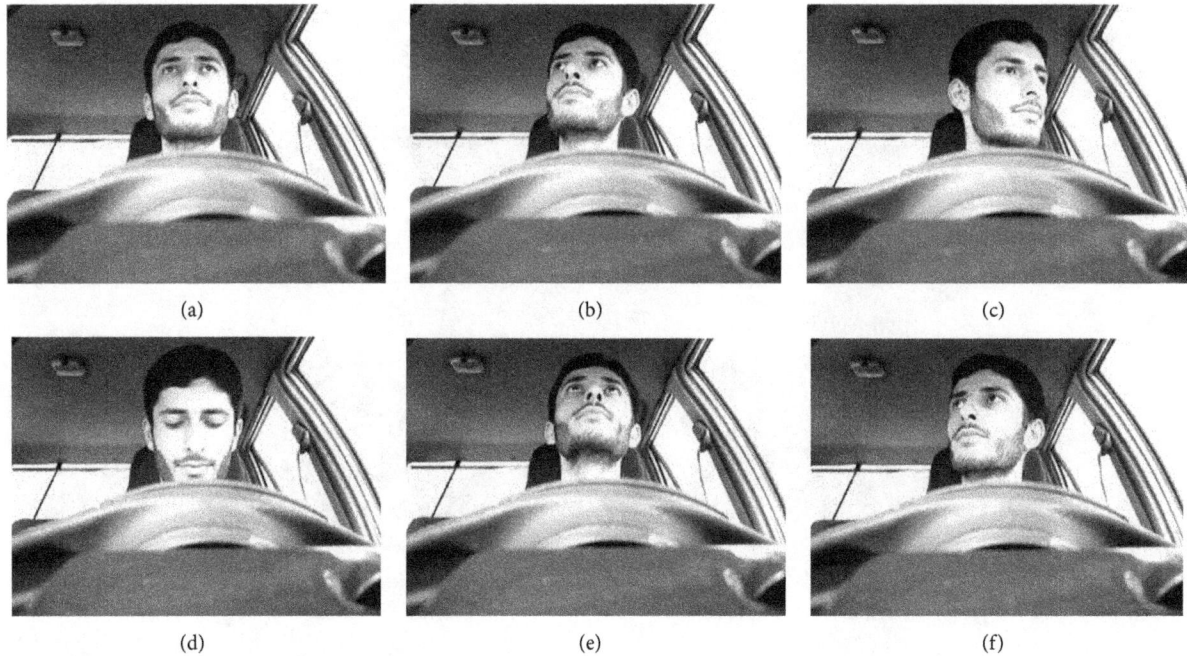

(a)   (b)   (c)

(d)   (e)   (f)

FIGURE 11: Some sample frames of a 2-minute-long sequence in which the driver rotated his head to different direction. (a) shows the head direction of driver in normal state.

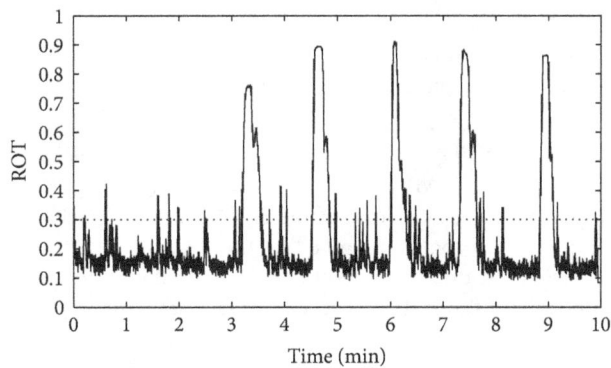

FIGURE 12: Results of head rotation detection for a sample video sequence depicted in Figure 11. Dashed horizontal line indicates the threshold for head rotation detection.

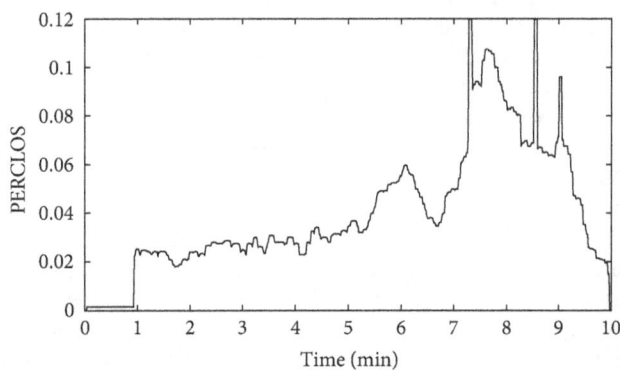

FIGURE 14: Curvature of CLOSNO for the sample sequence.

Table 5 shows the comparison result of the proposed method and the method presented in [19]. The comparison result shows that our method achieves higher precision rate.

5. Conclusions

In this paper, a new adaptive method for symptom extraction and driver state estimation was proposed for driver hypovigilance detection. Two types of symptoms were considered: symptoms related to eye region (including PERCLOS, ELSDC, and CLOSNO) and symptom related to face region (ROT). The proposed method extracts the symptoms related to eye region using horizontal projection of top-half segment without explicit eye detection; the symptom related to face region is extracted based on face template matching. Then, the normal value of the extracted symptoms

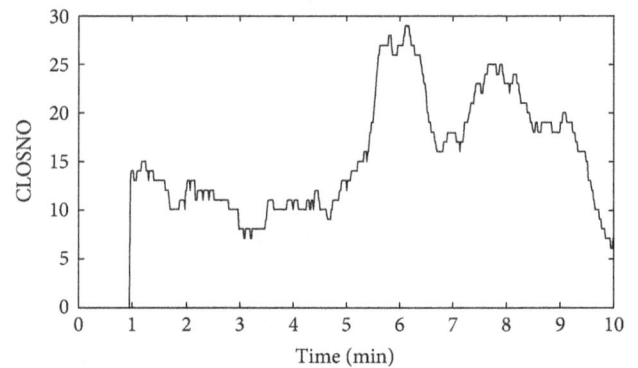

FIGURE 13: Curvature of PERCLOS for the sample sequence.

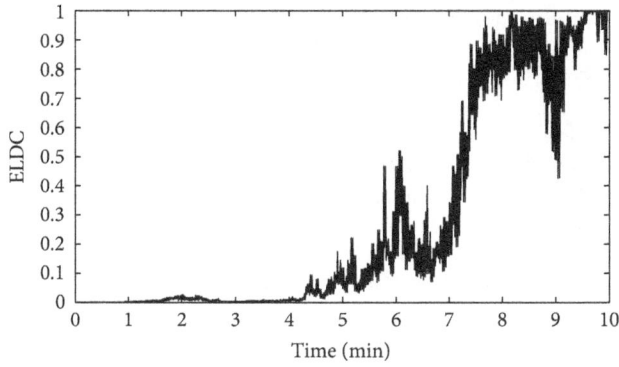

FIGURE 15: Curvature of ELDC for the sample sequence.

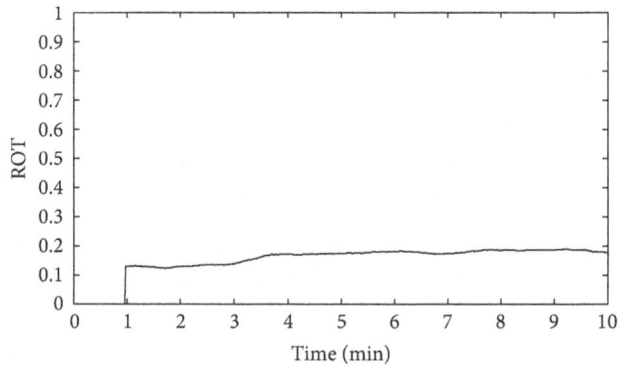

FIGURE 16: Curvature of ROT for the sample sequence.

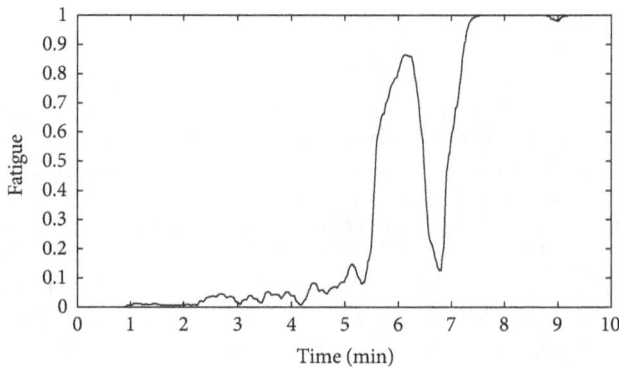

FIGURE 17: Estimated level of fatigue for the sample sequence.

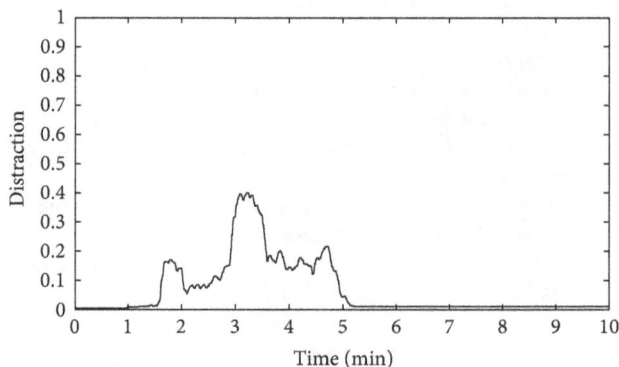

FIGURE 18: Estimated level of distraction for the sample sequence.

TABLE 1: Rules of the proposed fuzzy expert system for estimation of driver fatigue.

If PERCLOS is normal and ELDC is normal, then fatigue is normal

If PERCLOS is normal and ELDC is high, then fatigue is semisleepy

If PERCLOS is normal and ELDC is dangerous, then fatigue is sleepy

If PERCLOS is high and ELDC is normal, then fatigue is semisleepy

If PERCLOS is high and ELDC is high, then fatigue is semisleepy

If PERCLOS is high and ELDC is dangerous, then fatigue is sleepy

If PERCLOS is dangerous and ELDC is normal, then fatigue is sleepy

If PERCLOS is dangerous and ELDC is high, then fatigue is sleepy

If PERCLOS is dangerous and ELDC is dangerous, then fatigue is sleepy

TABLE 2: Rules of the proposed fuzzy expert system for estimation of driver distraction.

If CLOSNO is normal and ROT is normal, then distraction is normal

If CLOSNO is low and PERCLOS is normal and ELDC is normal, then distraction is semidistracted

If CLOSNO is dangerous and PERCLOS is normal and ELDC is normal, then distraction is distracted

If ROT is high and PERCLOS is normal and ELDC is normal, then distraction is semidistracted

If ROT is dangerous and PERCLOS is normal and ELDC is normal, then distraction is distracted

TABLE 3: Experimental results for eye closure detection.

	FPR	FNR
Normal state without glasses	7.7%	15.4%
Drowsy state without glasses	0.8%	56.1%
Normal state with glasses	10.3%	26.7%

TABLE 4: Comparison results of eye closure detection algorithms.

Algorithm	FPR	FNR
Batista [10]	9.5%	1.7%
Bergasa et al. [19]	?	6.9%
Smith et al. [21]	71%	15%
The proposed algorithm	7.7%	15.4%

TABLE 5: Comparison results of head rotation detection algorithms.

Algorithm	Precision
Bergasa et al. [19]	72.5%
The proposed algorithm	90.8%

is calculated during a short training phase. According to the normal value of the extracted features, an adaptive fuzzy expert system estimates the level of fatigue and distraction.

The short training phase makes the system robust and adaptive. In other words, the proposed system may be used efficiently for different individuals with different face and eyelid behaviors. Experiments show that the accuracy of the proposed method for extracting the symptoms of driver fatigue and distraction is very good. Additionally, the system can estimate the driver fatigue and distraction very well by subjective evaluation.

The proposed method was also tested on video sequences captured in visible spectrum, but the color information was not used in any part of the system. In other words, the proposed system operates in gray-level visible spectrum. Therefore, the system may operate in IR spectrum with a few changes. The main disadvantage of our system is the face tracking method which is inaccurate and very computationally complex. Adaptive filters such as Kalman filter may reduce the complexity and increase the processing speed and accuracy of the system.

References

[1] N. L. Haworth, T. J. Triggs, and E. M. Grey, *Driver Fatigue: Concepts, Measurement and Crash Countermeasures*, Human Factors Group, Department of Psychology, Monash University, 1988.

[2] C. T. Lin, L. W. Ko, I. F. Chung et al., "Adaptive EEG-based alertness estimation system by using ICA-based fuzzy neural networks," *IEEE Transactions on Circuits and Systems*, vol. 53, no. 11, pp. 2469–2476, 2006.

[3] T. V. Jan, T. Karnahl, K. Seifert, J. Hilgenstock, and R. Zobel, *Don't Sleep and Drive—VW's Fatigue Detection Technology*, Centre for Automotive Safety Research, Adelaide University, Adelaide, Australia, 2005.

[4] Q. Ji and X. Yang, "Real-time eye, gaze, and face pose tracking for monitoring driver vigilance," *Real-Time Imaging*, vol. 8, no. 5, pp. 357–377, 2002.

[5] T. Brandt, R. Stemmer, and A. Rakotonirainy, "Affordable visual driver monitoring system for fatigue and monotony," in *Proceedings of the IEEE International Conference on Systems, Man and Cybernetics (SMC '04)*, pp. 6451–6456, Hague, The Netherlands, October 2004.

[6] M. Bayly, B. Fildes, M. Regan, and K. Young, "Review of crash effectiveness of intelligent transport system," TRaffic Accident Causation in Europe (TRACE), 2007.

[7] H. Cai and Y. Lin, "An experiment to non-intrusively collect physiological parameters towards driver state detection," in *Proceedings of the SAE World Congress*, Detroit, Mich, USA, 2007.

[8] T. Nakagawa, T. Kawachi, S. Arimitsu, M. Kanno, K. Sasaki, and H. Hosaka, "Drowsiness detection using spectrum analysis of eye movement and effective stimuli to keep driver awake," *DENSO Technical Review*, vol. 12, pp. 113–118, 2006.

[9] M. H. Yang, D. J. Kriegman, and N. Ahuja, "Detecting faces in images: a survey," *IEEE Transactions on Pattern Analysis and Machine Intelligence*, vol. 24, no. 1, pp. 34–58, 2002.

[10] J. Batista, "A drowsiness and point of attention monitoring system for driver vigilance," in *Proceedings of the 10th International IEEE Conference on Intelligent Transportation Systems (ITSC '07)*, pp. 702–708, Seattle, Wash, USA, October 2007.

[11] S. Abtahi, B. Hariri, and S. Shirmohammadi, "Driver drowsiness monitoring based on yawning detection," in *Proceedings of the Instrumentation and Measurement Technology Conference*, Hangzhou, China, 2011.

[12] Y. Du, P. Ma, X. Su, and Y. Zhang, "Driver fatigue detection based on eye state analysis," in *Proceedings of the Joint Conference on Information Science*, Shen Zhen, China, 2008.

[13] W. B. Horng, C. Y. Chen, Y. Chang, and C. H. Fan, "Driver fatigue detection based on eye tracking and dynamic template matching," in *Proceedings of the IEEE International Conference on Networking, Sensing and Control*, pp. 7–12, Taipei, Taiwan, March 2004.

[14] P. Viola and M. Jones, "Rapid object detection using a boosted cascade of simple features," in *Proceedings of the IEEE Computer Society Conference on Computer Vision and Pattern Recognition*, pp. I511–I518, Cambridge, Mass, USA, December 2001.

[15] A. de la Escalera, M. J. Flores, and J. M. Armingol, "Driver drowsiness warning system using visual information for both diurnal and nocturnal illumination conditions," *EURASIP Journal on Advances in Signal Processing*, vol. 2010, Article ID 438205, 2010.

[16] T. Wang and P. Shi, "Yawning detection for determining driver drowsiness," in *Proceedings of the IEEE International Workshop on VLSI Design and Video Technology (IWVDVT '05)*, pp. 373–376, Suzhou, China, May 2005.

[17] A. Liu, Z. Li, L. Wang, and Y. Zhao, "A practical driver fatigue detection algorithm based on eye state," in *Proceedings of the 2nd Asia Pacific Conference on Postgraduate Research in Microelectronics and Electronics (PrimeAsia '10)*, pp. 235–238, Shanghai, China, September 2010.

[18] R. Grace, V. E. Byme, D. M. Bierman et al., "A Drowsy driver detection system for heavy vehicles," in *Proceedings of the 17th AIAA/IEEE/SAE Digital Avionics Systems Conference (DASC '98)*, pp. I36/1–I36/8, Washington, DC, USA, 1998.

[19] L. M. Bergasa, J. Nuevo, M. A. Sotelo, R. Barea, and M. E. Lopez, "Real-time system for monitoring driver vigilance," *IEEE Transactions on Intelligent Transportation Systems*, vol. 7, no. 1, pp. 63–77, 2006.

[20] M. J. Flores, J. M. Armingol, and A. D. l. Escalera, "Driver drowsiness detection system under infrared illumination for an intelligent vehicle," *IET Intelligent Transport Systems*, vol. 5, pp. 241–251, 2011.

[21] P. Smith, M. Shah, and N. da Vitoria Lobo, "Determining driver visual attention with one camera," *IEEE Transactions on Intelligent Transportation Systems*, vol. 4, no. 4, pp. 205–218, 2003.

[22] P. R. Tabrizi and R. A. Zoroofi, "Drowsiness detection based on brightness and numeral features of eye image," in *Proceedings of the 5th International Conference on Intelligent Information Hiding and Multimedia Signal Processing*, pp. 1310–1313, Kyoto, Japan, September 2009.

[23] Z. Zhang and J. S. Zhang, "Driver fatigue detection based intelligent vehicle control," in *Proceedings of the 18th International Conference on Pattern Recognition (ICPR '06)*, pp. 1262–1265, Hong Kong, China, August 2006.

[24] Y. Zheng and Z. Wang, "Robust and precise eye detection based on locally selective projection," in *Proceedings of the 19th International Conference on Pattern Recognition (ICPR '08)*, Tampa, Fla, USA, December 2008.

[25] D. Wenhui, Q. Peishu, and H. Jing, "Driver fatigue detection based on fuzzy fusion," in *Proceedings of the Chinese Control and Decision Conference (CCDC '08)*, pp. 2640–2643, Shandong, China, July 2008.

[26] J. C. McCall and M. M. Trivedi, "Facial action coding using multiple visual cues and a hierarchy of particle filters," in *Proceedings of the IEEE Conference on Computer Vision and Pattern Recognition Workshops (CVPRW '06)*, pp. 150–155, New York, NY, USA, June 2006.

[27] W. Dong and X. Wu, "Driver fatigue detection based on the distance of eyelid," in *Proceedings of the IEEE International Workshop on VLSI Design and Video Technology (IWVDVT '05)*, pp. 365–468, Suzhou, China, May 2005.

[28] M. H. Sigari, N. Mozayani, and H. R. Pourreza, "Fuzzy running average and fuzzy background subtraction: concepts and application," *International Journal of Computer Science and Network Security*, vol. 8, pp. 138–143, 2008.

7

Fuel Efficiency by Coasting in the Vehicle

Payman Shakouri,[1] **Andrzej Ordys,**[1] **Paul Darnell,**[2] **and Peter Kavanagh**[2]

[1] *School of Mechanical and Automotive Engineering, Kingston University of London, London SW15 3DW, UK*
[2] *Jaguar Land Rover, Banbury Road, Gaydon, Warwickshire CV35 ORR, UK*

Correspondence should be addressed to Payman Shakouri; paymash2006@yahoo.com

Academic Editor: Rakesh Mishra

This paper investigates the possibility of improving the fuel efficiency by decreasing the engine speed during the coasting phase of the vehicle. The proposed approach is stimulated by the fact that the engine losses increase with the engine speed. If the engine speed is retained low, the engine losses will be reduced and subsequently the tractive torque will be increased, enabling the vehicle to remain moving for longer duration while coasting. By increasing the time period of the coasting the fuel efficiency can be increased, especially travelling downhill, since it can benefit from the kinematic energy stored in the vehicle to continue coasting for a longer duration. It is already industry standard practice to cut fuel during coasting and refuel at low engine speed. The substantial difference proposed in this paper is the controlled reduction of engine speed during this phase and thus reduction in the engine losses, resulting in improved fuel economy. The simulation model is tested and the results illustrating an improvement to the fuel efficiency through the proposed method are presented. Some results of the experimental tests with a real vehicle through the proposed strategy are also presented in the paper.

1. Introduction

Fuel efficiency has been widely researched in the literature from different aspects. Barrand and Bokar [1], Tarpinian et al. [2] have investigated the influence of tyre rolling resistance on fuel saving. Lee et al. [3] have investigated the fuel efficiency through the powertrain and explained that it can be improved by almost locking up the torque converter clutch (TCC) when the vehicle is in coasting. The reason is that energy efficiency can increase up to 100% due to eliminating the slip between the turbine and the impeller. However, if the torque converter is fully locked-up at high speed and the driver lifts the foot from the accelerator pedal, the engine torque suddenly will fall to zero which causes the driver to feel the momentary shock at the moment of lift-up. Thus, the latter paper has proposed an adaptive antishock coasting control to reduce the shock without degrading the fuel economy. Another approach which is commonly used in Hybrid Electric Vehicle (HEV) is the engine start/stop system [4]. Based on this approach, the engine fuelling can be cut off when the engine operates in idle speed. This approach is significantly beneficial

for the vehicle in coasting, because fuelling the engine can be stopped during the coast phase of the vehicle. Lee [4] has studied the benefit of applying the "burn and coast" method in the fuel consumption. According to this method, the vehicle accelerates to the high velocity and then coasts to the lower velocity with the engine off instead of driving at the steady velocity. The "burn and coast" technique is significantly useful for HEV [5–8], because through this technique the vehicle can store high electric energy in the battery during the coast. Some other advanced autonomous systems have been introduced in the literature to control the vehicle speed with respect to the road grade such as Look Ahead Cruise Controller [9–12]. Furthermore, because the engine needs to use higher power to accelerate to high velocity in a short time, the engine brake thermal efficiency is increased. On contrary to [3] which has stated that the greatest fuel efficiency could be achieved by fully locking-up the torque converter, in this paper it has been demonstrated through the simulation that the fuel efficiency can be increased by controlling the slip between the components of the torque converter, that is, the impeller and turbine, through the torque converter slip control when the vehicle is in the coast phase. Based on the approach proposed in this paper, the coasting duration can be increased by reducing the friction losses in the engine, which requires that the torque converter get unlocked whenever the accelerator pedal is not applied (revised condition), even if the vehicle travels at higher speed. In normal condition, the torque converter is locked at higher vehicle speed about 40 mph (approximately 64.4 km/h) which is equivalent to the third or fourth gear of the transmission. While utilising the revised algorithm the engine can get disengaged from the rest of powertrain, and torque coming to the engine is controlled through the torque converter slip control. In this way, the engine friction losses can decrease which causes fuel efficiency to be improved. The vehicle equipped with revised algorithm can use the potential energy stored during driving downhill so as to increase the coasting duration. In addition, it is proposed that the engine fuelling must be stopped at high vehicle speed during the coasting and the engine must be refuelled at lower vehicle velocity in order to provide further improvement to the coasting duration. Since, at the lower vehicle speed, the torque backed to the engine from the road wheel is not adequate to keep the engine rotating (at engine idle speed), it is crucial to refuel the engine when the vehicle coasts to lower velocity (lower than 15 km/h) such that the engine idle speed can be maintained. Therefore, according to the method proposed in this paper, the engine speed is initially retained at idle level through the torque converter slip control while fuel is cut off, and when the vehicle velocity decreases to 15 km/h, fuelling is started.

Muller et al. [13] propose Stop/Start Coasting, that is, the Bosch concept, in which the fuel injection is turned off during the coasting; however, the combustion engine must be decoupled from the powertrain by opening the clutch or shifting into neutral gear in order to avoid negative engine torque during coasting. Disconnecting the engine from the powertrain, while turning off the fuel injection, prevents the engine from rotating. It has disadvantages as an additional power supply is required to supply systems such as power steering or air conditioning during the coasting condition. The concept proposed in this paper, that is, the Jaguar Land Rover (JLR) concept, is based on the controlled reduction of engine speed through the torque converter slip control during the coasting phase, and thus reduction in the engine frictional and pumping losses, resulting in less drag on vehicle and improved fuel economy over the conventional cases. In this way, unlike the Bosch concept, the combustion engine is not entirely disconnected from the powertrain and can be remained rotated at idle speed through the torque backed to the engine from the road wheels.

This paper is organized as follows. The longitudinal vehicle model is explained in Section 2. Section 3 proposes a new approach for increasing the coasting duration. Section 4 explains the parameters required for measuring fuel consumption and efficiency in the simulation. Section 5 illustrates the simulation results for the coasting and investigates the potential of the fuel efficiency through the approach proposed in this paper. Section 6 presents the results of the real implementation. Finally, Section 7 draws some conclusions and proposes future directions.

2. Vehicle Longitudinal Model

To carry out analysis in this paper, an intergraded simulation model of the vehicle has been developed using Simulink/MATLAB software. The detailed derivation of the vehicle simulation model is presented in the appendix.

The vehicle dynamics is classified into two separate categories [14]: (1) the dynamics of powertrain comprising of the engine, the torque converter, the gear box, the final drive, and the wheels; (2) the dynamics of the vehicle considering the external forces acting on the vehicle. These include aerodynamic drag force, gravitational force, and rolling resistance force.

In the powertrain model, two states of the torque converter, that is, locked and unlocked, are simulated. The transmission between these two states is implemented in the simulation by devising a switching algorithm which depends on the vehicle velocity and the throttle opening position.

The torque converter can get locked up when the vehicle travels at speed higher than 40 mph (approximately 17.88 m/s) or equivalently at third or fourth gear; at this speed the throttle is slightly opened. When torque converter locksup, the impeller and turbine will get entirely engaged so as to provide the one to one drive from the engine to the gear box input shaft without any slippage. Thus, the engine speed will be correlated to the vehicle speed directly through gear and final drive ratios.

In order to take the friction and pumping losses of the engine (for brevity in this paper, these two terms are lumped and denoted by the engine friction) into consideration, a map corresponding to 5-litre naturally aspirated V8 engine (provided by JLR) has been used. This map is illustrated in Figure 1, which introduces the engine friction versus engine rotation speed for the air mass flow of 42%. As it is observed, the engine friction increases along with the engine speed. The engine friction also depends on other characteristics of the engine such as the compression ratio and the engine sizing.

3. Revised Algorithm

When the torque converter enters the lockedup state, the engine speed becomes directly related to the vehicle velocity through gear box and the final drive; therefore the engine speed will remain higher for the longer period of time. An algorithm needs to be devised to prevent the system from staying locked-up during coasting, which can help the engine speed to be reduced promptly. The lower engine speed results in the lower engine friction (see Figure 1). This avoids generating the opposing torque on the wheel due to the engine friction losses during coasting, which consequently helps the vehicle coast for longer duration.

An integrated simulation model of the vehicle dynamics and powertrain considering an automatic transmission and incorporating the unlocked and locked states of the torque converter as well as a switching logic to perform transition between those states is illustrated in Figure 2. In the simulation, it requires two engine models for presenting different conditions of operation: the engine with an unlocked torque converter (Engine 1) and the engine with a locked torque

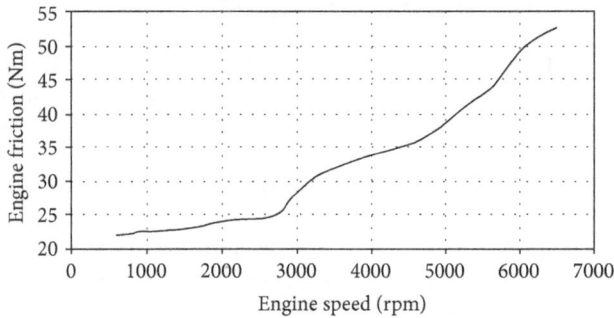

FIGURE 1: The engine friction map for 5-litre naturally aspirated V8 engine.

converter (Engine 2). When the torque converter is unlocked, the operation of system is represented by Engine 1 together with the torque converter model. However, when the torque converter is locked the model does not need to consider the torque converter, since it may be assumed that the engine crank shaft is directly connected to the transmission input shaft. Therefore, the operation of system is only represented by Engine 2, without considering the torque converter. In reality, in our simulation, we use the model which combines those two models together with a switching algorithm (see Figure 2).

In order to investigate the effect of the locked and the unlocked states of the torque converter on the coast duration and fuel efficiency through simulation, an algorithm is presented based on a simple switching algorithm which is triggered by the throttle opening position and the vehicle speed. To investigate the impact of new approach proposed in this paper on the coasting and fuel efficiency, two conditions are, namely, introduced: normal condition and revised condition. During normal condition the switching logic only uses the vehicles velocity in order to transit between locked and unlocked states; regardless of the position of the throttle opening, that is, whether throttle opening is zero or not. In this way the torque converter is locked up when the vehicle travels at speed approximately higher than 17.88 m/s (40 mph) while it is unlocked up at the vehicle speed lower than 17.88 m/s. However, based on the revised condition the torque converter can get unlocked up once the throttle opening position has become zero, even though the vehicle speed is high, which can provide an improvement to the coasting duration. A switching algorithm is utilized in the simulation for performing transition between the locked and unlocked states of the torque converter. Depending on the position of the switches, different values of the engine speed and transmission input torque will be used for the computation. Thus, if either the vehicle velocity is less than 17.88 m/s or the throttle opening is zero (indicating the unlocked state of the torque converter), the engine speed (N_e) calculated by Engine 1 (see Figure 2) and the turbine torque (T_{in1}) calculated by the torque converter model will be used for the engine speed (also as an initial engine speed for the next instance of computation) and transmission input torque (T_{in}), respectively. Otherwise, the transmission input speed (N_{in}) and the torque (T_{in2}) calculated by Engine 2 will be used.

The engine has to be rotated at idle speed during coasting, rather than stopping or rotating fully, to allow ancillary components to operate, that is, power steering, air conditioning, alternator, and so forth and also to improve speed of response should the driver press the accelerator pedal. According to the new concept, the torque converter gets unlocked during coasting and no fuel is injected to the engine. In this condition the engine speed will fall to idle, that is, approximately 610 rpm. In order to keep the engine rotating at the idle speed, the following two approaches can be proposed, depending on the vehicle speed.

3.1. The Torque Converter Slip Control. Based on this new approach, the fuel is not injected into the engine during the coasting phase; thus, the engine speed drops to lower than the idle speed. Therefore, in order to hold the idle engine speed, sufficient connection across the torque converter needs to be generated to provide enough torque back to the engine from the road wheel. This can be achieved through the torque converter slip control.

3.2. Refueling the Engine at Lower Vehicle Speed. At the lower vehicle speed, the torque backed to the engine from the road wheel will not be sufficient to keep the engine rotating at the idle speed. Therefore, it is crucial to refuel the engine when the vehicle coasts to the lower velocity such that the engine idle speed can be maintained. Thereby, refueling the engine at lower vehicle speed provides further improvement on the coasting duration.

To present the two approaches explained above in the simulation, two Proportional-Integral-Derivative (PID) controllers are used. Both controllers try to keep the engine idle speed; one controls the torque coming back to the engine from torque converter, that is, impeller torque, when the vehicle coasts at high speed; other one controls the throttle opening position at low vehicle speed. These algorithms are depicted in Figure 3, which corresponds to the engine model for the unlocked state of the torque converter (Engine 1 depicted in Figure 2).

In the simulation, the torque converter slip control has been virtualized by use of a PID controller (PID controller 1) that controls the impeller torque in order to help the engine speed reach to its set-point. The engine idle speed is introduced as the set-point. Depending on the switching conditions, either the magnitude of the impeller torque calculated by the controller (T_{ic}) or that generated by torque converter (T_i) will be sent to the engine model. The magnitude calculated by the controller (T_{ic}) will be used, if the following three conditions hold at the same time:

(1) the throttle signal is zero;

(2) the vehicle velocity is higher than 15 km/h (approximately 4.16 m/s);

(3) the engine reaches to its idle speed, that is, approximately 610 rpm.

Otherwise the torque generated by the torque converter (T_i) will be used.

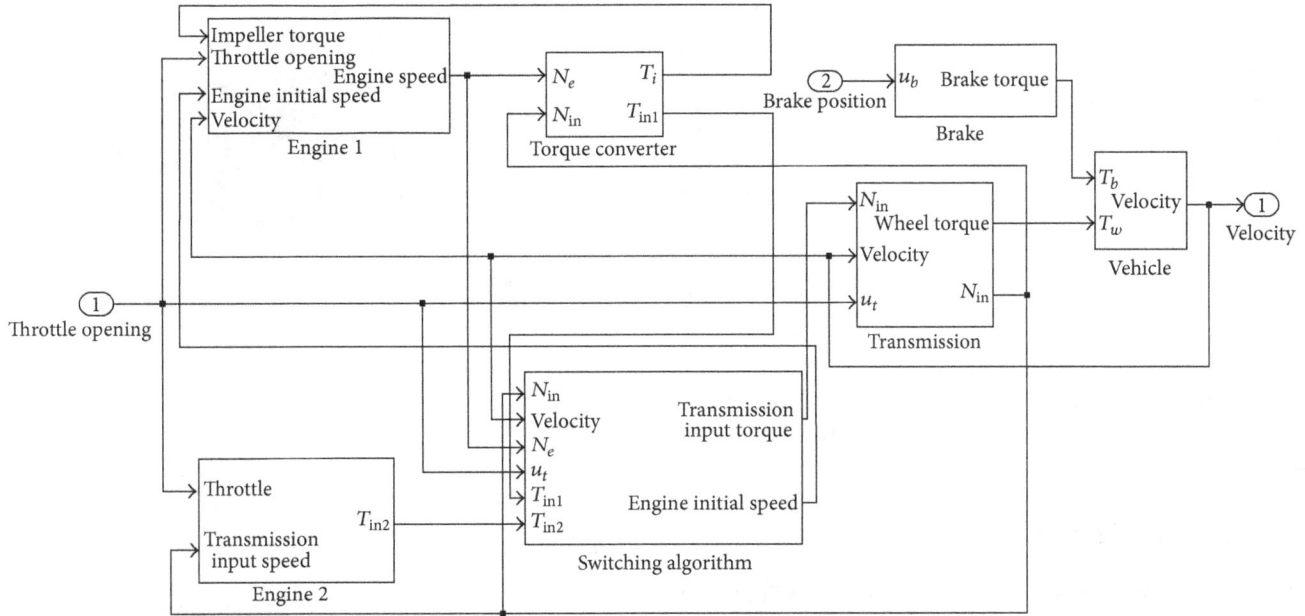

FIGURE 2: An integrated simulation model of the vehicle dynamics and powertrain considering an automatic transmission—both locked and unlocked states of the torque converter are modelled.

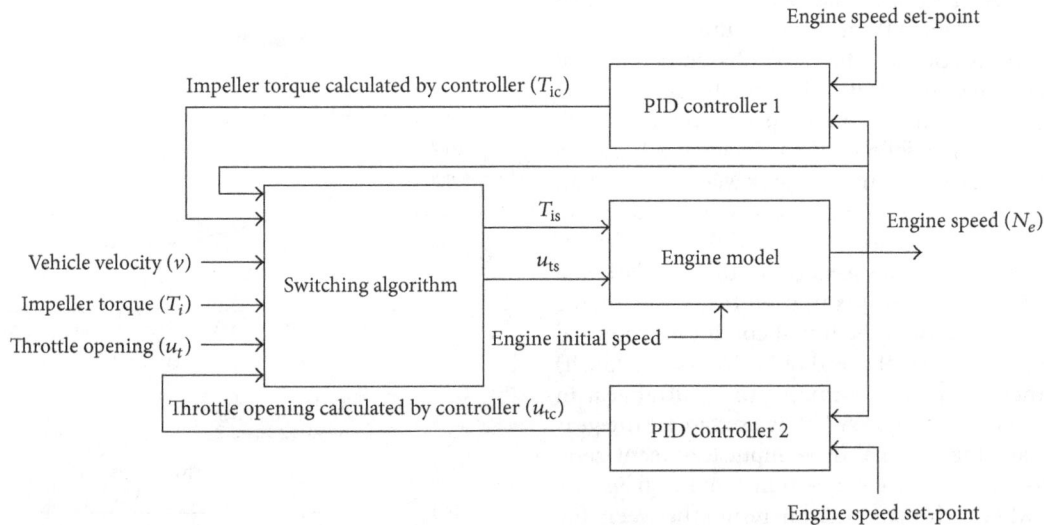

FIGURE 3: Two PID controllers used to keep the idle engine speed (610 rpm) during coasting—the torque backed to the engine from torque converter (impeller torque) is controlled at the higher vehicle speed, while the throttle is controlled to hold the idle engine speed at the lower vehicle speed.

Refueling of the engine at the low vehicle speed is modeled by another PID controller (PID controller 2) which regulates the throttle opening position. The throttle position is adjusted such that the idle engine speed is retained. In this way, the throttle position slightly opens to produce enough torque for keeping the engine at idle speed. Depending on the switching conditions, either the throttle opening magnitude calculated by the controller (u_{tc}) or that predefined as an input to the vehicle simulation model (u_t) will be sent to the engine model. In order for the magnitude calculated by the controller to be used, the following two conditions must hold at the same time:

(1) throttle signal is zero;

(2) the vehicle speed is less than 15 km/h.

Otherwise the pre-defined throttle opening (u_t) will be used.

4. Fuel Efficiency and Fuel Consumption

A term to investigate fuel efficiency is the Brake Specific Fuel Consumption (BSFC). It may be determined by the following equation:

$$\text{BSFC} = \frac{\dot{m}_f}{P_{eb}}. \tag{1}$$

Here $\dot{m}_f$ denotes fuel consumption rate (g/h) and P_{eb} is engine brake power (kW). To calculate the fuel consumption rate in the simulation, a look-up table giving the brake specific fuel consumption (BSFC) versus brake mean effective pressure bmep (Pa) and engine speed (rpm) is used. The look-up table is not presented in this paper due to the confidentiality issue. The bmep can be calculated from the engine brake torque T_{eb} [15]:

$$\text{bmep} = \frac{n\pi T_{eb}}{V_d}, \tag{2}$$

where V_d denotes the displacement volume of the engine (m^3), n denotes the number of the time the piston moves during one cycle of the engine, that is, the engine stroke, and T_{eb} denotes the engine brake torque (Nm).

5. Simulation Results

The vehicle used for illustration of the proposed approach (through simulation) is a Range Rover Sport L320 with 5-litre naturally aspirated V8 engine including a ZF HP28 six-speed automatic transmission (Figure 4). Hence, the appropriate engine map (provided by JLR) has been used. The torque converter data was obtained from the literature [3].

The simulation results and the results based on the real test data are compared to establish that the dynamics of the vehicle during coasting has correctly been simulated. The real data was collected as follows: the vehicle was driven to 135 km/h on the flat road and neutral gear was selected with the vehicle allowed to coast with no steering input down to 15 km/h. Recording of the data was started from 125 km/h. In order to resemble the same scenario in the simulation, a cruise control (CC) model [16] is utilized to get the engine speed and vehicle velocity to the initial conditions required for the test; thus, the set-point speed of 34.72 m/s (125 km/h) is chosen. Furthermore, for determining the neutral gear in the powertrain model, a simple switch is used to set the gear box output torque, that is, final drive input torque, at zero. In this condition, there is no torque transmission from the engine to the wheel. The comparison results between the simulation and real test data are illustrated in Figure 5. The simulation results are almost identical with the real test data. This indicates that dynamics of the vehicle during coasting is correctly presented by the simulation model.

As it was explained, in normal condition, the torque converter only unlocks up when the speed is less than 64.4 km/h (40 mph). However, according to the new approach the torque converter can get unlockedup whenever the throttle is closed regardless of the vehicle speed. It helps the engine speed decrease faster which in turn reduces the engine friction and the fuel efficiency can be obtained. In order to investigate the influence of the revised algorithm on the coasting duration along with fuel efficiency two scenarios based on the road condition are determined: flat road and inclined road. These two scenarios are presented as follows.

First scenario considers the flat road. For this test, the cruise control (CC) is initially used to make the vehicle velocity track the reference velocity set at 144 km/h. The

FIGURE 4: Range Rover Sport L320 with 5-Litre naturally aspirated V8 engine including a ZF HP28 six-speed automatic transmission.

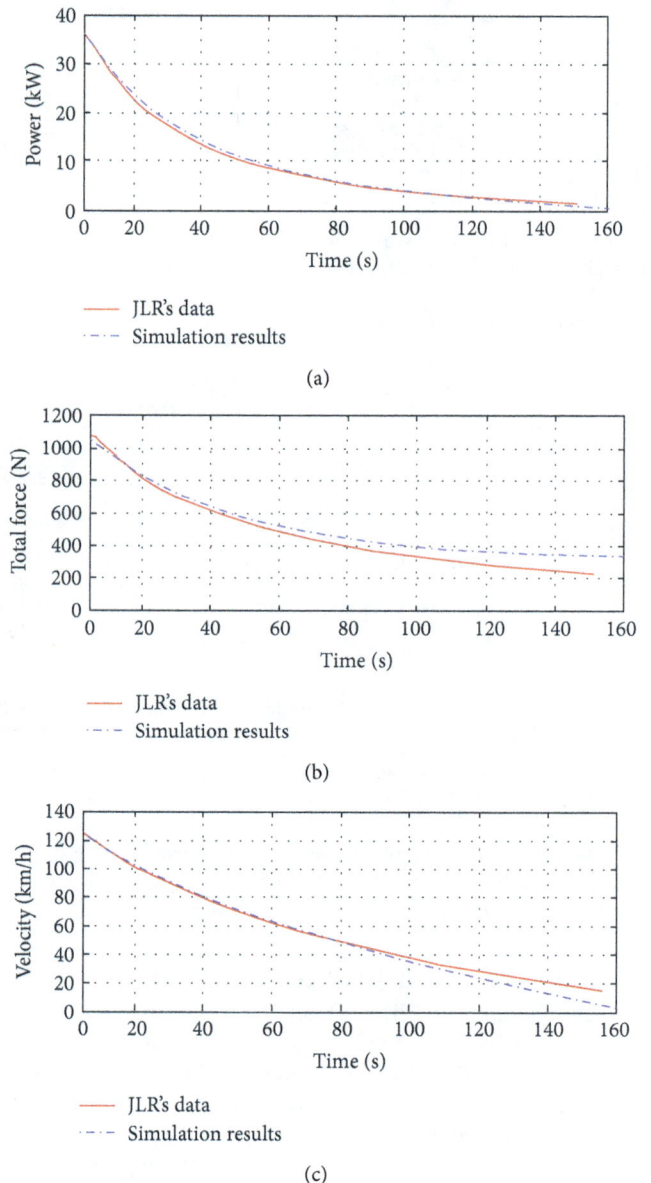

(a)

(b)

(c)

FIGURE 5: (a) Total amount of the external forces acting on the vehicle during coasting—the result of the simulation compared with data obtained from the test on a real vehicle, (b) power presents the reduction of the stored energy over the time as the vehicle coasts down, and (c) comparison of the vehicle's velocities obtained from the simulation model and from the test vehicle.

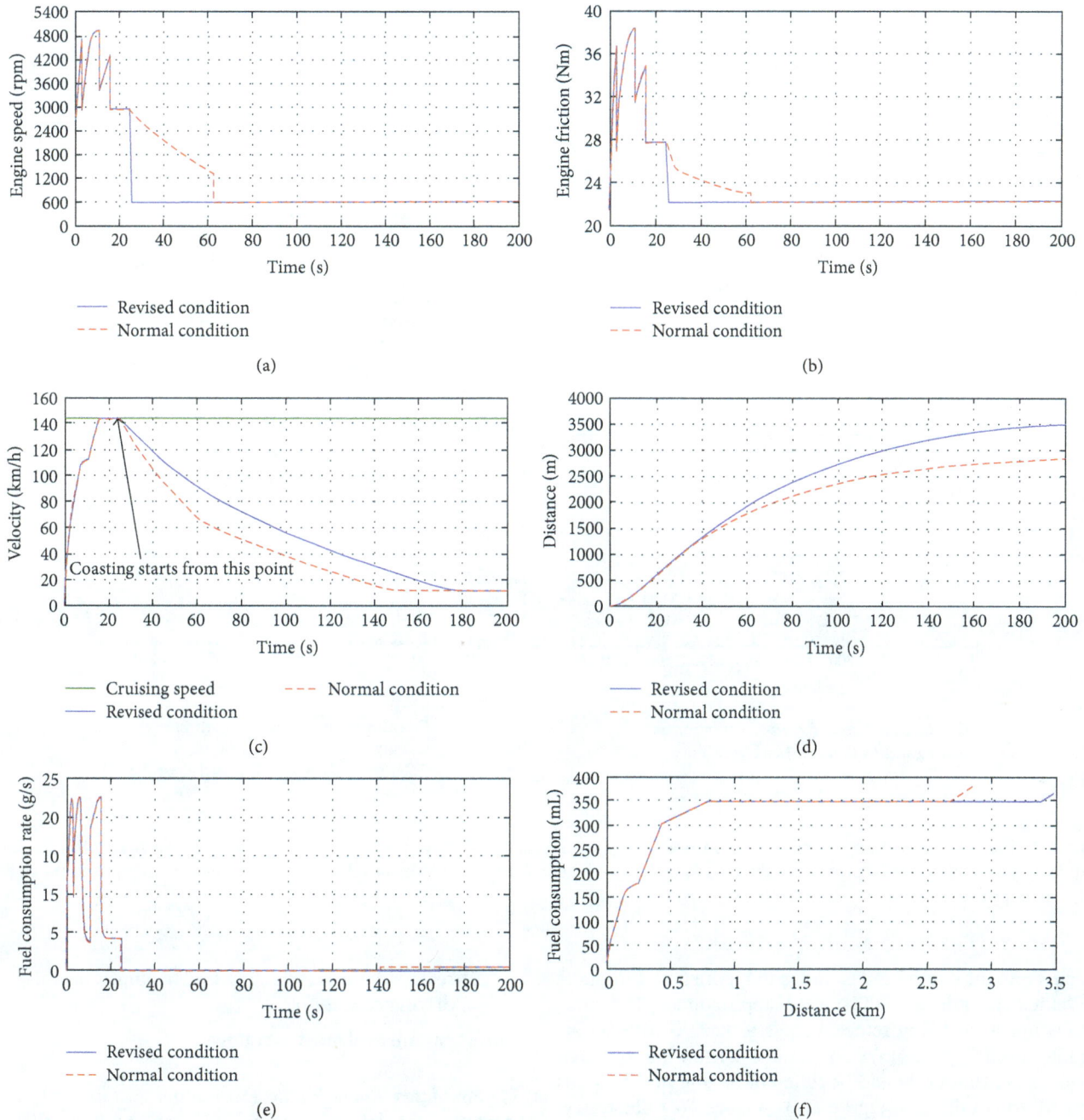

FIGURE 6: (a) The engine speed obtained for the normal and revised conditions—the lower engine speed obtained by the revised algorithm. (b) The engine friction obtained for two conditions—the amount of friction losses is lower via revised algorithm. (c) Vehicle velocity during coasting—longer coasting duration achieved through the revised algorithm. (d) Longer distance covered by the revised algorithm during coasting. (e) Comparison of the fuel consumption rates. (f) Fuel consumption versus distance for two conditions.

CC is switched off to set both the brake and the throttle inputs at zero and to let the vehicle coast, which happens after 25 s. During the normal condition the torque converter stays lockedup at the speed higher than 64.4 km/h resulting in high engine speed (Figure 6(a)) and consequently higher engine friction (Figure 6(b)). High engine friction reduces the overall tractive force which causes the reduction of the coasting duration. However, the revised algorithm does not allow the torque converter to remain lockedup and helping

the engine speed to be reduced quicker, hence longer coasting duration (Figure 6(c)) and coasting distance (Figure 6(d)) can be achieved. The comparisons between two conditions (revised and normal conditions) are illustrated in Figure 6. When using revised algorithm, the torque converter gets unlocked and stays at this state after 25 s, while without the revised algorithm (normal condition) the torque converter stays unlocked between 25 and 62 s and becomes unlocked after 62 s onward. When the vehicle is coasting, no fuel is

(a)

- · - Cruising velocity
— Coasting without revised algorithm
— Coasting with revised algorithm
— CC with dead-zone and revised algorithm

(b)

— CC with dead-zone and revised algorithm
-·- Coasting with/without revised algorithm

(c)

— Coasting without revised algorithm
— Coasting with revised algorithm
— CC with dead-zone and revised algorithm

(d)

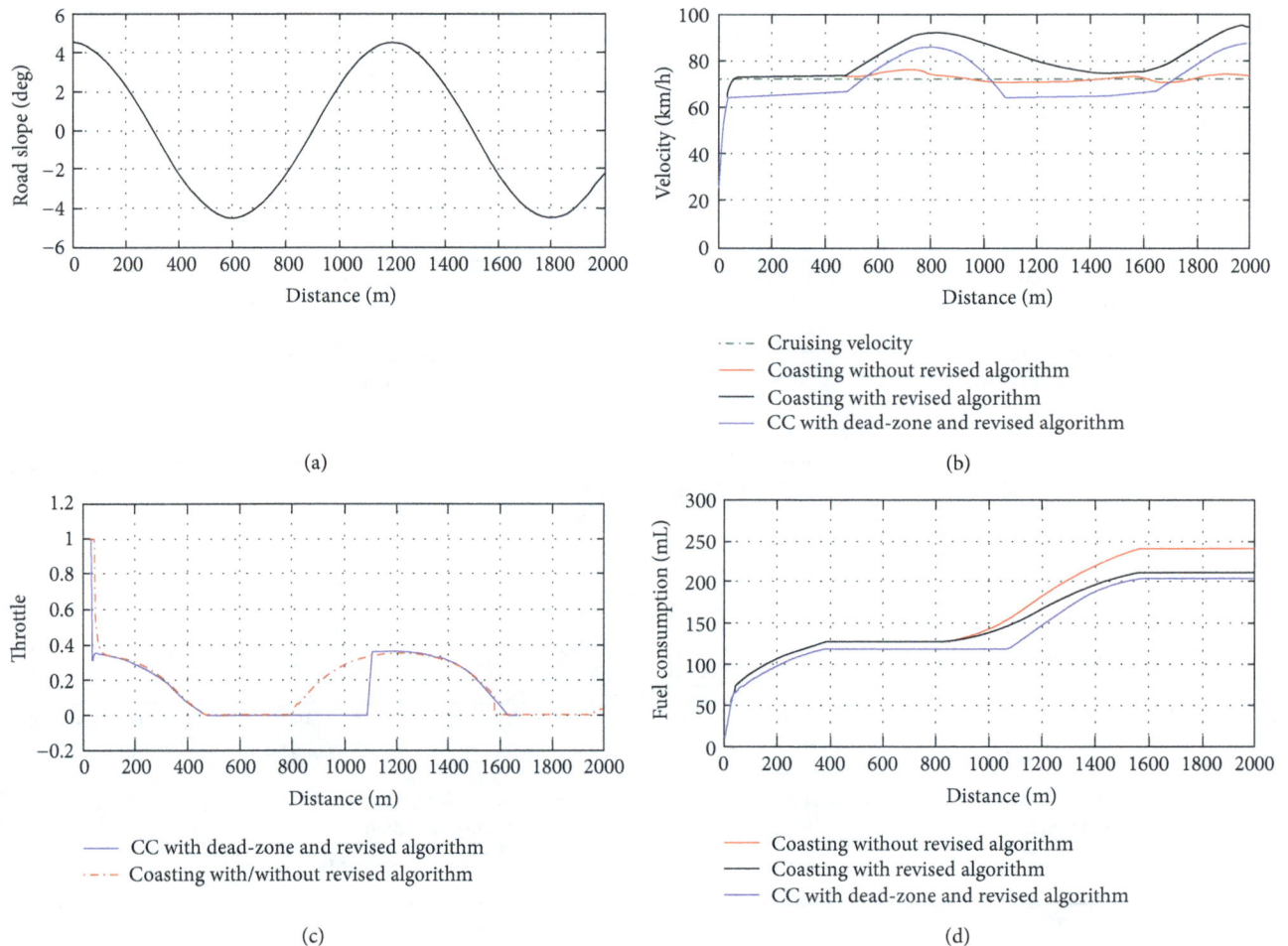

FIGURE 7: The influence of revised algorithm on both fuel consumption and travelled distance is investigated on the inclined road by introducing a simulated road profile (sinusoidal inclination)—(a) road slope. (b) Vehicle velocity. (c) Throttle opening positions. and (d) fuel consumption.

injected into the engine as it is shown in Figure 6(e). But once the vehicle slows down to the speed of approximately 15 km/h, the engine starts being refueled so as to keep it rotating at the idle speed (Figure 6(a)). The time instance when the fuel injection is restarted would be different for the normal and the revised conditions (Figure 6(e)). Figure 6(e) illustrates that the refueling starts at 140 s for normal condition, while it occurs at about 165 s for revised condition. It is due to the vehicle slows down faster in the normal condition (Figure 6(c)). The results demonstrate (Figure 6(f)) that fuel consumption can be reduced about 7.9% through the proposed approach. By consuming the same amount of fuel in both conditions, revised and normal conditions, the travelled distance (Figure 6(d)) could increase about 17.14% through the longer coast duration.

It is believed that the most fuel economy saving can be obtained when the vehicle travels downhill. Thus, the next set of two scenarios (scenarios 1 and 2) is determined to test the new approach (revised algorithm) on the inclined road. The two following tests are carried out:

(i) simulated road profile (sinusoidal inclination),

(ii) real road profile, which is used to emulate a real-world driving condition.

Those tests are explained as follows.

5.1. Simulated Road Profile (Sinusoidal Inclination). Three scenarios are defined here to evaluate the fuel efficiency gained through the new approach (Figure 7) for the described road profile, which is based on the sinusoidal variation of the road slope as shown in Figure 7(a). This test is carried out for the three following cases.

5.1.1. Coasting without Revised Algorithm. The test is started with applying the cruise control (CC) algorithm to the road profile illustrated by the road slope in Figure 7(a), so as to maintain the constant speed (72 km/h) as shown in Figure 7(b) after applying CC algorithm and analyzing the results, it is realized that in this particular case the brake is almost zero and braking is not essential. Therefore, it is decided not to consider the brake, especially that the brake is not important from the point of view of fuel consumption. Thus, the CC algorithm is run with the brake set at zero and

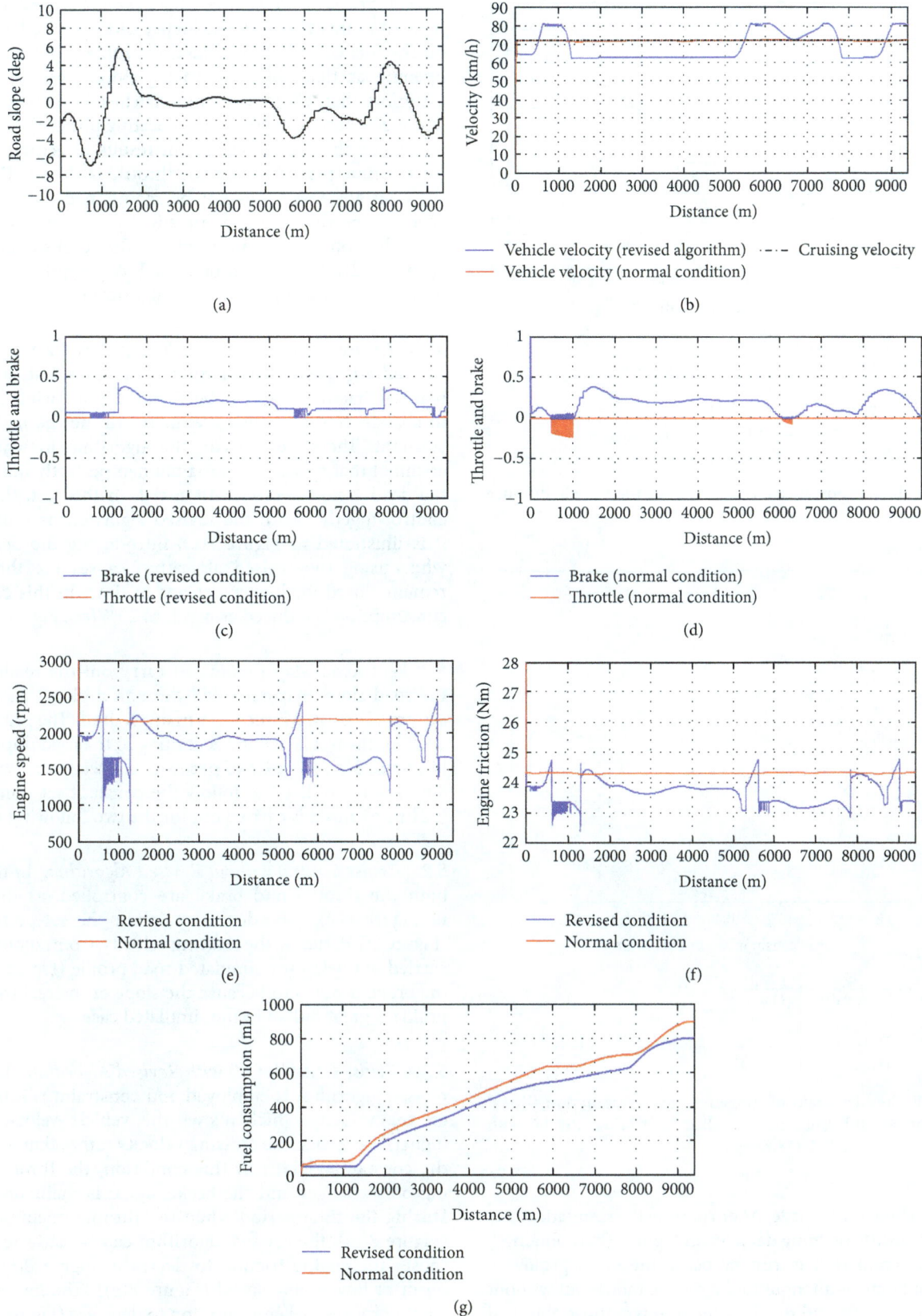

FIGURE 8: (a) Road slope. (b) Velocity obtained with and without revised algorithm. (c) Control action resulted from revised condition. (d) Control action resulted from normal condition, (e) comparison of the engine speeds. (f) Comparison of the engine frictions, and (g) comparison of the fuel consumptions obtained for two conditions.

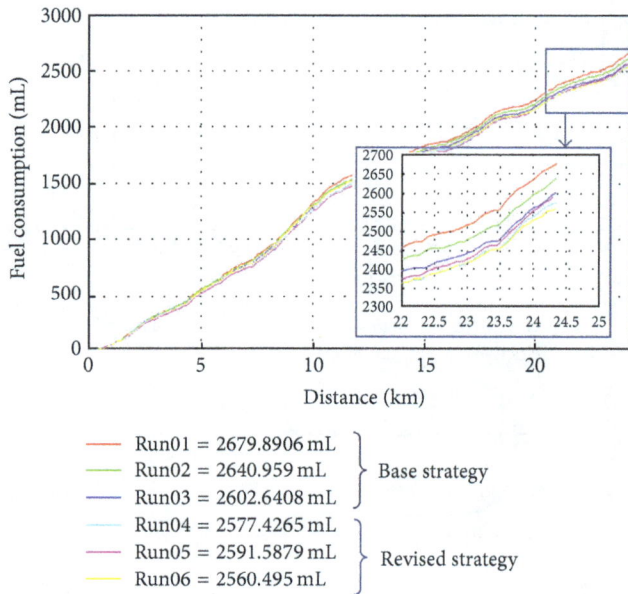

FIGURE 9: The accumulated fuel used versus the vehicle distance travelled.

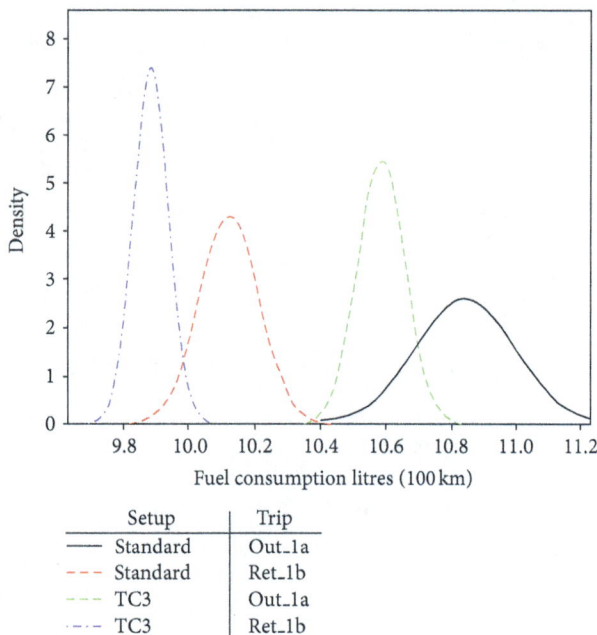

FIGURE 10: The histogram of measured fuel consumption (litres/ 100 km) for two different journeys called "Out_1a" & "Ret_1a", with standard and revised (TC3) strategies.

with only using the throttle. After running the simulation, the profile of throttle opening depicted in Figure 7(c) is obtained. Then, the simulation is run by using the same profile of throttle opening as an input to the vehicle model but without utilizing the CC algorithm. The same results as those obtained using the CC, that is, the vehicle velocity follows the cruising velocity, are expected and achieved. The throttle opening obtained from this test (see Figure 7(c)) is zero within two distances (from 475 to 800 m and from 1600 to 1950 m).

5.1.2. Coasting with Revised Algorithm. This test is implemented to investigate the influence of the revised algorithm on the fuel consumption. Here, the same profile of throttle opening as that obtained from previous test is used but assuming that when the throttle is closed, the revised algorithm gets activated during the coasting phase of the vehicle. Activating the revised algorithm results in getting less loss for the time periods when the throttle is closed. Therefore, the constant speed will not be maintained but rather the speed of the vehicle will go up slightly; however, we can get the reduction of fuel consumption. The results demonstrate (Figure 7(d)) that a reduction in fuel consumption of 12% can be achieved by using the revised algorithm.

5.1.3. Cruise Control (CC) including Dead Zone and the Revised Algorithm. Based on the previous test, it can be realized that the saving can be obtained by allowing the speed to change; that is, cruising velocity will not be maintained constant. Therefore, we run the algorithm in which it is assumed that speed is not maintained perfectly but there is the dead zone (1.5 m/s) with speed. In this test, the cruise control together with the revised algorithm is utilized. As it is illustrated in Figure 7(c), introducing the dead zone while using the revised algorithm causes the throttle to remain closed for longer periods of time. In this case, fuel consumption is reduced as much as 20% (see Figure 7(d)).

5.2. Real Road-Map Profile. To carry out the realistic test, the road slope map recorded through a global positioning system (GPS) in Munich-Germany is used. The road profile used for the test is defined according to the road slope shown in Figure 8(a). In order to perform this test, a cruise control algorithm is utilized to follow the constant set-point speed (72 km/h) and it is carried out for the two following cases.

5.2.1. Cruise Control without Revised Algorithm. In this case, both the throttle and brake are controlled, as illustrated in Figure 8(d), in order to maintain the set-point speed (Figure 8(b)), that is, the cruising velocity. Contrary to the test carried out using the simulated road profile (Figure 7), here the brake is not zero because the slope of the real road-map profile is greater than in the simulated case.

5.2.2. Cruise Control (CC) with Revised Algorithm. Here, the revised algorithm is employed and constraint is introduced on the velocity which allows the vehicle velocity to go higher/lower than the cruising velocity rather than following the constant velocity. In this condition, the throttle opening only changes and the brake signal is maintained zero. During the time period when the throttle opening is zero (Figure 8(c)), the revised algorithm can be activated which causes the engine friction to decrease (Figure 8(f)) as the result of low engine speed (Figure 8(e)). The lower engine friction causes fuel consumption to decrease (Figure 8(g)).

Here, fuel consumption is reduced up to 12% through the revised condition. The fact that the vehicle does not have to always track a fixed cruising velocity helps to mitigate the loss of energy due to braking. It also allows the vehicle to increase speed (Figure 8(b)) when going downhill which enables the vehicle to store more potential energy to cope with travelling uphill without having to consume so much fuel. Furthermore, in this condition the throttle can be closed for longer period which causes the revised algorithm to be activated and consequently engine friction to decrease.

6. Real Implementation

Some real tests have also been implemented in a vehicle utilising the revised strategy. This testing was conducted on a Range Rover Sport (Figure 4). The transmission torque converter control was modified to replicate the strategy. The testing is limited, but does show a statistically significant improvement in fuel economy. This modification was not optimal, and significant additional improvements in fuel consumption will be possible with an optimal controller in place. The results of the testing are shown in Figure 9. The plot shows the accumulated fuel used versus the vehicle distance travelled. The plots indicated by Run01 to Run06 were all captured from driving the vehicle on the identical route with similar traffic conditions, and there were no special driving styles adopted. However, there were the small variations in driver style and traffic. The plots depicted for Run01 to Run03 are for the tests for the base strategy, that is, does not use the revised strategy, while those depicted from Run04 to Run06 are for the tests when the revised strategy is used.

Figure 10 shows a histogram of data for two different journeys, called "Out_1a" & "Ret_1a", with standard and revised (TC3) strategies. Each journey is repeated several times with the original and new revised control system. The curves show the variation in measured fuel consumption (litres/100 km) used for each of these journeys and each control system. For example, journey "Ret_1b" has an average of 10.12 litres/100 km fuel consumption with original strategy, and new strategy achieves 9.89 litres/100 km. The data was analysed to confirm that this improvement was statistically significant (i.e., the separation between the curves is greater than test to test variability).

7. Conclusions

In this paper, the feasibility study on the fuel efficiency during coasting was carried out through the simulation. Both locked-up and unlocked-up modes of the torque converter and the switching between these two modes of operation were investigated in the simulation. The accuracy of the integrated simulation model of the longitudinal vehicle dynamics was validated by comparing the results against the Jaguar Land Rover (JLR) data. The paper has presented modeling of the following situations:

(i) vehicle coasting down on the flat-road with the torque converter locked-up (in normal operating mode),

(ii) vehicle coasting down on the flat-road with the torque converter unlocked-up,

(iii) to present a realistic scenario, the tests have been implemented on the inclined road. These tests have been carried out by utilizing the simulated road profile (sinusoidal inclination) and a real road profile, which is used to emulate a real-world drive cycle, with and without using the revised algorithm.

In all above cases, two virtual PID controllers have been utilized to keep the engine idle speed depending on the vehicle speed and throttle opening position. As a result of simulation investigation, longer coasting duration can be achieved or less fuel can be consumed through reducing the engine speed. Reduction of the engine speed can be achieved by controlling the torque converter operation which is called as the revised algorithm in this paper. Therefore, incorporating the revised algorithm in the typical power-train system increases the fuel economy saving. The test results indicated that the coasting by employing the revised algorithm could reduce fuel consumption and increase the travelled distance. The results have demonstrated that fuel consumption can be reduced about 7.9% through the proposed approach when the vehicle coasts on the flat road. In other word, the travelled distance could increase about 17.14%, that is, the longer coast duration. Fuel saving can be increased up to 12% when coasting on the inclined road. The losses in the torque converter during reengaging the engine over the repeated costing/acceleration schedule will have only a small effect in the overall fuel consumption value, since the total time elapsed during this phase is relatively small compared to the whole drive cycle. Of course, further work will be undertaken to assess all efficiencies and other factors in more detail, which will be subject of future research. Unconstrained vehicle speed increases during cruise control coasting would not be suitable for production. Some balance between speed deviation and fuel economy would need defining. Furthermore, even with no speed deviation, a fuel economy benefit is realized due to longer coast duration. In order to take the most advantage of the revised algorithm, the optimal velocity trajectory for a particular road profile needs to be calculated by utilizing a suitable algorithm such as dynamic programming (DP) and Pontryagin's Minimum Principle. This investigation would be carried out in the future research work.

Appendix

A. Derivation of the Vehicle Simulation Model

A.1. Powertrain Model. Figure 11 shows the schematic diagram of a power-train which demonstrates the transmission of the torque and velocity from the engine to the wheels. The automatic transmission is considered in the power-train model. Therefore, the torque produced by the engines is transmitted to the gear box through the torque converter [17–19]. The torque converter consists of three essential parts, that is, the impeller, the turbine, and the reactor. The impeller is connected to the crankshaft that transmits the power of the engine to the turbine through the hydraulic oil inside the torque converter. In turn the turbine is connected to the output shaft of the converter which is coupled to the input shaft of the gear box. The torque getting through the gear

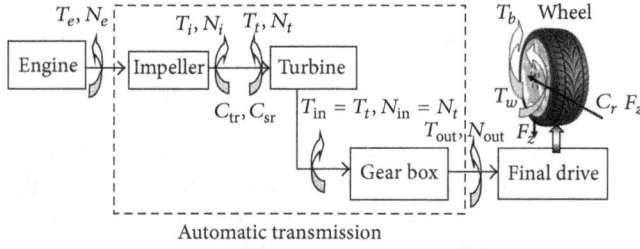

Automatic transmission

FIGURE 11: Schematic diagram of the powertrain, demonstrating the transmission of the torque and velocity from the engine to the wheels.

box varies depending on the gear ratio. Finally the output torque from the gear box is transmitted to the wheels after passing through the final drive. The torque and rotation speed of the wheel are affected by the brake torque and the external forces exerted on the vehicle including rolling resistance force resulted from the interaction between the tyre and the road, gravitational force, and aerodynamic drag force. The powertrain model is simulated by considering the two states of the torque converter-locked and unlocked states.

A.1.1. Model Presenting Unlock State of the Torque Converter. If the torque converter is unlocked-up, the torque produced via engine is transmitted by the fluid flowing through the impeller, turbine, and stator; this is the so-called fluid coupling. In order to develop the simulation model which presents unlocked state of the torque converter, the interaction between the engine and the torque converter must be taken into consideration. The engine model corresponding to the unlocked state of the torque converter can be defined by following equation [15, 20]:

$$I_{ei}\dot{N}_e = T_{eb},$$ (A.1)

where N_e denotes the engine rotation speed (rpm), I_{ei} denotes the summation of moment of inertia of the engine and the impeller, and T_{eb} denotes the engine brake torque which is determined by considering all the losses in the engine as follows:

$$T_{eb} = T_{ei} - T_f - T_P - T_l,$$ (A.2)

where T_{ei} denotes the engine indicated torque, which depends both on the value of the engine rotation speed (N_e) and the throttle opening position (u_t) within the range [0, 1]. T_f and T_p denote, respectively, the friction loss and pumping loss (for brevity in this paper, these two terms are lumped and denoted by the engine friction). T_l indicates the additional acting load on the engine and here, the impeller torque (T_i) is only considered as an additional acting load on the engine; that is, $T_l = T_i$.

The engine model determined by the above equations together with the two PID controllers explained in Section 3 presents Engine 1, illustrated in Figure 4. The engine indicated torque, and the engine friction are obtained by using two look-up tables in the simulation. The value of the engine friction depends on the engine rotation speed and air mass flow.

The magnitudes of the impeller torque and throttle opening are acquired by the switching algorithm (see Figure 4).

Parameters playing an important role in the performance of a torque converter are the speed ratio ($C_{sr} = w_t/w_i$), the torque ratio ($C_{tr} = T_t/T_i$), the efficiency ($\eta_e = C_{tr}C_{sr}$), and the capacity factor or K-factor (K_{tc}), where T_i is the impeller torque (converter input torque), w_i is the impeller speed (converter input speed), T_t is the turbine torque (converter output torque), and w_t is the turbine speed (converter output speed). The capacity factor shows the ability of the converter to absorb or transmit the torque [17, 21].

By assuming that the impeller speed (N_i) is identical to the engine speed (N_e), the impeller torque can be obtained as follows:

$$T_i = \left(\frac{N_e}{K_{tc}}\right)^2.$$ (A.3)

Knowing the torque ratio (C_{tr}) and speed ratio (C_{sr}) enables us to calculate the turbine torque as follows:

$$T_t = C_{tr}T_i.$$ (A.4)

The torque ratio (C_{tr}) and K-factor (K_{tc}) can be interpolated from the torque converter characteristic maps against the speed ratio (C_{sr}) [3]. These maps are depicted in Figure 12. By neglecting distortion and damping effect of the rotating part, the turbine torque and its speed will be identical to input torque (T_{in1}) and input speed (N_{in}) of the transmission. T_{in1} indicates the transmission input torque while the torque converter is unlocked.

A.1.2. Model Presenting Locked-Up State of the Torque Converter. Considering the torque converter to be locked up, the engine torque and its speed will be entirely transmitted to gear box input shaft; therefore, the torque converter can be disregarded in the equations presenting that condition. The equation presenting locked-up state of the torque converter is determined as follows:

$$I_{et}\dot{N}_e = T_{ei} - T_f - T_p - T_{in2},$$ (A.5)

where N_e denotes the engine rotation peed (rpm). I_{et} denotes the summation of moment of inertia of the engine, torque converter, and gear box. T_{ei} denotes the engine indicated torque. T_f and T_p denote the friction and pumping losses, respectively. T_{in2} denotes the transmission input torque while torque converter is locked. By neglecting distortion and damping effect of the rotating part, the engine speed (N_e) will be identical to the transmission input speed (N_{in}). Therefore, the transmission input torque (T_{in2}) can be calculated as follows:

$$T_{in2} = T_{ei} - T_f - T_p + I_{et}\dot{N}_{in}.$$ (A.6)

The engine indicated torque (T_{ei}) and friction (T_f) and pumping (T_p) losses are obtained from the same lookup tables as those used for the unlocked state of the torque converter, however, here the transmission input speed (N_{in})

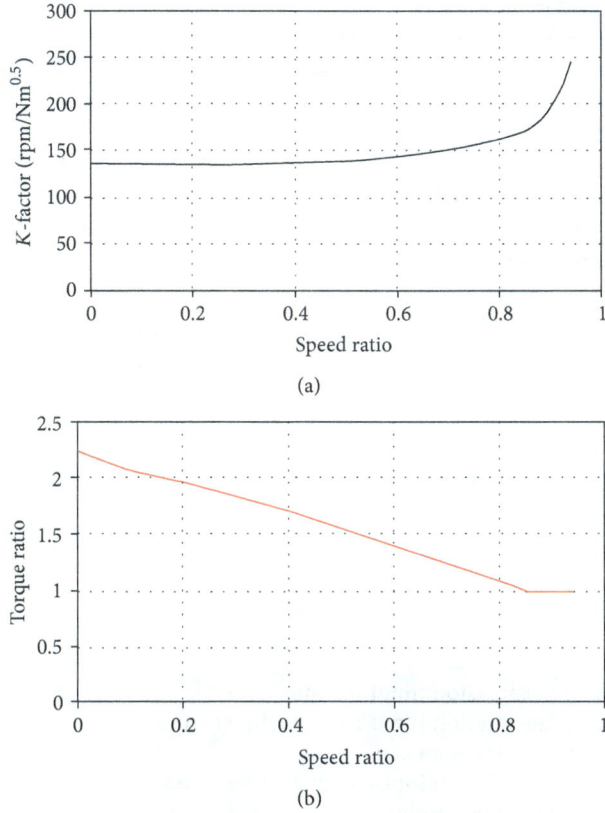

FIGURE 12: Performance characteristic of a torque converter—(a) K-factor and (b) torque ratio.

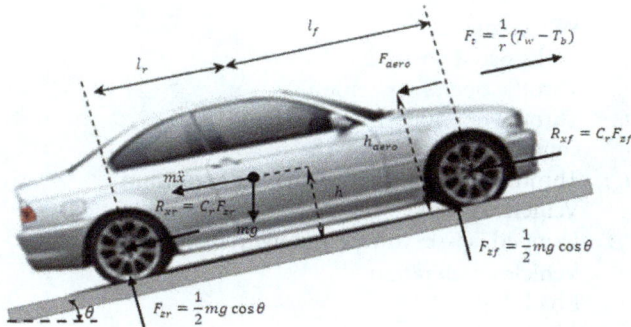

FIGURE 13: Exerting forces on the vehicle during travelling on inclined road.

is used in place of the engine speed (N_e) in the interpolation. (A.6) presents Engine 2, depicted in Figure 2.

Transition may occur between the two states of the torque converter; locked and unlocked states. This transition is implemented through a switching algorithm in the simulation. For normal condition, the switch is only triggered according to the velocity of the vehicle. Therefore, the torque coveter is locked-up at the speed higher than 17.88 m/s, and subsequently the model presented by (A.6) is utilized.

A.1.3. Common Elements in the Power-Train Model for Locked and Unlocked States of the Torque Converter. The torque and

rotation speed transmitted to the wheel through gear box and the final drive can be obtained as follows:

$$T_w = R_{tr}R_{fd}T_{in},$$

$$N_w = \frac{N_{in}}{R_{tr}R_{fd}}, \tag{A.7}$$

where T_{in} and N_{in} are the torque and the speed on the input side of the transmission, respectively. R_{tr} and R_{fd} are the gear ratio and the final drive ratio, respectively. T_w and N_w indicate the torque and rotation speed of the wheel.

A.2. Brake Model. The torque generated by the brake system can be obtained by the following equation [22, 23]:

$$T_b = K_bP_b, \tag{A.8}$$

where K_b denotes the lumped gain for entire brake system and P_b, the amount of pressure produced behind the brake disk. This pressure is described by the following dynamic equation:

$$P_b = 150K_cu_b - \tau_bP_b, \tag{A.9}$$

where K_c denotes the brake pressure gain, τ_b the lumped lag obtained by combining two lags relating to the dynamics of the servo valve and the hydraulic system, and u_b is the brake pedal position within the range $[-1, 0]$.

A.3. Vehicle Dynamics Model. Utilizing the second Newton's law which takes balancing of the forces exerted on the vehicle consisting of rolling resistance force (R_z), aerodynamic force (F_{aero}), gravitational force (F_g) and the forces generated on the wheel by the brake torque (T_b) and the engine torque (T_w) (Figure 13), the acceleration of the vehicle (a) can be obtained as follows:

$$Ma = \frac{1}{r}(T_w - T_b) - \underbrace{\frac{1}{2}\rho ACdv^2}_{F_{aero}} - \underbrace{C_rmg\cos(\theta)}_{R_z} - \underbrace{mg\sin(\theta)}_{F_g}. \tag{A.10}$$

Here ρ denotes the air density, r the wheel radius, C_r the rolling resistance coefficient, C_d the drag coefficient which depends on the vehicle's body shape, v vehicle velocity, A the frontal cross area of the vehicle, and g the gravitational acceleration. θ indicates the road slope. In order to take into account the effect of rotating part on the vehicle dynamics, the moment of inertial of the wheels has to be lumped with the vehicle mass:

$$M = \frac{n_wI_w}{r^2} + m, \tag{A.11}$$

where I_w denotes the moment of inertia of a wheel, n_w indicates the number of the wheels, m the mass of the vehicle, and r the wheel radius. Consequently By integrating the acceleration, the velocity of the vehicle will be obtained. The values of the parameter used in the equation are given in Table 1.

TABLE 1: Values of the parameters used in the equations.

Parameter	Description	Numerical value
m	Vehicle mass	2270 [kg]
r	Wheel radius	0.326 [m]
R_{tr}	Gear ratio	3, 2.34, 1.85, 1.45, 1.00, 0.68
R_{fd}	Final driver ratio	3.28
I_{ei}	Moment of inertia of engine and torque converter	0.224 [kgm^2]
I_{et}	Moment of inertia of engine, torque converter and transmission	0.226 [kgm^2]
I_w	Moment of inertia of the wheel	1.7 [kgm^2]
$\rho \cdot A \cdot c_d$	Aerodynamic force coefficient	1.2 [kg/m]
C_r	Rolling resistance coefficient	0.015
n	Engine stroke	4
V_d	Engine displacement volume	0.005 [m^3]
g	Gravitational acceleration	9.8 [m/s^2]
τ_b	Lumped lag-servo valve and the hydraulic system	0.2
K_c	Pressure gain	1
K_b	Lumped gain for entire brake system	20 [Nm/bar]
n_w	Number of the wheels	4

Abbreviations

A: Frontal cross-section area of the vehicle
C_d: Aerodynamic drag coefficient
C_r: Rolling resistance coefficient
C_{sr}:   Speed ratio
C_{tr}:   Torque ratio
F_{aero}:   Aerodynamic force
F_g: Gravitational force
g: Gravitational acceleration
I_{ei}:   Summation of moment of inertia of the engine and the impeller
I_{et}:   Summation of moment of inertia of the engine, torque converter, and gear box
I_w: Moment of inertia of a wheel
K_b: Lumped gain for entire brake system
K_c: Brake pressure gain
K_{tc}:   Capacity factor or K-factor
m: Mass of the vehicle
n: Engine stroke
N_e: Engine rotation speed
N_i: Impeller speed
N_{in}:   Transmission input speed
N_t: Turbine torque
N_w: Wheel rotation speed
n_w: Number of the wheels
P_b: Pressure behind the brake disk
r: Wheel radius
R_{fd}:   Final drive ratio
R_{tr}:   Gear ratio
R_z: Rolling resistance force
T_b: Brake torque
T_{eb}:   Engine brake torque
T_{ei}:   Engine indicated torque
T_f: Friction loss
T_i: Impeller torque
T_{ic}:   Impeller torque calculated by PID controller

T_{in}:   Transmission input torque
T_{in1}:   Transmission input torque while torque converter is locked
T_{in2}:   Transmission input torque while torque converter is unlocked
T_{is}:   Impeller torque from switching algorithm
T_l: Acting load
T_p: Pumping loss
T_t: Turbine torque
T_w: Wheel torque
u_b: Brake pedal position
u_t: Throttle opening position
u_{tc}:   Throttle opening calculated by PID controller
u_{ts}:   Throttle opening from switching algorithm
v: Vehicle velocity
τ_p: Lumped lag-servo vale and hydraulic system
a: Vehicle acceleration
θ: Road slop
ρ: Air density.

Conflict of Interests

The authors do not have any financial relationship with the company Mathworks (MATLAB software).

Acknowledgments

The authors would like to thank Peter Kock from the Central Engineering Division of MAN (Truck & Bus) Nutzfahrzeuge AG in Munich, Germany, for providing the real road-map profile. This work has been carried out as a collaborative project between Kingston University of London and Jaguar Land Rover (JLR) in the UK.

References

[1] J. Barrand and J. Bokar, "Reducing tire rolling resistance to save fuel and lower emissions," *SAE International Journal of Passenger Cars*, vol. 1, no. 1, pp. 9–17, 2009.

[2] H. D. Tarpinian, G. H. Nybakken, and J. Mishory, "A fuel saving passenger tire," *SAE Technical Paper*, no. 790726, 1979.

[3] D. Y. Lee, H. H. Ju, J. S. Rhee, S. H. Lee, and H. S. Lee, "Adaptive anti-shock coasting lock-up control of the torque converter clutch," *World Academy of Science, Engineering and Technology*, vol. 15, pp. 97–102, 2006.

[4] J. Lee, *Vehicle Inertia Impact on Fuel Consumption of Conventional and Hybrid Electric Vehicle Using Acceleration and Coast Driving Strategy [Ph.D. thesis]*, Virgina Polytechnic Institute and State University, Blacksburg, Virgina, USA, 2009.

[5] T. van Keulen, G. Naus, B. de Jager, R. van de Molengraft, M. Steinbuch, and E. Aneke, "Predictive cruise control in hybrid electric vehicles," *World Electric Vehicle Journal*, vol. 3, no. 1, 2009.

[6] T. Van Keulen, B. De Jager, D. Foster, and M. Steinbuch, "Velocity trajectory optimization in Hybrid Electric trucks," in *Proceedings of the American Control Conference (ACC '10)*, pp. 5074–5079, Marriott Waterfront, Baltimore, Md, USA, July 2010.

[7] A. Rousseau, S. Pagerit, and D. W. Gao, "Plug-in hybrid electric vehicle control strategy parameter optimization," *Journal of Asian Electic Vehicles*, vol. 6, no. 2, pp. 1125–1133, 2008.

[8] D. Sinoquet, G. Rousseau, and Y. Milhau, "Design optimization and optimal control for hybrid vehicles," *Optimization and Engineering*, vol. 12, no. 1-2, pp. 199–213, 2011.

[9] E. Hellstrom, *Explicit use of road topography for model predictive cruise control in heavy truck [M.S. thesis]*, Linkopings universite, Likopings, Sweden, 2005.

[10] E. Hellström, M. Ivarsson, J. Åslund, and L. Nielsen, "Look-ahead control for heavy trucks to minimize trip time and fuel consumption," *Control Engineering Practice*, vol. 17, no. 2, pp. 245–254, 2009.

[11] E. Kozica, *Look Ahead Cruise Control: Road Slop Estimation and Control Sensitivity [M.S. thesis]*, Master's Degree Project, Stockholm, Sweden, 2005.

[12] P. Kock, H.-J. Welfers, B. Passenberg, S. Gnatzig, O. Stursberg, and A. W. Ordys, "Saving energy through predictive control of longitudinal dynamics of heavy trucks," *VDI Berichte*, no. 2033, pp. 53–67, 2008.

[13] N. Muller, S. Straub, S. Tumback, and A. Christ, "Coasting-next generation start/stop systems," *MTZ*, vol. 72, no. 9, pp. 14–19, 2011.

[14] P. Shakouri, A. Ordys, M. Askari, and D. S. Laila, *Longitudinal Vehicle Dynamics Using Simulink/Matlab*, UKACC International conference on CONTROL, Coventry, 2010.

[15] J. B. Heywood, *Internal Combustion Engine Fundamentals*, McGraw-Hill, 1988.

[16] P. Shakouri, A. Ordys, and D. S. Laila, "Adaptive cruise control system: comparing gain-scheduling PI and LQ controllers," in *Proceedings of the 18th IFAC World Congress*, Milano, Italy, August 2011.

[17] J. Y. Wang, *Theory of Ground Vehicle*, Wiley-IEEE, 3rd edition, 2001.

[18] D. Geuns, *Discrption Automatic Transmission VT1F*, ZF Getriebe Sint-Truiden, 2003.

[19] C. H. Yao, *Automotive Transmissions: Efficiently Transferring Power from Engine to Wheels*, Discovery Guides, Pro Quest, 2008.

[20] H. Heisler, *Advanced Vehicle Technology*, College of North West London, 2nd edition, 2002.

[21] MathWorks, *Using Simulink and Stateflow in Automotive Application*, Automatic transmission control, 1998.

[22] M. Short, M. J. Pont, and Q. Huang, *Simulation of Vehicle Longitudinal Dynamic*, Embedded System Laboratory University of Leicester, safety and reliability of Distributed Embedded Systems, Leicester, 2004.

[23] J. C. Gerdes and J. K. Hedrick, "Vehicle speed and spacing control via coordinated throttle and brake actuation," *Control Engineering Practice*, vol. 5, no. 11, pp. 1607–1614, 1997.

8

Improving Energy Conversion Efficiency by means of Power Splitting in Dual Drive Train EV Applications

Michael A. Roscher, Roland Michel, and Wolfgang Leidholdt

imk automotive GmbH, Annaberger Straße 73, 09126 Chemnitz, Germany

Correspondence should be addressed to Michael A. Roscher; michael.roscher@rwth-aachen.de

Academic Editor: Lingyang Song

The limited amount of energy stored on board of battery electric vehicles (BEV) spurs research activities in the field of efficiency optimization for electric drive train applications in order to achieve an enhanced mileage. In this work a control method for BEV applications with two drive trains (e.g., one at the front and one at the rear axle) is presented. Herein, a simple optimization algorithm is introduced enabling to operate the two drives with different torque values, depending on the instantaneous operation point, leading to a reduction of apparent power losses on board. Simulations on a virtual BEV yield a decrease in the cumulated energy consumptions during typical BEV operation, leading to an increase in the achievable mileage.

1. Introduction

In recent years, the need for zero emission transportation spurred the broad market introduction of battery electric vehicles (BEV). BEVs can play a key role on the way to the environmentally conscious society [1]. Especially in mega cities where air pollution is a critical, the emission free vehicles can contribute to improve the health-related quality of life. Moreover, if the energy for battery charging comes from renewable sources (wind, solar, hydro power, etc.), the vehicles operate almost 100% CO_2 neutral.

However, one main drawback of battery powered vehicles is the very limited range, due to the very limited amount of energy stored in the battery. On the one hand, the efficiency of energy conversion of electric power trains is 2-3 times higher ($\eta > 80\%$) than the efficiency of combustion engines ($\eta = 20$–40%). On the other hand, the nowadays commercial battery cells comprise energy densities of maximum 250 Wh/kg [2], which is less than one fortieth of the theoretical energy density of conventional fuels (petrol gas approx. 12.000 Wh/kg). Moreover, the energy density of common battery systems is reduced about almost one-half, due to efforts for housing, cooling, and integrated electrics/electronics.

An electric drive train substituting the combustion engine is one possible vehicle concept. Besides, various approaches

(e.g., four separate in-wheel drives) are possible. In the following a vehicle architecture including two drives (i.e., one at the front and one at the rear axle ($\rightarrow$ 4 WD); see Figure 1) is exemplarily considered.

In the illustrated BEV architecture the HVAC or at least the integrated compressor is supplied directly from the battery. To supply additional aggregates, the battery voltage is transferred into, for example, 12 V by a DC/DC converter. During vehicle acceleration and even speed, the drive train components (power converters and electric motors) convert the chemical energy stored in the battery in mechanical work. During regenerative breaking, the battery is recharged by the converters. The 4 WD concept enables a very efficient recuperation, because regenerative breaking proceeds on both axles. Through incorporating a recuperative breaking, the possible range can be increased about up to several 10% [3, 4]. However, the battery state-of-charge successively decreases during driving due to the apparent losses (wind, friction), differences in the level (track profile), and the energy consumption of the ancillary components [5].

From the literature several approaches are known referring to range prediction algorithms for HEV applications (e.g., in [6]) and similarly for BEVs [7] incorporation track profile information. Further, Zhang et al. [8] presented methods for driving range prediction of BEVs considering

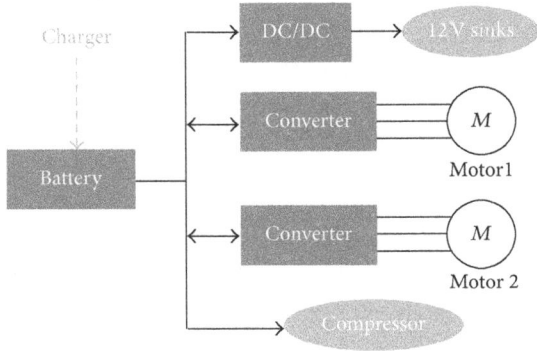

FIGURE 1: Power supply architecture of a two motor BEV concept.

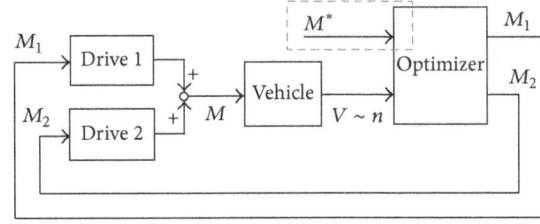

FIGURE 2: Control structure to optimize the energy consumption in a two motor EV power train concept.

environmental aspects (wind) and the driver's habits during driving, too. Besides, model based and expert systems specific energy management systems using stochastic approaches are known [9]. According to control strategies designed in order to optimize the energy efficiency on board of electrified vehicles, several references exist on the one hand focusing on managements systems of vehicles including at least two different energy storage systems (application with battery and supercap storage [10] or battery and fuel cell [11]) based on climatic, track profile, and operation point information.

On the other hand, energy saving strategies are known focusing on optimized control strategies for distinct vehicle components, for example, the air-conditioning system [12], by means of occasionally switching ancillary components according to the driving situation (battery temperature, power demand of the drive train, etc.). Optimizing the efficiency of electric motors in boost and regeneration operation is addressed in the literature references for HEV applications [13] and also HEV applications with two individual electric power trains in addition to the combustion engine, by optimizing the instantaneous operation points of the three individual drive trains [14]. Referring to the aforementioned approaches, a control strategy for the two-motor application in BEVs aiming an energy saving through optimized torque and speed control of the individual motors is presented in the following section.

2. Power-Saving Operation

In BEV applications with two individual power trains (on one axle or applied on two different axles), each of the power trains can be controlled independently [14]. During normal operation, the BEV's speed varies according to drive cycle and the driver's habits, but most of the time the speed can be assumed to remain almost constant over several seconds, except for sections of fast acceleration or regenerative breaking. During these periods of even speed with varied torque the overall mechanical power demand can be supplied by the both of the power trains or only by one power train, if possible. With only one motor supplying the torque (herein named *individual operation*), the electric power input can be reduced in comparison to two motors in synchronous operations, which is the cases if the power conversion with one motor is

more efficient, due to the magnetization and switching losses (idle losses) [15, 16]. The instantaneous efficiency of power conversion of converter, and motor (electric to mechanical power, i.e., motor operation) depends on the rotation speed n and the torque M [17, 18]; for example, given with (1), where P_{mech} is the mechanical power at the axles, P_{el} is the electric power input, η_{pc} is the actual efficiency of the power converter and η_{mot} is the efficiency of the motor.

The general condition to meet in order to achieve the optimum efficiency is given with (2), where M_1 is the torque supplied from one of the drives (drive 1) and M_2 is the torque supplied from the other drive (drive 2). The overall torque supply M correlates with the accelerator position operated by the vehicle driver as follows:

$$\eta_{el} = \frac{P_{mech}}{P_{el}} = \eta_{pc} \cdot \eta_{mot} = f(n, M), \tag{1}$$

$$P_{el} = \frac{n \cdot M_1}{\eta(n, M_1)} + \frac{n \cdot M_2}{\eta(n, M_2)} \overset{!}{=} \text{Min}. \tag{2}$$

A proposed control structure is illustrated in Figure 2. The overall motor torque demand M^* (correlating with the accelerator position) and actual rotation speed n (proportional to the vehicle's velocity V) are the input value for the optimizer unit.

From M^* and n, the optimum set points M_1 and M_2 are derived, which are the torque control set point the both drives. The optimizer unit considers the efficiencies of the two possible operation cases, as stated above (1) and (2). The efficiencies in both cases, depending on the torque M and the rotation speed n, are extracted form point maps ($\eta = f(n, M^*)$), see (1) of the drive trains, using, for example, look-up tables or polynomials. Taking into account the actual efficiencies of the two drive trains in the two operation modes a distinction of cases is proposed according (3)

$$M_1 = \begin{cases} M^* & \text{for } \eta(n, M^*) > \eta\left(n, \frac{1}{2}M^*\right) \\ \frac{M^*}{2} & \text{for } \eta(n, M^*) < \eta\left(n, \frac{1}{2}M^*\right) \end{cases}$$

$$M_2 = \begin{cases} 0 & \text{for } \eta(n, M^*) > \eta\left(n, \frac{1}{2}M^*\right) \\ \frac{M^*}{2} & \text{for } \eta(n, M^*) < \eta\left(n, \frac{1}{2}M^*\right). \end{cases} \tag{3}$$

Accordingly, both drive trains supply equal torques, that is, one-half of the overall torque demand M^*, if the efficiency

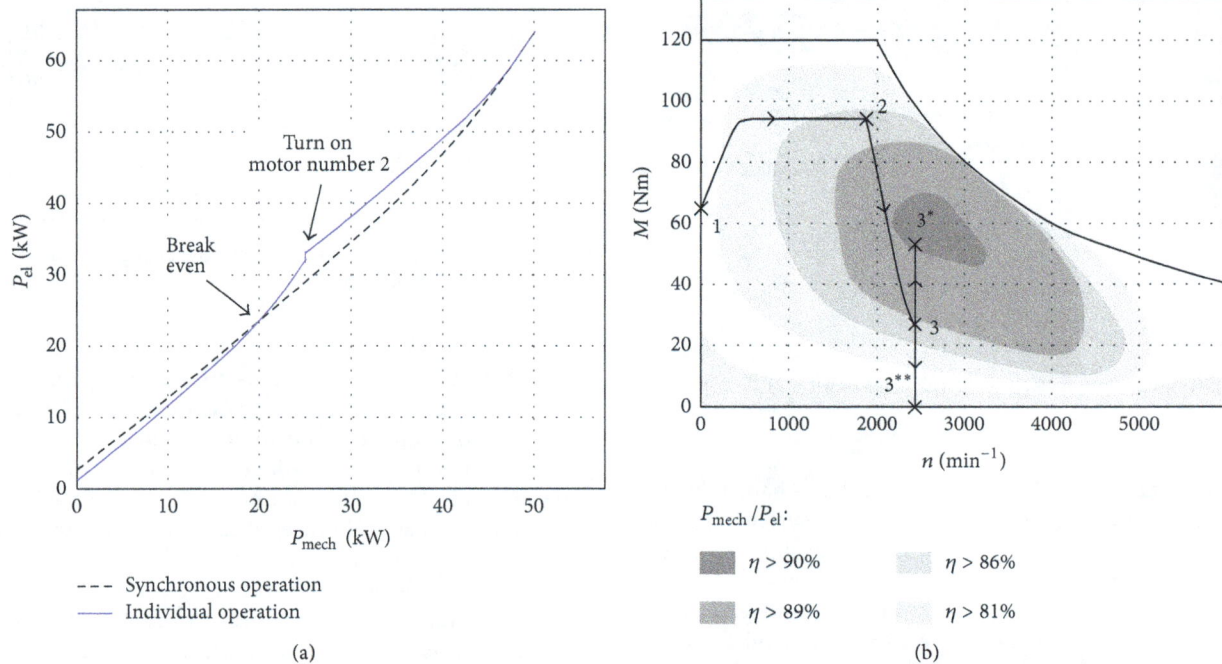

FIGURE 3: Comparison of the electric power demand at constant speed of two electric motors in synchronous and individual operation plotted versus the actual torque (a) and the possible trajectory for acceleration and subsequent even speed operation, with one drive train switched off after acceleration (b).

at $1/2 \cdot M^*$ is higher than the efficiency at M^* (constant speed n assumed). In the contrary case, the control set point of one motor is set to M^*, and the set point of the second motor is set to zero (i.e., turned off second drive train).

To elucidate a detailed control strategy of the drive trains itself (e.g., [15, 19]), considering the variable battery voltage, temperature, load limitations and so forth, lies beyond the scope of this work and therefore is left out here.

With this approach the instantaneous power losses during even speed periods can be minimized. A comparison of both mentioned operation modes is illustrated in Figure 3(a), where the electric power effort is plotted over the instantaneous power at the drive side.

The dotted line indicates the power demand of two motors supplying the same torque. The blue solid line reflects the case where firstly one power train is active (the other converter is turned off) until the maximum power is reached. Towards higher torques, the second drive is active too (indicated in Figure 3(a) with turn on, motor number 2). At the point where the second drive is turned-on the power demand increases approximately about the amount of the idle losses. Obviously, at high total torque demand in case of operating the two power trains with different torques, the energy conversion is less efficient in comparison to two drives with equal torque operation. Therefore, an operation with both power trains supplying equal torques is to prefer (synchronous operation). But at lower torques the operation with one active motor can be more efficient, approximately until the saturation losses compensate the magnetization/swichting losses in idle running. The point where the efficiencies of

synchronous and individual operations are equal is indicated in Figure 3(a) with "break even."

Hence, the intention is to operate with two active drive trains in high power demand operation and to switch-off one of the drives at low mechanical power demand, where the individual operation is more efficient. A possible trajectory during a driving situation with acceleration and even speed operation is depicted in Figure 3(b). Herein, the vehicle is accelerating (point $1 \rightarrow 2$). After accelerating greatly, the velocity is slightly increased for a certain period (point $2 \rightarrow 3$), and after that the vehicle speed is constant. In this operation (in the neighbourhood of point 3 in Figure 3(b)) it is more efficient to propel the axle with only one of the drives. Hence, one of the drives (converter and motor) is turned off, and two different operation points exist (points 3^* and 3^{**}). Due to the second motor being turned off, no magnetization losses are apparent in one motor, and no switching losses appear in one converter (point 3^{**}). In the shown exemplary case the overall efficiency is increased about approximately 1.3% at an electric power input of approximately 14.5 kW. The instantaneous power saving is in a range of 180–200 W.

The intention to turn-on and turn-off one of the drives, if it is preferable form energetic point of view, bears drawbacks during operation scenarios where the actual operation point lies in neighbourhood of the break-even point. In this case, on one hand it might be necessary to alternately switch between synchronous and individual operation, which can be detrimental in terms of mechanical stress on the components (gears, axles, etc.) and performance, due to the reaction time. On the other hand, the energy saving realized through

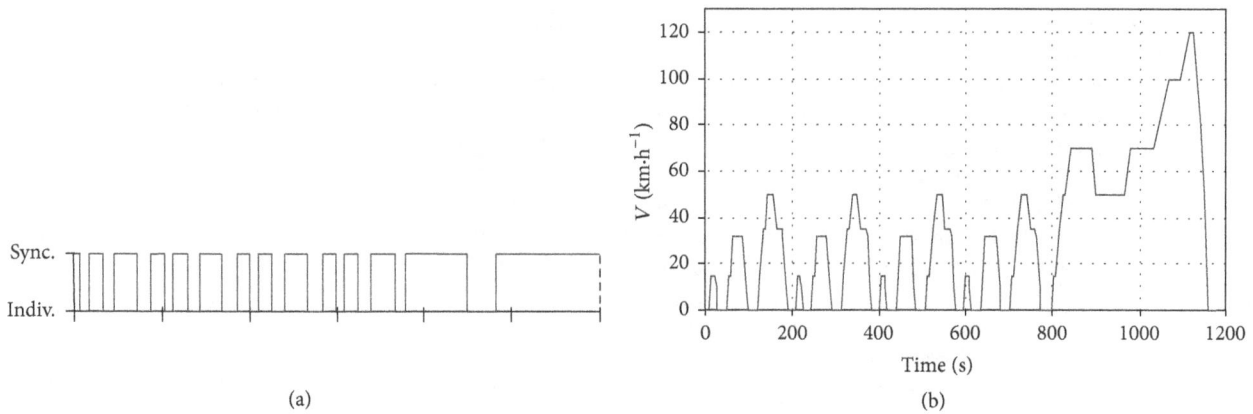

FIGURE 4: NEDC speed profile (b) and the periods where the drive trains operate in synchronous and individual operation (a).

operation mode switch is marginal in the close near of the break-even point, and therefore it is preferable to consider a locking function in order to avoid a switch from synchronous and individual mode too often in the mentioned case. A possible solution would be a hysteresis two-step control based on the efficiency values or the actual power saving which can be calculated instantaneously. Another approach includes a time based locking function to avoid the switch between the modes for certain periods.

The outlined strategy (η-optimization with locking function) is easy to be implemented in a control unit of a BEV. Simulation results carried out in use of the presented method are given and discussed in the following section.

3. Simulation and Discussion

The proposed approach to optimize the efficiency of power conversion of a battery electric vehicle is implemented in the control strategy of a virtual BEV in MATLAB/Simulink. Herein, the model of the BEV includes two individual drives (including PSM drives with 25 kW nominal power each) referring to Figure 1, where both motors are mechanically connected to one and the same differential gear at the rear axle. By this way the both drives always operate with equal speed and the torques add. In order to avoid the excessive switching from synchronous to individual operation and vice versa, a time based lock is included (lock duration = 5 s). The time constant for locking was chosen in accordance with the reaction time of the power train components. If the power demand exceeds the power which can be applied from one drive (compare to Figure 3(a)), the locking function is neglected, and the second drive instantly becomes active.

The incorporated BEV model includes a 20 kWh lithium-ion-based battery, and the total weight of the vehicle is assumed to be 1100 kg with a drag coefficient $c_d = 0.28$ (front cross section area 2.5 m^2), coefficient of rolling friction $c_R = 0.025$. For validation various standardized drive cycles were taken into account. In the following the results from simulations incorporating the common NECD ("New European Drive Cycle") load cycle are illustrated in detail (preferable the NEDC reflects highway and urban use; investigations on

different load cases yield similar results and therefore are left out here). The speed profile of the NEDC is depicted in Figure 4(b). One NEDC tasks 1180 seconds with a maximum speed of 120 km/h and the covered distance is approximately 11 km.

Besides the speed profile also the periods where both motors are active (sync.) and the periods where only one drive is operating (indiv.) are illustrated in Figure 4(a). Obviously, most of the time both drive trains are active. The individual operation mainly occurs during low speed driving with almost constant speed. The electric energy demands are given in Table 1, where a comparison is given between the modes with event-oriented operation in individual and synchronous operations (sync./indiv.) and an operation where the two drives always supply the same torque (only sync.). The given values refer to the energy consumption during one NEDC cycle, the electric energy required for 100 km mileage under NEDC load and the theoretical range under NEDC load achievable with a fully charged 20 kWh Li-ion battery (additional components are assumed to have a total power consumption of 200 W).

Noticeably, the energy consumption of the BEV model incorporating the proposed control strategy is reduced in comparison to an operation without splitting of torque. As visible in Figure 4, the individual mode is active mainly in sections with low, even speed, that is, low/moderate power demand. Even if the energy saving during low power operation is in a range of several tens or up to a few hundreds Watt, the saved energy cumulates over longer driving periods. The relative energy saving for the investigated case is approximately 0.6%. This value does not seem to be very much, but the energy can be saved without any hardware effort, only by using the proposed control strategy. Anyhow, the saving of

TABLE 1: Energy consumption during one NEDC cycle and corresponding maximum range.

Operation	1 × NEDC	100 km (NEDC)	max. range (20 kWh)
sync./indiv.	1.567 kWh	14.22 kWh	140.6 km
only sync.	1.578 kWh	14.32 kWh	139.7 km

0.6% of the consumed energy extends the maximum range in the presented case about almost 1 km or is equal to a weight reduction of the vehicle about approximately 12 kg, respectively. Furthermore, in the investigated case two similar drives are considered, including two converters and two motors. A further energy saving may be possible by using drives optimized for different ranges of operation. This way the break even point can be shifted towards higher power values, and the amount of saved energy can be increased. In case of using two individual current-excited motors, the relative energy saving can be increased due to the higher idle losses. Herein, during individual operation the rotor losses of one motor can be eliminated completely.

However, the presented investigation results are carried out on a BEV simulation model and therefore may differ from results obtained from tests conducted on real vehicles. Hence, the implementation of the proposed methods in the energy management system of a BEV and the subsequent evaluations on vehicle handling, performance, and energy consumption are challenging topics of our further research and development efforts. Moreover, the application of two (or more) drives at different axles may be more critical and needs detailed investigations concerning the vehicle handling and performance in case of torque splitting among the two axles, and therefore those investigations are also a topic of future work.

4. Conclusion

This work focuses on the optimization of power conversion on board of a battery electric vehicle including two drive trains. Due to the characteristics of the drive trains in case of low or moderate constant speed with varying torque operating, only one drive instead of two active drive trains may yield a higher efficiency in power conversion. Accordingly, a control structure is derived and is implemented in the energy management control unit of a virtual BEV. During operation (i.e., NEDC load) a reduction of energy consumption about 0.6% can be achieved without any additional hardware effort. This energy saving corresponds to a reduction of the vehicle's weight about approximately 12 kg.

References

[1] M. A. Roscher, J. Vetter, and D. U. Sauer, "Influence of cathodes technology on the power capability and charge acceptance of lithium ion batteries," in *Proceedings of the 24th International Battery, Hybrid and Fuel Cell Electric Vehicle Symposium (EVS '09)*, Stavanger, Norway, May 2009.

[2] Panasonic, "NNP series—NCR18650A," Technical Datasheet, 2010, http://www.panasonic.com/.

[3] K. Imai, T. Ashida, Y. Zhang, and S. Minami, "Theoretical performance of EV range extender compared with plugin hybrid," *Journal of Asian Electric Vehicles*, vol. 6, no. 2, pp. 1181–1184, 2008.

[4] Y. Yang, J. Liu, and T. Hu, "An energy management system for a directly-driven electric scooter," *Energy Conversion and Management*, vol. 52, no. 1, pp. 621–629, 2011.

[5] Y. Ota, H. Taniguchi, T. Nakajima, K. M. Liyanage, J. Baba, and A. Yokoyama, "Autonomous distributed V2G (vehicle-to-grid) considering charging request and battery condition," in *Proceedings of the IEEE PES Innovative Smart Grid Technologies Conference Europe (ISGT Europe '10)*, November 2010.

[6] S. Kermani, S. Delprat, R. Trigui, and T. Guerra, "Predictive energy management of hybrid vehicle," in *Proceedings of the IEEE Vehicle Power and Propulsion Conference (VPPC '08)*, vol. 20, no. 1, pp. 1–6, September 2008.

[7] A. Dardanelli, M. Tanelli, S. M. Savaresi, and M. Santucci, "Active energy management of electric vehicles with cartographic data," in *Proceedings of the IEEE International Electric Vehicle Conference (IEVC '12)*, Greenville, SC, USA, March 2012.

[8] Y. Zhang, W. Wang, Y. Kobayashi, and K. Shirai, "Remaining driving range estimation of electric vehicle," in *Proceedings of the IEEE International Electric Vehicle Conference (IEVC '12)*, Greenville, SC, USA, March 2012.

[9] C. Dextreit, F. Assadian, I. Kolmanovsky, J. Mahtani, and K. Burnham, "Hybrid electric vehicle energy management using gametheory," SAE Technical Paper 2008-01-1317, 2008.

[10] A. A. Ferreira, J. A. Pomilio, G. Spiazzi, and L. de Araujo Silva, "Energy management fuzzy logic supervisory for electric vehicle power supplies system," *IEEE Transactions on Power Electronics*, vol. 23, no. 1, pp. 107–115, 2008.

[11] J. Schiffer, O. Bohlen, R. W. De Doncker, D. U. Sauer, and K. Y. Ahn, "Optimized energy management for fuelcell-supercap hybrid electric vehicles," in *Proceedings of the IEEE Vehicle Power and Propulsion Conference (VPPC '05)*, pp. 716–723, September 2005.

[12] M. A. Roscher, W. Leidholdt, and J. Trepte, "High efficiency energy management in BEV applications," *International Journal of Electrical Power and Energy Systems*, vol. 37, no. 1, pp. 126–130, 2012.

[13] H. Wu, L. Li, B. Kou, and Z. Ping, "The research on energy regeneration of permanent magnet synchronous motor used for hybrid electric vehicle," in *Proceedings of the IEEE Vehicle Power and Propulsion Conference (VPPC '08)*, pp. 1–4, September 2008.

[14] L. Jishun, L. Jun, W. Qingnian, W. Jiaxue, and S. Jinhu, "Study on mechanism of energy saving for double motor configuration hybrid electric vehicle," in *Proceedings of the International Conference on Mechatronic Science, Electric Engineering and Computer (MEC '11)*, pp. 2596–2602, August 2011.

[15] J. Lee, K. Nam, S. Choi, and S. Kwon, "A lookup table based loss minimizing control for FCEV permanent magnet synchronous motors," in *Proceedings of the IEEE Vehicle Power and Propulsion Conference (VPPC '07)*, pp. 175–179, September 2007.

[16] J. J. C. Gyselinck, L. Vandevelde, D. Makaveev, and J. A. A. Melkebeek, "Calculation of no load losses in an induction motor using an inverse vector Preisach model and an eddy current loss model," *IEEE Transactions on Magnetics*, vol. 36, no. 4, pp. 856–860, 2002.

[17] B. J. Chalmers and I. Musaba, "Performance characteristics of permanent-magnet and reluctance machines to meet EV requirements," in *Proceedings of the IEE Colloquium on Machines and Drives for Electric and Hybrid Vehicles*, Digest No: 1996/152, August 2002.

[18] F. Deng, "Improved analytical modeling of commutation losses including space harmonic effects in permanent magnet in brushless DC motors," in *Proceedings of the IEEE International Electric Machines and Drives Conference (IEMDC '97)*, May 1997.

[19] J. Lee, K. Nam, S. Choi, and S. Kwon, "Loss-minimizing control of PMSM with the use of polynomial approximations," *IEEE Transactions on Power Electronics*, vol. 24, no. 4, pp. 1071–1082, 2009.

9

River Flow Lane Detection and Kalman Filtering-Based B-Spline Lane Tracking

King Hann Lim,[1] Kah Phooi Seng,[2] and Li-Minn Ang[3]

[1] Electrical and Computer Department, School of Engineering, Curtin University Sarawak, CDT 250, Sarawak, 98009 Miri, Malaysia
[2] School of Computer Technology, Sunway University, No. 5, Jalan Universiti, Bandar Sunway, Selangor, 46150 Petaling Jaya, Malaysia
[3] Centre for Communications Engineering Research, Edith Cowan University, Joondalup, WA 6027, Australia

Correspondence should be addressed to King Hann Lim, glkhann@curtin.edu.my

Academic Editor: T. A. Gulliver

A novel lane detection technique using adaptive line segment and river flow method is proposed in this paper to estimate driving lane edges. A Kalman filtering-based B-spline tracking model is also presented to quickly predict lane boundaries in consecutive frames. Firstly, sky region and road shadows are removed by applying a regional dividing method and road region analysis, respectively. Next, the change of lane orientation is monitored in order to define an adaptive line segment separating the region into near and far fields. In the near field, a 1D Hough transform is used to approximate a pair of lane boundaries. Subsequently, river flow method is applied to obtain lane curvature in the far field. Once the lane boundaries are detected, a B-spline mathematical model is updated using a Kalman filter to continuously track the road edges. Simulation results show that the proposed lane detection and tracking method has good performance with low complexity.

1. Introduction

Automation of vehicle driving is being developed rapidly nowadays due to the vast growth of driver assistance systems (DASs) [1]. In conjunction with the development of low-cost optical sensors and high-speed microprocessors, vision-based DASs become popular in the vehicular area to detect apparent imaging cues from various road scenes for visual analysis and therefore warn a driver of an approaching danger and simultaneously perform autonomous control to the vehicle's driving. Of all fatal errors happened, driver's inattention and wrong driving decisions making are the main factors of severe crashes and casualties on road [2]. The deviation of a vehicle from its path without a signal indication has threatened the nearby moving vehicles. As a consequence, vision-based lane detection and tracking system becomes an important mechanism in vehicular autonomous technology to alert a driver about road physical geometry, the position of the vehicle on the road, and the direction in which the vehicle is heading [3].

In the last few decades, a lot of vision-based lane detection and tracking techniques [4–8] have been developed in order to automatically allocate the lane boundaries in a variety of environmental conditions. It can broadly be divided into three major categories, that is, region-based method, feature-driven method, and model-driven method. Region-based method [9–14] basically classifies the road and non-road pixels using color or texture information. Although it has simple algorithm, it may suffer from color inconstancy and illumination problem. Feature-driven method [15–18] extracts the significant features such as lane markings from road pixels to identify the lane edges. This method is highly dependent on feature detection methods such as edge detection which are sensitive to occlusion, shadow, or other noises. On the other hand, model-driven method builds a mathematical function such as linear-parabolic [19, 20], hyperbola [21, 22], or spline-based [23, 24] methods to mimic the lane geometry. Due to its comprehensive learning and curvature flexibility, this method has been widely used in the lane detection and tracking system.

As stated in [19], Jung and Kelber manually cropped the effective road region to obtain the possible lane edges detection. At the same time, they applied a fixed threshold to split near- and far-field segment for the lines prediction using linear-parabolic model. On the other hand, Wang et al. [24] have proposed a lane detection and tracking technique using B-snake lane model, which measures dual external forces for generic lane boundary or marking. Initially, lane boundaries are detected using Canny/Hough estimation of vanishing points (CHEVPs). It is followed by constructing B-snake external force field for lane detection iteratively based on gradient vector flow (GVF) [25]. Nevertheless, the above-mentioned methods have encountered some flaws in the process. Manual cropping to obtain the effective lane region may not be an efficient way in the automation. In addition, fixed near-far-field threshold or segmented lines are not always applicable to all on-road conditions to determine the edges. Moreover, edge detector and Hough transform (HT) may easily be affected by shadow cast or weather change. The lane boundaries in the far-field range would be gradually undetectable using HT. Furthermore, CHEVP method is sensitive to numerous thresholding parameters initialization. Significant numbers of iterations are required to obtain GVF in the lane-tracking process.

Motivated by the above-mentioned problems, a new system composed of lane detection and tracking is presented in Figure 1. Horizon localization is first applied to a set of traffic scene image sequence to automatically segment the sky and the road region. Road region is then analyzed to further separate nonroad and road pixels. Subsequently, an adaptive line segment is computed using multiple edge distribution functions to monitor the change of road geometry. The portion below the adaptive line threshold is estimated with 1D HT method, while the upper part is determined using a low-processing method with the concept of river flow topology. After the lane edges are successfully located, they are passed over to lane tracking for reducing the computational time. Possible edge scanning is applied to seek for the nearby lane edges with respect to the estimated lane line. Control points are determined in order to construct a B-spline lane model. In assistance with Kalman filtering, B-spline control points are updated and predicted for the following frame's lane curve.

The proposed lane detection and tracking method has offered some advantages over the method [19, 24]. The performance of edge detection and HT are always distorted by shadow effect. Therefore, a regional dividing line is first applied to discard disturbance from sky region. The elimination of shadow effect is achieved by using an adaptive statistical method. Instead of using fix line to segment near and far-field, an adaptive line segment is proposed to monitor the change of angles along the lane boundaries. The concept of river flow is proposed for the lane detection system to follow the road in the far-field region. Furthermore, Kalman filter plays a twofold-role: (i) to correct B-spline control points for the current image frame and (ii) to predict the lane model for the consecutive frames. Unlike [24], less parameter tuning or thresholding values are required in the proposed system. Moreover, it has no camera

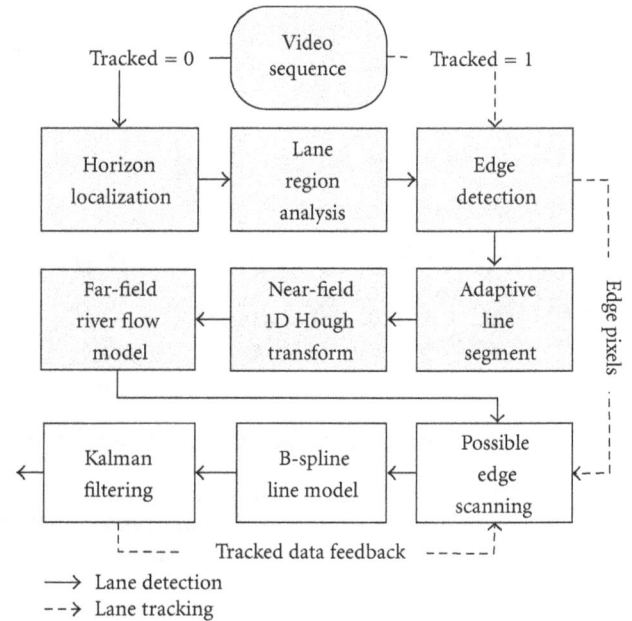

FIGURE 1: Block diagram of the proposed lane detection and tracking system.

parameters involved in the system determination. Overall, it gives a better performance with promising computational speed. This paper is organized as follows. Section 2 discusses the proposed lane detection, and Section 3 explains the Kalman filtering-based B-spline lane tracking technique. Some simulation results are shown in Section 4 and followed by conclusion and future works.

2. Lane Detection

Lane detection is a crucial task to estimate the left-right edges of driving path on a traffic scene automatically. In this section, four-stage lane boundary detection is proposed, that is, (i) horizon localization, (ii) lane region analysis, (iii) adaptive line segment, and (iv) river flow model. In the traffic scene, road region is the main focus for lane detection. First, horizon localization splits the traffic scene into sky and road region. Then, lane region analysis is applied adaptively based on the surrounding environment in order to remove mostly road pixels and keep the lane mark pixels. Adaptive line segment is therefore used to analyze the road edge curvature and separate the road region into near and far-field. In the near-field, a 1D HT is applied to draw a near-field line. At the same time, a river flow method is applied to obtain far-field edges which are hardly to be estimated using common line detection.

2.1. Horizon Localization. To eliminate the disturbances from sky segment, horizon localization [26] is performed to partition an $M \times N$ image into sky and road region, whereas M and N are the image row and column, respectively. First, a minimum pixel value filter, with a 3×3 mask, is applied on the image $I(x, y)$ as depicted in Figure 2(a) to enlarge the effect of low intensity around the horizon line. Subsequently,

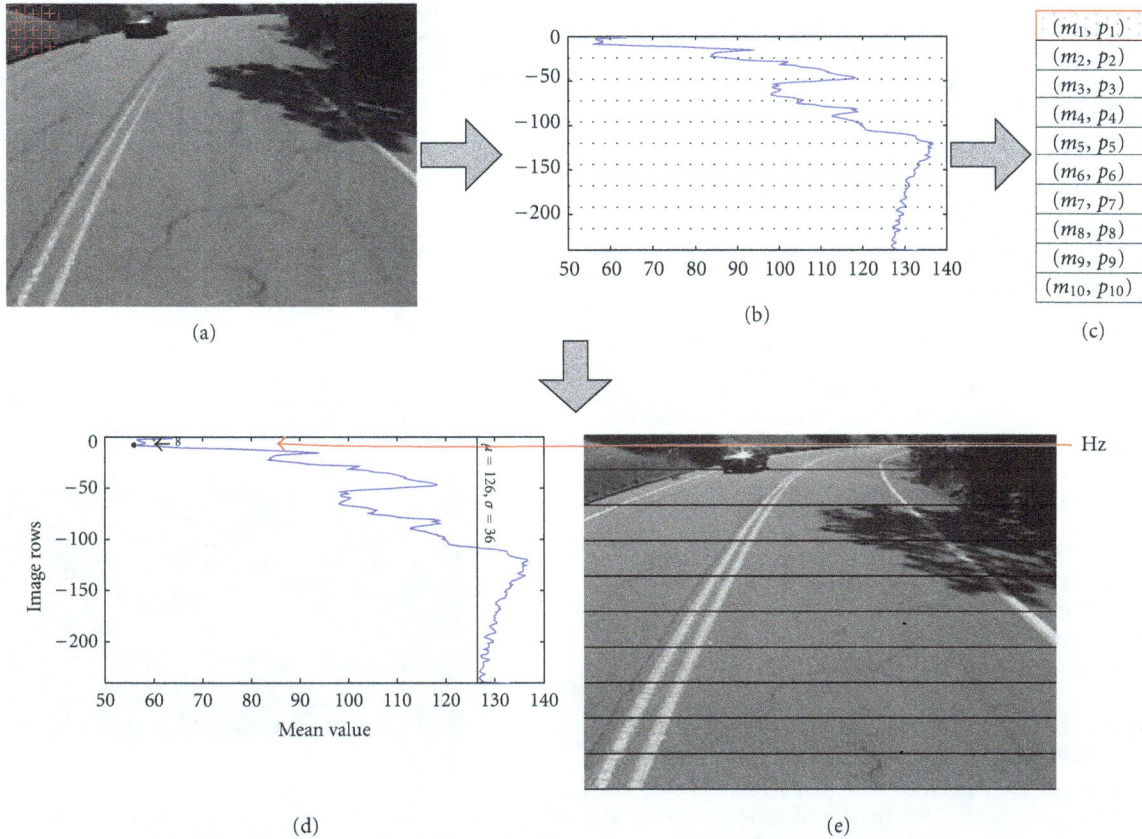

FIGURE 2: (a) Minimum pixel value filter, (b) vertical means distribution segmentation, (c) regional minima, (d) selection of the horizon line threshold, and (e) separation of the sky and road region.

a vertical mean distribution is computed as plotted in Figure 2(b) by averaging every row of gray values on the blurry image. Instead of searching for the first minimum value along the upper curve, the plot curve is then divided into β segments to obtain minima, whereas β represents the number of dividing segments. In this context, β is chosen to be 10 throughout the experiments. All regional minima are recorded to the subset A as follows:

$$A = \{(m_i, p_i) : m_i \in [0, 255] \mid p_i \in [1, M] \mid 1 \leq i \leq \beta\}, \tag{1}$$

where m_i is the magnitude of row pixel mean, and p_i is the row index where the minimum point occurred as shown in Figure 2(c). Naturally, sky region always appears on top of the road image. Therefore, m_1 is taken as a reference minimum point for condition comparison to prevent local minima occurred in the sky portion. Additionally, mean value (μ) of the entire image is calculated to determine an overall change in intensity. The regional minimum point search for the horizon localization is written as follows:

$$Hz = \begin{cases} p_1, & \text{if } (m_1 < m_{i+1}) \cap (m_1 < \mu) \cap (p_1 > \varepsilon_{Hz}), \\ p_i, & \text{else if } [(m_i < m_{i-1}) \cup |m_{i-1} - m_i| < \Delta m] \\ & \cdots \cap (m_i < m_{i+1}) \cap (m_i < m_1), \\ & \cdots \cap (m_i < \mu) \cap |m_{i+1} - m_i| > \Delta m, \\ \lambda, & \text{if others,} \end{cases} \tag{2}$$

where ε_{Hz} is a small integer to prevent a sudden drop from the top of image; Δm is the minor variation of mean value change, whereas $\Delta m = \pm 2$ intensity value, and λ is a user-defined value in case the minimum point cannot be found in the plot. As illustrated in Figure 2(d), the adaptive value of regional dividing line is obtained throughout the regional minimum search, whereas $Hz \in A$ is denoted as the horizon line. The notion of getting the horizon line is because the sky usually possesses higher intensity than road pixels, and it might have a big difference of intensity as the sky pixels approach the ground. Nevertheless, the horizon line often happens at neither the global minimum nor the first local minimum appeared from the upper curve. Hence, a regional minimum search by obtaining minimum points from the segments is proposed to ensure the correct localization for the horizon line dividing the sky and road region, accurately. In Figure 2(e), the horizon line threshold is applied to separate the sky and road region and a road image R_{roi} is generated where all vertical coordinates below the Hz value are discarded.

2.2. Lane Region Analysis.

Lane region analysis is performed with an adaptive road intensity range to further classify road and nonroad pixels with regards to the variation of surrounding environment. The lane region analysis steps are described as follows.

Step 1. Select κ rows of pixels for lane region analysis where κ is the number of pixel rows to be selected. The selected rows are started at δ number of rows from the bottom of road image to avoid the likely existence of interior part of a vehicle at the image edge.

Step 2. The intensity voting scheme is carried out on every selected row and each maximum vote of selected row, is defined as v_i, while g_i is the grey value whereas the maximum vote occurs, assuming that

$$B = \{(g_i, v_i) : g \in [0, 255] \mid v \in \mathfrak{R} \mid 1 \le i \le \kappa\}. \quad (3)$$

Subset B contains all maximum votes and gray value pixels for ith selected row. Hence, the vote, noted as v_m, and the greatest gray value threshold, noted as g_m, are selected as

$$(g_m, v_m) = \max(B), \quad (4)$$

where $\max(B)$ is a function to get a maximum value from the set B. In addition, the most frequent gray level of the selected region is recorded as g_s, where the global maximum vote occurs for the entire selected pixels. The standard deviation of selected rows is marked as σ_s.

Step 3. Define the adaptive road intensity range as $[g_s - \sigma_s; g_m + \sigma_s]$. Pixels that fall within the range are denoted as possible road pixels, and a binary map (R_{bin}) is formed as depicted in Figure 3(a). This region of interest could be further analyzed to investigate the road conditions. However, in our concern, lane marks are main features to identify the direction of lane flow. Only high intensity values are considered as the lane marks, whereas the pixel values being greater than $g_m + \sigma_s$ are denoted in R_{plm} as "1". This processing step may get rid of shadow problem since shadow is usually in low intensity.

Step 4. By summing up each row of R_{plm}, the values being greater than a threshold (T_1) are discarded to remove the possible high intensity of a vehicle at the frontal view of the image. T_1 is obtained by averaging nonzero rows of R_{plm}.

Step 5. Finally, a difference map (D_{map}) is generated by multiplying R_{roi} and R_{plm} maps. The remaining binary pixels are possible lane mark pixels as shown in Figure 3(b).

2.3. Adaptive Line Segment. Lane markings are the salient features on the road surface, and they are often used to define the boundaries of road region. Initially, the gradient magnitude (∇D_{map}) and the orientation (θ) are denoted as

$$\left| \nabla D_{\mathrm{map}} \right| \approx |D_x| + |D_y|,$$
$$\theta = \tan^{-1}\left(\frac{D_y}{D_x}\right), \quad (5)$$

where D_x is the horizontal edge map, and D_y is the vertical edge map. In order to monitor the changes of lane direction, an adaptive line segment is proposed to split the near- and far-field regions using edge distribution functions (EDFs) [19].

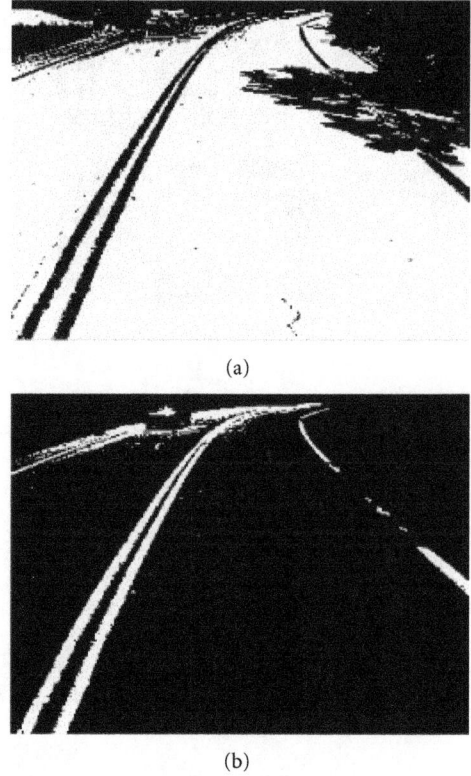

(a)

(b)

FIGURE 3: (a) Extracted road region using the adaptive road intensity range; (b) remaining binary pixels are the possible lane markings.

The edge map is first partitioned into small segments for every T_2 rows, whereas T_2 is the number of rows to be grouped for each partition. Multiple EDFs are applied to these partitions to observe the local change of lane orientation based on the strong edges as denoted in Figures 4(a)– 4(c). With reference to the gradient map and its corresponding orientation, multiple EDFs are constructed, with its x-axis is the orientation in the range of, $[-90°; 90°]$, and its y-axis is the accumulated gradient value of each orientation bin. The maximum peak acquired on the negative and positive angles denotes the right and left boundary angles, respectively. Peak values that go below a certain threshold T_3 are discarded, whereas T_3 is the mean EDF value of each partition. As shown in Figure 4(c), there is no detection for right angle since it has no significant edge points existed in the image illustrated in Figure 4(d).

Subsequently, EDFs' grouping is done by joining those angles that are approximately equal into the same array. They are grouped from bottom to top based on the difference of each partition to its previous partition within ±5°. The observation of orientation variation is that it will have an instant change of lane geometry when it comes to the far field. Assume that there are nth groups of lane orientation. If $n = 1$, it is a linear lane boundary. If $n \ge 2$, $(n - 1)$th of EDF bins are combined to obtain a global lane orientation (θ_{nf}) for near field, while the nth group refers to the far-field orientation (θ_{ff}). Although shadows deviate the lane edges,

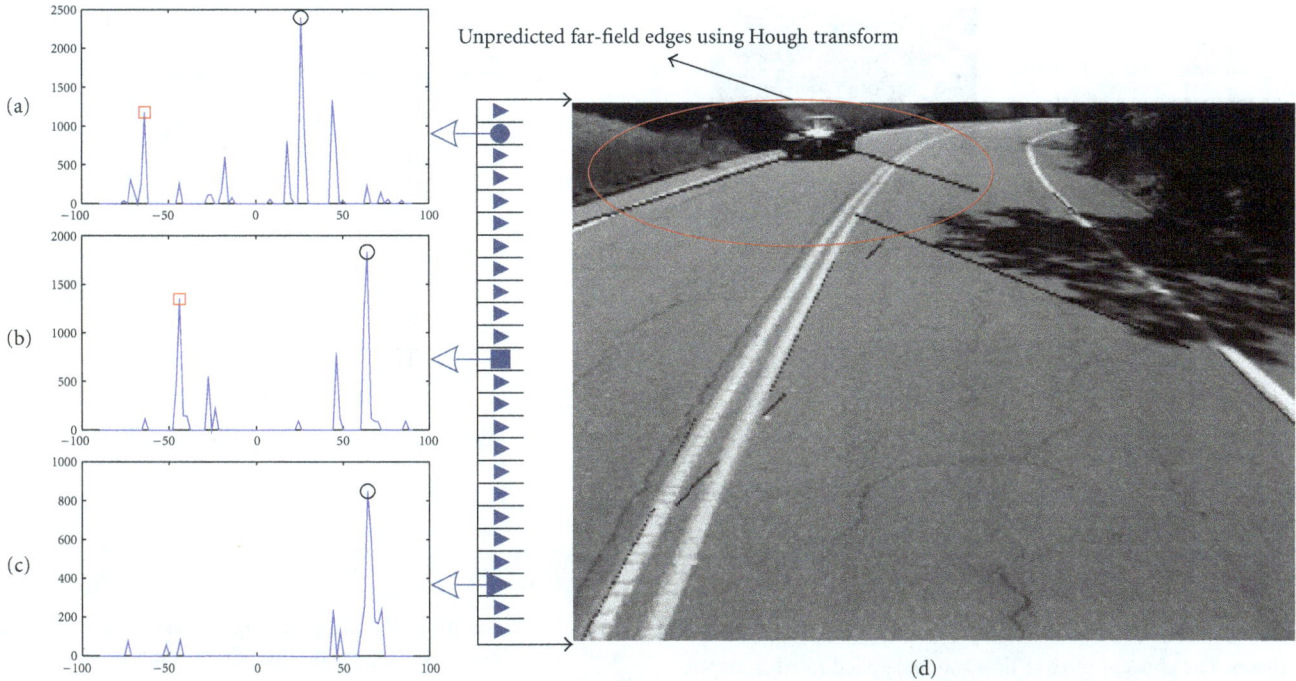

FIGURE 4: (a)–(c) Multiple EDFs on segmented regions, (d) results of lane detection after EDF partitions.

other partitions of EDF with more frequent angle value may correct the orientation of the road path. However, the angle θ_{ff} may not be successfully estimated for the far-field edges because the left-right boundaries may deflect into the same direction.

Eventually, a 1D-weighted gradient HT [19] is applied to near field to determine the lane's radius $r(\theta_{nf})$ based on $r(\theta_{nf}) = x_o \cos \theta_{nf} + y_o \sin \theta_{nf}$, with known θ_{nf} values. The voting bins for each radius are accumulated with the gradient edge values, and the maximum vote is selected. Figure 5 demonstrates the left-right lane boundaries in the near field constructed by the measured angles and radii.

2.4. River Flow Edge Detection. At the far-field region, lane edges flow is hard to be estimated using HT because the lane geometry has become unpredictable in the road scene. As noticed in Figure 4(d), the circle points out the failure of far-field edges detection using EDF and HT. This is because far-field lane edges may have irregular road curvature, and both lane edges may turn into the same direction. Two peaks of angles may fall onto the same side of EDF plot. Therefore, a river flow method is proposed to handle the far-field edge detection. From another perspective, lane flow detection has the same topology of a river flow. When there is an existing path, river will flow along the path. In this context, the path refers to the strength of edges for the lane boundaries. It starts flowing from the closest edges to the adaptive line segment and continues on to the vanishing point regarding to the connectivity and the strongest edge points as shown in Figure 6. With the 3×2 mask provided in Figure 6(b), the selection of the edge will be based on the surrounded large neighboring edge pixel clockwise.

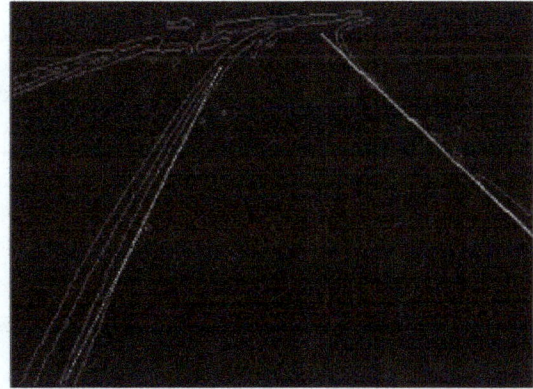

FIGURE 5: Detected lane boundaries by combining $(n - 1)$th EDF voting bins.

In Figure 6(c), white pixels represent the estimated line from near-field region, while grey pixels represent the edge pixels given by the edge operator. It stops when it has no connectivity or flows in reverse direction.

Assuming the pixels map shown in Figure 7 is an image after edge operator, the river flow is operated in such a manner: check for the nearby high edge pixels and link up all the possible high edge values. The edge pixel having higher value than its neighboring pixels constructs the flow pathway. Initially, the starting flow point has to be allocated in the image before start flowing in the map. In this context, the starting point is allocated by using the previous near-field edge detection technique. As we know, the salient edge points would have higher intensity value than nonsalient pixels. By

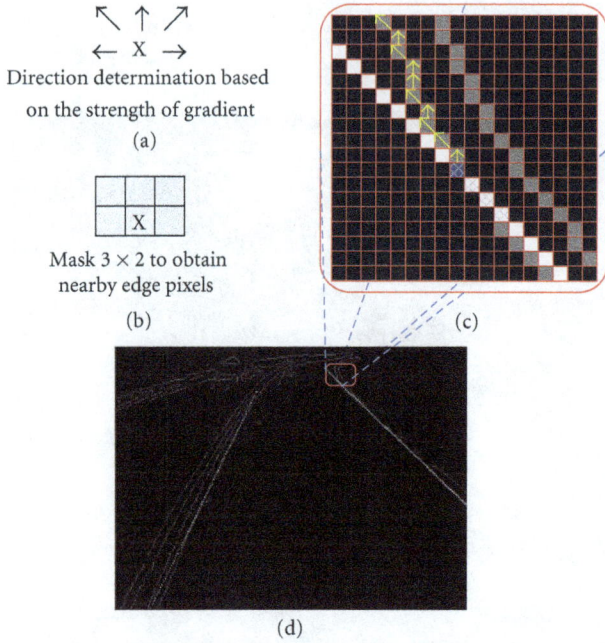

FIGURE 6: The concept of river flow model applied to edge map to detect the most significant edges in the far-field region.

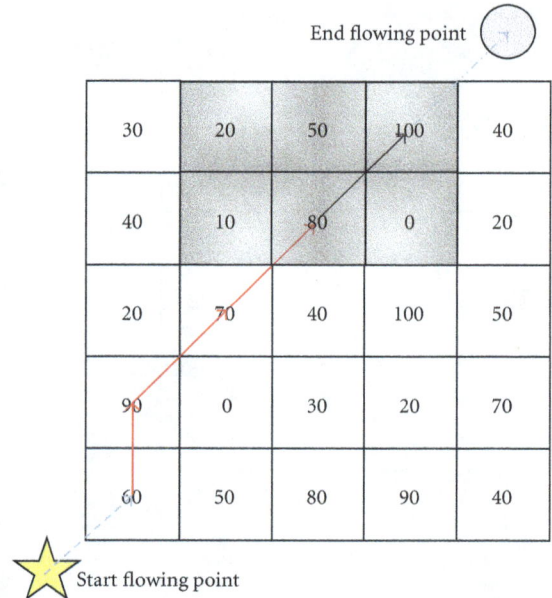

FIGURE 7: The example of flowing path on the edge pixels, where the "star" indicates the starting point; the "circle" indicates the ending point.

moving the 3×2 mask upwards as shown in the Figure 7, the maximum neighboring pixel is chosen in a clockwise manner indicating the connected path. The flowing process might halt if two or more same higher pixel values existed in the next flow or there is no connectivity detected in the next pixel. Due to the application of lane detection, reversion of the edge point flow is prohibited. The lane flow is either moving forward or moving in the same row of current detected pixel. Finally, the lane edges for left and right road boundaries are detected for lane tracking system.

3. Lane Tracking

After the lane detection is generated in Figure 8, lane tracking system is implemented to restrict the edges searching area on the subsequent frames and simultaneously estimate a lane model to follow the road boundaries. Lane tracking system has three stages, that is, possible edge scanning, B-spline lane modeling, and Kalman filtering. The nearest edges to the estimated line are determined with the possible edge scanning. A B-spline lane model is constructed with three control points to estimate lane edges in the current frame. Kalman filter corrects the lane model's control points and use them to predict the next frame lane edges. These predicted lane edges will be passed to the next frame for edge scanning again, and this will reduce the overall lane detection process.

3.1. Possible Edge Scanning.
The detected lines are used to scan for the nearby edge pixels on the consecutive image after an edge detector, and these lines would be updated iteratively from the previously estimated lane model. The closest pixels to the estimated lane model are considered as the possible

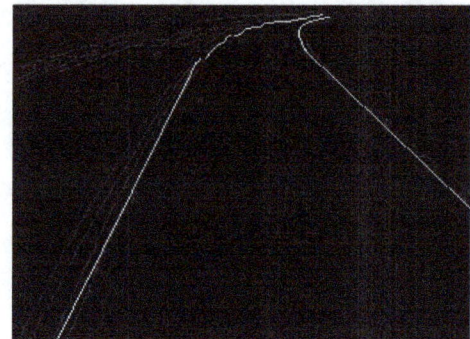

FIGURE 8: The detected edges after river flow model.

lane edges. The detected possible edges are very important for the lane model estimation.

3.2. B-Spline Lane Modeling.
An open cubic B-spline [28] with $t+1$ control points $\{Q_0, Q_1, \ldots, Q_t\}$ consists of $(t-2)$th connected curve segments, where $t \geq 3$. Each curve segment $\mathbf{g}_i(s)$ is a linear combination of four control points, and it can be expressed as

$$\mathbf{g}_i(s) = \mathbf{S}_i \mathbf{W}_i \mathbf{Q}_i, \quad i = -1, 0, 1, 2, \ldots, t,$$

$$\mathbf{S}_i = \begin{bmatrix} s^3 & s^2 & s & 1 \end{bmatrix}, \quad 0 \leq s \leq 1,$$

$$\mathbf{W}_i = \begin{bmatrix} \mathbf{w}_1 & \mathbf{w}_2 & \mathbf{w}_3 & \mathbf{w}_4 \end{bmatrix} = \frac{1}{6} \begin{bmatrix} -1 & 3 & -3 & 1 \\ 3 & -6 & 3 & 0 \\ -3 & 0 & 3 & 0 \\ 1 & 4 & 1 & 0 \end{bmatrix}, \quad (6)$$

$$\mathbf{Q}_i = \begin{bmatrix} Q_{i-1} & Q_i & Q_{i+1} & Q_{i+2} \end{bmatrix}^T,$$

where $\mathbf{S}_i$ is the knot vector which is uniformly distributed from 0 to 1, $\mathbf{W}_i$ is the spline basis functions, and $\mathbf{Q}_i$ is the control points with 4×4 dimension. According to [24], three control points are found to be efficient to describe the lane shapes, that is, Q_0, Q_1, and Q_2. The Q_0 and Q_2 are the first and last points of the detected edges, while the Q_1 is defined as follows

$$Q_1 = \frac{3}{2}P_1 - \frac{1}{4}(Q_0 + Q_2), \tag{7}$$

where P_1 is the adaptive line segment point. Next, Q_0 and Q_2 are tripled to ensure the line completely passing through the control points. For further prediction, control points have to be rearranged into state $\mathbf{x}$ and the observation model matrix $\mathbf{H}$:

$$\mathbf{x} = \begin{bmatrix} Q_0 & Q_1 & Q_2 \end{bmatrix}^T, \tag{8}$$

$$\mathbf{H} = \begin{bmatrix} \mathbf{S}_{-1}(\mathbf{w}_1 + \mathbf{w}_2 + \mathbf{w}_3) & \mathbf{S}_{-1}\mathbf{w}_4 & \mathbf{0} \\ \mathbf{S}_0(\mathbf{w}_1 + \mathbf{w}_2) & \mathbf{S}_0\mathbf{w}_3 & \mathbf{S}_0\mathbf{w}_4 \\ \mathbf{S}_1\mathbf{w}_1 & \mathbf{S}_1\mathbf{w}_2 & \mathbf{S}_1(\mathbf{w}_3 + \mathbf{w}_4) \\ \mathbf{0} & \mathbf{S}_2\mathbf{w}_1 & \mathbf{S}_2(\mathbf{w}_2 + \mathbf{w}_3 + \mathbf{w}_4) \end{bmatrix}. \tag{9}$$

3.3. Kalman Filtering. Kalman filtering method [6] can be used to predict the control points for left-right edges in consecutive frames. The linear state and measurement equations are defined as

$$\mathbf{x}_{k|k-1} = \mathbf{F}_{k|k-1}\mathbf{x}_{k-1|k-1} + \boldsymbol{\varphi}_k,$$
$$\tilde{\mathbf{y}}_k = \mathbf{H}_k\mathbf{x}_{k|k-1} + \mathbf{e}_k, \tag{10}$$

where the state space $\mathbf{x}$ is the control points of B-spline lane model defined in (8); $\mathbf{F}_{k|k-1}$ is the transition matrix bringing state $\mathbf{x}$ from time $k-1$ to k; $\boldsymbol{\varphi}_k$ is known as process noise; $\tilde{\mathbf{y}}_k$ is the measurement output; $\mathbf{H}_k$ is the observation model that maps the true state space to the observed space; $\mathbf{e}_k$ is the measurement noise. In this context, $\mathbf{F}_{k|k+1} = \mathbf{I}_{3\times3}$ with the assumption of zero external forces. The state is then corrected using Kalman filtering as defined in the following

$$\mathbf{P}_{k|k-1} = \mathbf{F}_{k|k-1}\mathbf{P}_{k-1|k-1}\mathbf{F}_{k|k-1}^T + \mathbf{v}_k,$$

$$\boldsymbol{\alpha}_k = \mathbf{y}_k - \mathbf{H}_k\mathbf{x}_{k|k-1},$$

$$\mathbf{K}_k = \mathbf{P}_{k|k-1}\mathbf{H}_k\left(\mathbf{H}_k\mathbf{P}_{k|k-1}\mathbf{H}_k^T + \mathbf{z}_k\right)^{-1}, \tag{11}$$

$$\mathbf{x}_{k|k} = \mathbf{x}_{k|k-1} + \mathbf{K}_k\boldsymbol{\alpha}_k,$$

$$\mathbf{P}_{k|k} = (\mathbf{I} - \mathbf{K}_k\mathbf{H}_k)\mathbf{P}_{k|k-1},$$

where $\mathbf{P}_{k|k-1}$ is a priori estimate error covariance; $\mathbf{P}_{k|k}$ is a posteriori estimate error covariance; $\boldsymbol{\alpha}_k$ is the error between the output $\mathbf{y}_k$ obtained from the possible edge scanning and the lane model; $\mathbf{K}_k$ is the Kalman gain. Hence, Figure 9 shows the Kalman filtering-based B-spline lane model which will be used for detecting the possible edges in the next frame.

4. Simulation Results

All results were generated using Matlab 2007a in a machine with core-2 duo processor at 1.8 GHz with 1 GB RAM. An

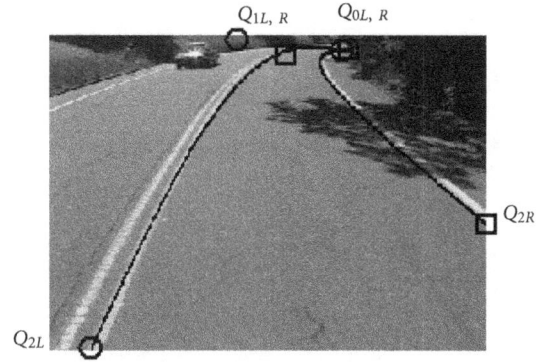

FIGURE 9: The Kalman filtering-based B-spline lane model with control points indication.

image sequence was downloaded from [29], and it was used to compare with the proposed method and the method [24] in terms of performance and computational complexity. Additional video sequences were captured for further experimental evaluation on the proposed lane detection and tracking method. The threshold values of proposed system were initialized as $\varepsilon_{Hz} = 7$, $\lambda = \kappa = 30$, and $T_2 = 10$, where all values were obtained throughout the experiment.

4.1. Performance Evaluation. The performance of the proposed system was evaluated by comparing the estimated lane model with regard to the nearest edges. To calculate an average pixel error, $d = 100$ random points were picked from the estimated lane model, and hence, the error was measured as follows:

$$\text{err} = \frac{\sum \left|\overline{x}_j - x_j\right| + \left|\overline{y}_j - y_j\right|}{d}, \quad j = 1, \ldots, d, \tag{12}$$

where $(\overline{x}_j, \overline{y}_j)$ is denoted as the coordinate of nearest edges, and (x_j, y_j) was the estimated lane model pixel. Figure 10 showed average error plots for left-right edges. The proposed method obtained lower average pixels error rate per frame which were 3.34 and 2.19 for left-right lane tracking than the method [24], which were 7.30 and 6.05, respectively. This was because the performance of method [24] was highly dependent on the CHEVP detection and the threshold value to terminate the measurement of ΔQ and Δk. Examples of using method [19, 24] with its limitations were pointed in Figure 11.

Some lane detection results were provided in Figure 12 where the first line indicated the horizon, and the second line was the near-far-field adaptive line. Figure 12(a) obtained continuous lines on left and right edges, while Figure 12(b) contained dashed lines as the lane marks. The white line was the successfully detected lane edges on the ground using the proposed lane detection method.

Figure 13 showed the simulation results for the proposed method where all continuous frames were detected and tracked successfully. Moreover, more tested results using the proposed method were demonstrated in Figure 14 with two on-road videos. Video no. 1 was recorded in the rural

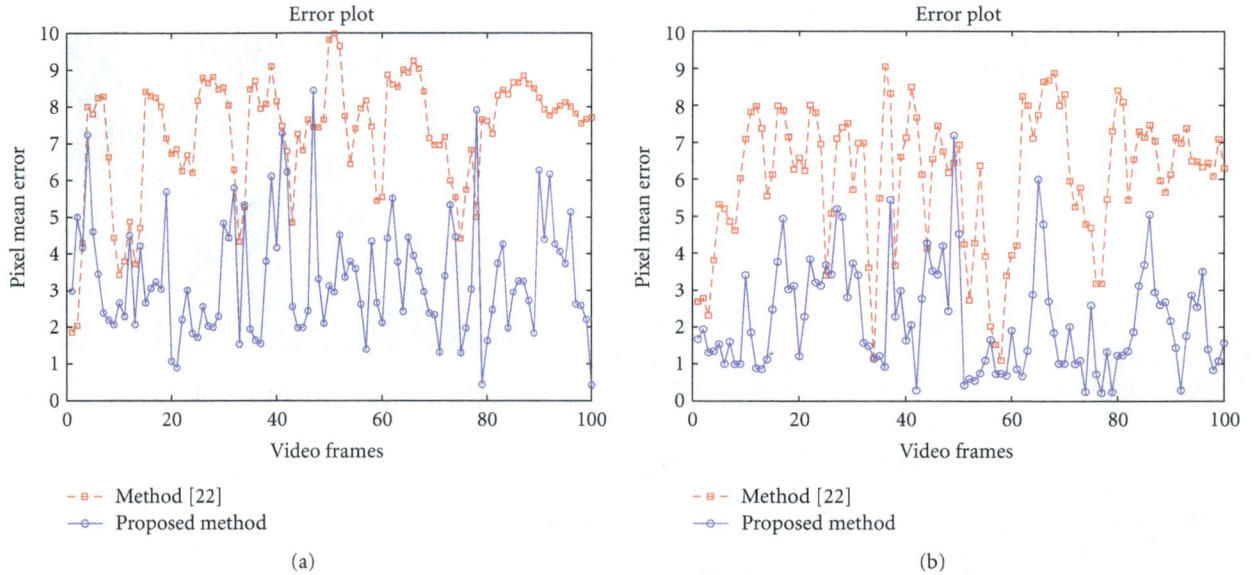

FIGURE 10: The error plots for (a) left lane tracking, (b) right lane tracking.

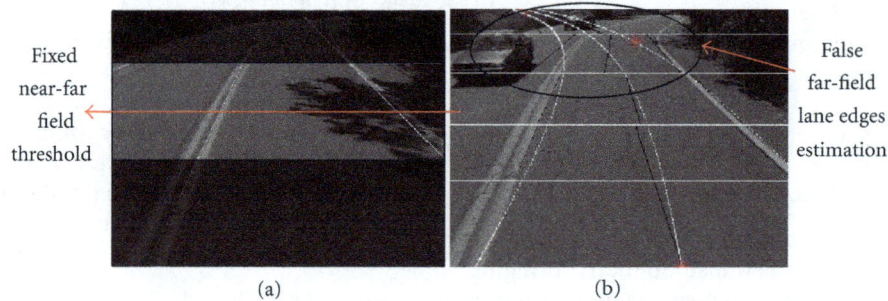

FIGURE 11: Examples of using methods (a) [19] and (b) [24].

area with continuous lines, and it was successfully detected although there was an existing frontal vehicle. Video no. 2 was shot in the highway with dotted dash line as shown in Figure 14(b), and the line was successfully estimated. A demo video has been uploaded to [30] for further understanding. However, the proposed method may still have a failure case in some aspect. When there was an overtaking vehicle which blocks the frontal road edge, it might not detect and predict the lane edges. Massive traffic flow might cause the misdetection too. At the same time, the proposed system may suffer on the scene without lane marks.

4.2. Computational Time. Complexity-wise, the proposed method achieved faster computational time than the method [24]. This was because the complexity of method [24] was greatly dependent on the number of detected lines in the HT process for computing the vanishing points and the GVF iterations which was more complex compared to the proposed method. A summary of computational time between the proposed method and the method [24] was presented in Table 1.

TABLE 1: Complexity comparison in average time base (sec.).

Method	Lane detection		Lane tracking
Method [24]	<4		<0.5
Proposed method	Horizon localization	0.19	0.78
	Lane region analysis	0.34	
	Adaptive line segment	0.79	
	River flow method	0.06	

5. Conclusion

A river flow lane detection and Kalman filtering-based B-spline tracking system has been presented to identify lane boundaries in image sequence. The advantages of horizon localization are to limit the searching region on ground and get rid of noises from the sky region. Moreover, lane region analysis eliminates shadow while maintains the lane markings. Meanwhile, an adaptive line segment with multiple EDFs is proposed to monitor the change of lane orientation from near to far-field. A 1D HT is applied to estimate the

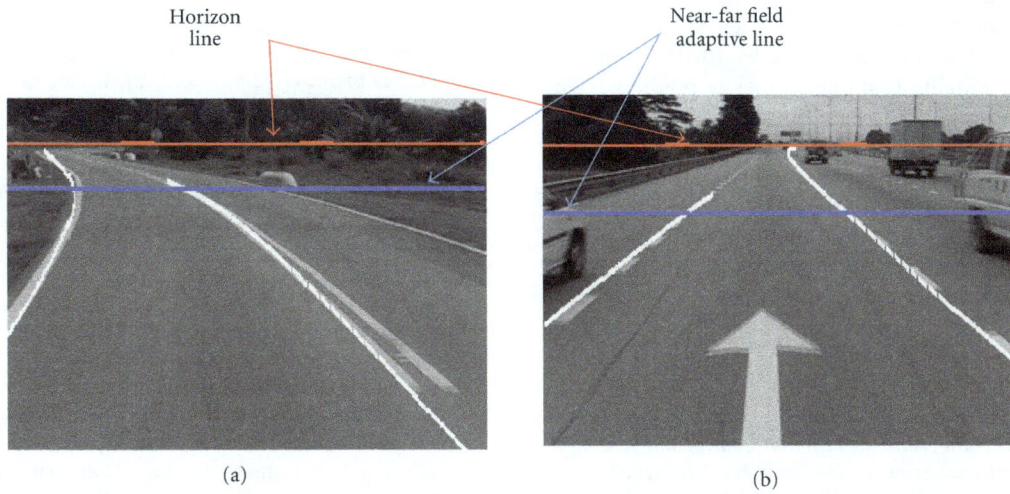

FIGURE 12: The results for the proposed lane detection on (a) rural area, (b) highway.

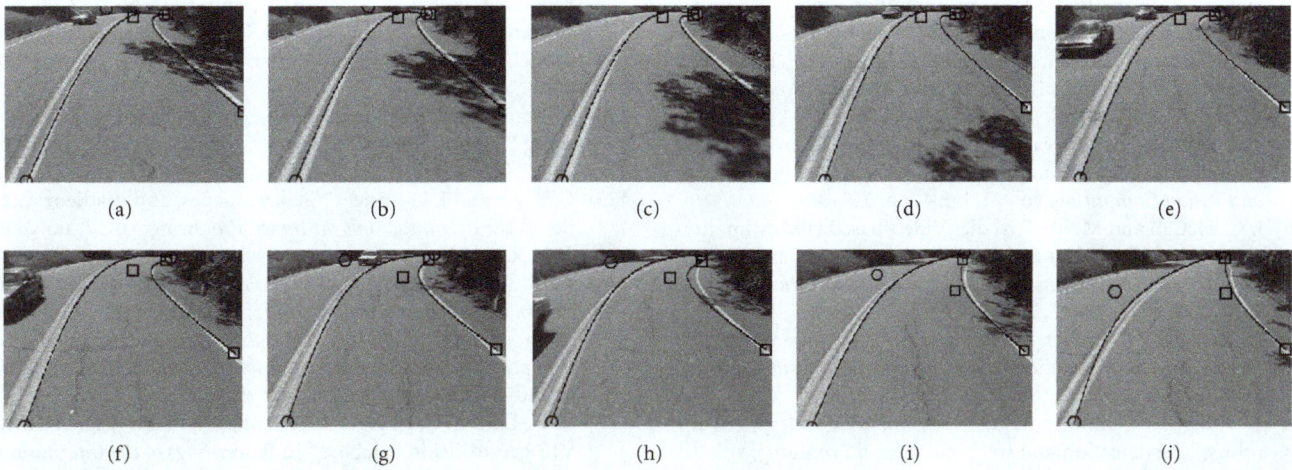

FIGURE 13: Continuous images with the proposed tracking method, where "○"s and "□"s are left-right control points. (c)–(e) Q_{1L} is located outside the image.

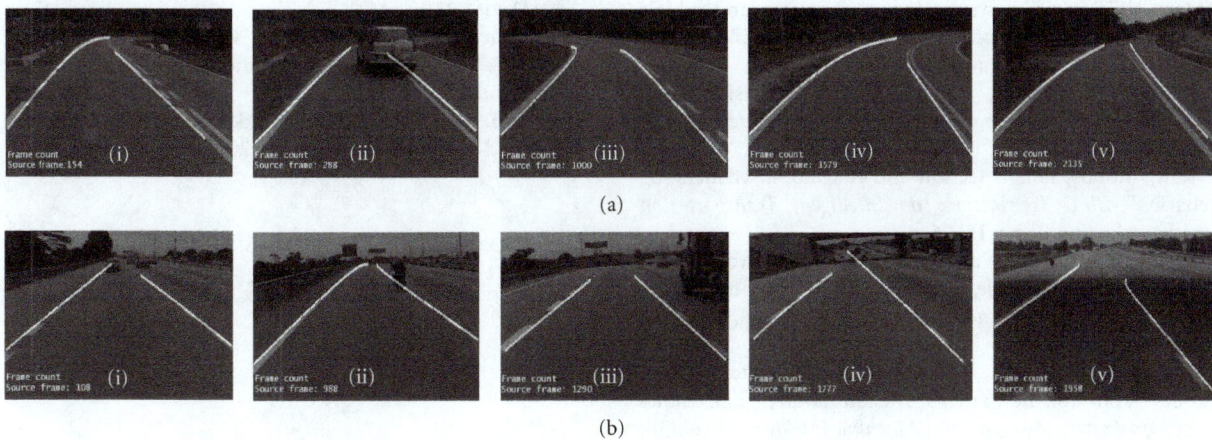

FIGURE 14: Random video samples extracted from the UNMC-VIER AutoVision [27] video clips: (a) video no. 1: rural area and (b) video no. 2: highway.

linear model in the near field. At the same time, river flow model is presented to detect far-field edges, and it could continuously detect and track lane edges for the following frame in future. Finally, B-spline model is predicted with a Kalman filter to follow the lane boundaries continuously. The proposed lane detection and tracking system will be further improved to suit more road scenarios. Further evaluation on river flow method will be investigated in the future. The concept of river flow method could be further extended to attain macroscopic and microscopic optimized mathematical mobility models in the future.

References

[1] L. Li and F. Y. Wang, *Advanced Motion Control and Sensing for Intelligent Vehicles*, Springer, New York, NY, USA, 2007.

[2] J. M. Armingol, A. de la Escalera, C. Hilario et al., "IVVI: intelligent vehicle based on visual information," *Robotics and Autonomous Systems*, vol. 55, no. 12, pp. 904–916, 2007.

[3] Y. Zhou, R. Xu, X. F. Hu, and Q. T. Ye, "A robust lane detection and tracking method based on computer vision," *Measurement Science and Technology*, vol. 17, no. 4, pp. 736–745, 2006.

[4] M. Bertozzi, A. Broggi, A. Cellario, A. Fascioli, P. Lombardi, and M. Porta, "Artificial vision in road vehicles," *Proceedings of the IEEE*, vol. 90, no. 7, pp. 1258–1270, 2002.

[5] V. Kastrinaki, M. Zervakis, and K. Kalaitzakis, "A survey of video processing techniques for traffic applications," *Image and Vision Computing*, vol. 21, no. 4, pp. 359–381, 2003.

[6] J. C. McCall and M. M. Trivedi, "Video-based lane estimation and tracking for driver assistance: survey, system, and evaluation," *IEEE Transactions on Intelligent Transportation Systems*, vol. 7, no. 1, pp. 20–37, 2006.

[7] A. Bar Hillel, R. Lerner, D. Levi, and G. Raz, "Recent progress in road and lane detection: a survey," *Machine Vision and Applications*. In press.

[8] Y. Wang, N. Dahnoun, and A. Achim, "A novel system for robust lane detection and tracking," *Signal Processing*, vol. 92, no. 2, pp. 319–334, 2012.

[9] J. D. Crisman and C. E. Thorpe, "SCARF. A color vision system that tracks roads and intersections," *IEEE Transactions on Robotics and Automation*, vol. 9, no. 1, pp. 49–58, 1993.

[10] M. A. Turk, D. G. Morgenthaler, K. D. Gremban, and M. Marra, "VITS–a vision system for autonomous land vehicle navigation," *IEEE Transactions on Pattern Analysis and Machine Intelligence*, vol. 10, no. 3, pp. 342–361, 1988.

[11] K. Kluge and C. Thorpe, "The YARF system for vision-based road following," *Mathematical and Computer Modelling*, vol. 22, no. 4–7, pp. 213–233, 1995.

[12] Z. Kim, "Robust lane detection and tracking in challenging scenarios," *IEEE Transactions on Intelligent Transportation Systems*, vol. 9, no. 1, pp. 16–26, 2008.

[13] S. Baluja, "Evolution of an artificial neural network based autonomous land vehicle controller," *IEEE Transactions on Systems, Man, and Cybernetics B*, vol. 26, no. 3, pp. 450–463, 1996.

[14] C. Thorpe, M. H. Hebert, T. Kanade, and S. A. Shafer, "Vision and navigation for the Carnegie-Mellon navlab," *IEEE Transactions on Pattern Analysis and Machine Intelligence*, vol. 10, no. 3, pp. 362–373, 1988.

[15] D. Pomerleau, "RALPH: rapidly adapting lateral position handler," in *Proceedings of the Intelligent Vehicles Symposium*, pp. 506–511, September 1995.

[16] S. K. Kenue and S. Bajpayee, "LaneLok: robust line and curve fitting of lane boundaries," in *Proceedings of the 7th Mobile Robots*, pp. 491–503, Boston, Mass, USA, November 1992.

[17] D. J. LeBlanc, G. E. Johnson, P. J. T. Venhovens et al., "CAPC: a road-departure prevention system," *IEEE Control Systems Magazine*, vol. 16, no. 6, pp. 61–71, 1996.

[18] C. Kreucher and S. Lakshmanan, "LANA: a lane extraction algorithm that uses frequency domain features," *IEEE Transactions on Robotics and Automation*, vol. 15, no. 2, pp. 343–350, 1999.

[19] C. R. Jung and C. R. Kelber, "Lane following and lane departure using a linear-parabolic model," *Image and Vision Computing*, vol. 23, no. 13, pp. 1192–1202, 2005.

[20] C. R. Jung and C. R. Kelber, "An improved linear-parabolic model for lane following and curve detection," in *Proceedings of the 18th Brazilian Symposium on Computer Graphics and Image Processing (SIBGRAPI'05)*, pp. 131–138, October 2005.

[21] Y. Wang, L. Bai, and M. Fairhurst, "Robust road modeling and tracking using condensation," *IEEE Transactions on Intelligent Transportation Systems*, vol. 9, no. 4, pp. 570–579, 2008.

[22] L. Bai, Y. Wang, and M. Fairhurst, "An extended hyperbola model for road tracking for video-based personal navigation," *Knowledge-Based Systems*, vol. 21, no. 3, pp. 265–272, 2008.

[23] Y. Wang, D. Shen, and E. K. Teoh, "Lane detection using spline model," *Pattern Recognition Letters*, vol. 21, no. 8, pp. 677–689, 2000.

[24] Y. Wang, E. K. Teoh, and D. Shen, "Lane detection and tracking using B-Snake," *Image and Vision Computing*, vol. 22, no. 4, pp. 269–280, 2004.

[25] C. Y. Xu and J. L. Prince, "Snakes, shapes, and gradient vector flow," *IEEE Transactions on Image Processing*, vol. 7, no. 3, pp. 359–369, 1998.

[26] K. H. Lim, K. P. Seng, and L. M. Ang, "Improvement of lane marks extraction technique under different road conditions," in *Proceedings of the 3rd IEEE International Conference on Computer Science and Information Technology (ICCSIT'10)*, pp. 80–84, Chengdu, China, July 2010.

[27] K. H. Lim, A. C. Le Ngo, K. P. Seng, and L.-M. Ang, "UNMC-VIER AutoVision database," in *Proceedings of the International Conference on Computer Applications and Industrial Electronics (ICCAIE '10)*, pp. 650–654, Kuala Lumpur, Malaysia, December 2010.

[28] K. I. Joy, "Cubic uniform B-spline curve refinement," in *On-Line Geometric Modeling Notes*, University of California (Davis), 1996.

[29] Carnegie-Mellon-University, "CMU/VASC image database, 1997–2003," http://vasc.ri.cmu.edu//idb/html/road/may30_90/index.html.

[30] K. H. Lim, "River flow lane detection and Kalman filtering based B-spline lane tracking," 2011.

10

Multiple-Observation-Based Robust Channel and Doppler Estimation in High Mobility Applications

Md. Jahidur Rahman[1] and Jiaxin Yang[2]

[1] *Department of Electrical and Computer Engineering, The University of British Columbia, Vancouver, Canada V6T 1Z4*
[2] *Department of Electrical and Computer Engineering, McGill University, Montreal, Canada H3A 0E9*

Correspondence should be addressed to Md. Jahidur Rahman; jrahman@ece.ubc.ca

Academic Editor: Shinsuke Hara

Channel estimation is a challenging task, especially in high mobility applications due to the rapid variation of the propagation environment. This paper presents a new technique that exploits past channel impulse responses (CIRs) in order to trace and compensate Doppler frequency in mobile applications, enabling robust estimation of time-varying channel. Based on the fact that channel taps at different time instants can be fitted with a sinusoidal wave, a joint estimator is proposed to estimate the channel parameters. Therefore, the efficiency of the channel estimation can be improved and stringent delay requirements for the communication systems can also be satisfied. Simulation results show that system performance in terms of bit error rate (BER) is significantly improved with the proposed algorithm.

1. Introduction

Channel estimation is of crucial importance for reliable coherent detection in high mobility applications, such as vehicular-to-vehicular (V2V) or vehicular-to-infrastructure (V2I) communications [1, 2]. It is complicated due to the large amount of Doppler frequency experienced by moving vehicles, and, in turn, the rate of channel variation can be sufficiently high, leading to poor channel estimation performance. Therefore, channel estimation performance in high mobility applications is generally poor. In an effort, past observations of channel impulse responses (CIRs) can be exploited to improve the channel estimation performance. However, in highly mobile environments, the direct combination of multiple observed CIRs is not effective due to the phase rotation introduced by Doppler frequency, and, hence the combination is noncoherent. Therefore, the estimation of Doppler frequency is essential to improve the channel estimation performance in mobile communication. If the total amount of Doppler frequency or change in Doppler frequency can be estimated exploiting observed past CIRs,

the channel can be tracked even after the elapse of the channel coherence time, which will make communication more robust in mobile applications.

One of the first studies reported in the literature for the estimation of Doppler frequency is [3], where the authors proposed estimation techniques using both complex and real envelop of the received signal. Another method to estimate the Doppler frequency in the presence of speckle and receiver noise was presented in [4]. This estimation technique is based on the correlation of the signal power spectra with an arbitrary weighting function and specifically tailored for Synthetic Aperture Radar (SAR) data. On the other hand, an estimation technique for OFDM-based transmission that relied on finding the autocorrelation function of time domain channel estimates was studied in [5]. Besides, a good review of Doppler estimation techniques can be found in [6]. However, most of these techniques are based on obtaining the autocorrelation function (ACF) of the received signals or channel estimates. Once the ACF is estimated, different methods can be used to calculate the Doppler spread as discussed in [6]. However, it requires precise estimation of

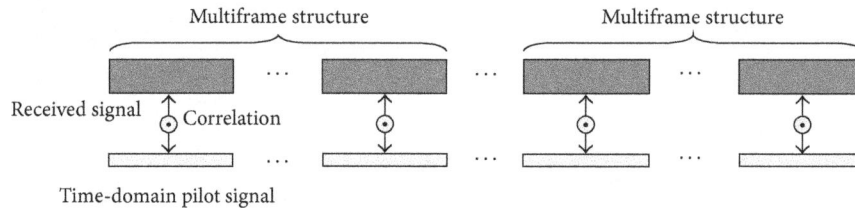

FIGURE 1: Multiframe based observation of channel impulse responses.

the time lag which often becomes erroneous due to the presence of line-of-sight component, nonisotropic scatterers, and so forth [6].

In this paper, we have studied a joint CIR and Doppler frequency estimator that exploit multiple observations of the past CIRs, which can be used to track the CIR in the current estimation period. At first, a multiframe time-domain correlation technique is used to obtain the time-domain estimates of the past CIRs, as shown in Figure 1. Thereafter, these observations of immediate past CIRs are exploited to model the variation of the Doppler frequency in the current estimation period, as described in Section 3. Comparing it to the sequential estimation, a joint estimator is more efficient as it takes into account the relation between the parameters and reduces the overhead necessary for achieving a certain system performance [7]. This is especially important in burst transmission systems, where the data burst is relatively short and the overhead of the preamble and pilot data must be kept to a minimum as in vehicular communications. Because of the joint estimation of the channel parameters, this technique can also satisfy the delay requirements for vehicular communications [8]. In our analysis, we have considered that multipath components in a channel profile may experience different Doppler shifts and may or may not vary from one observation instant to another. Therefore, the research problem is to estimate the CIR and Doppler frequency jointly, based on multiple observations of past CIRs. It should be noted that time-domain implementation improves the estimator performance because of the reduced number of channel parameters to be estimated and the exact modeling of the phase rotation suffered by the Doppler frequency [9]. Therefore, the data reception performance can be improved by estimating the variation of the channel model in a mobile environment.

Note that the proposed method does not rely on the estimation of ACF as in conventional techniques, that is, prone to erroneous estimation of Doppler due to model imperfections such as presence of line-of-sight component and nonisotropic scattering. In addition, the proposed method uses local time-domain pilot signals to obtain multiframe observations of the channel, therefore free of model imperfections, that is, a major drawback for the conventional techniques [6]. If we compare the computation complexity of the proposed algorithm against the conventional schemes, the proposed algorithm is much simpler to implement. In conventional algorithm, the estimated channel needs to be correlated, that is, obtaining the ACF in order to determine the time lag that is necessary to estimate the Doppler frequency. In the proposed

algorithm, on the other hand, we can simply use the channel parameters observed from past CIRs in order to determine the Doppler for the current estimation period. Therefore, the proposed algorithm provides an accurate estimation of Doppler frequency as seen from our simulation results with much lower computation complexity than the conventional schemes.

The rest of the paper is organized as follows. Section 2 describes the multiframe-based channel observation technique. The estimation of channel parameters is described in Section 3. The initializations of the parameters are described in Section 4. The channel estimation technique with varying Doppler for the past observations is discussed in Section 5. Simulation results for this study are presented in Section 6. Finally, the paper is summarized in Section 7.

2. Multiframe Channel Observation

Let us consider a multipath channel as shown in Figure 2. Each path is affected by a different Doppler frequency at different time instant and can be modeled as follows [10].

$$h(t) = \sum_{l=0}^{L-1} a_{t,l} e^{j(2\pi\zeta_{t,l}t + \gamma_{t,l})}, \qquad (1)$$

where $a_{t,l}$, $\gamma_{t,l}$, and $\zeta_{t,l}$ denote the time-dependent path amplitude, initial phase, and Doppler frequency for the lth tap at time instant t, and L is the total number of paths for any given CIR.

Over a short interval of time, for example, when observation times t_1 and t_2 are very close, we can assume that the magnitudes of the multipath components do not vary significantly. Considering that the multipath amplitudes do not vary significantly over a short observation of the channel, it can be written that $\Gamma(a_{t_1,l}) \approx \Gamma(a_{t_2,l})$, where Γ is the magnitude operation and $a_{t_1,l}$ and $a_{t_2,l}$ are the amplitudes of the multipath components at time instants t_1 and t_2. In first part of our analysis, we have considered that the Doppler frequency and initial phase do not change over the observation period.

A multiframe technique is used to obtain multiple observations of the past CIRs [11]. This technique is useful to obtain multiple observations of the channel at the same time and exploit these observations to estimate Doppler frequency and channel in the current estimation period.

In order to obtain the reference CIR, we have used time-domain channel estimation for the OFDM system,

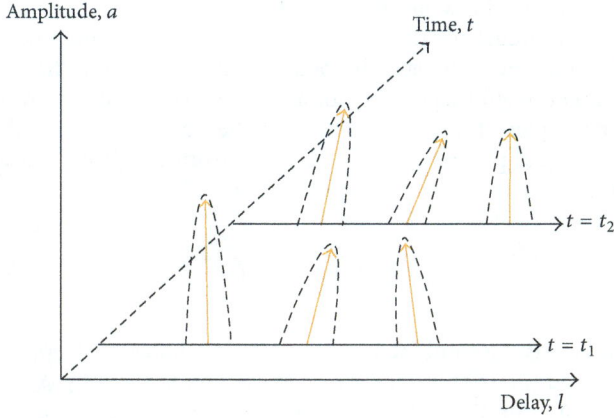

FIGURE 2: Typical multipath channel impulse responses observed at different time instants.

tailored for DVB-T system [12]. Let us consider that the time domain data and pilot signals are denoted by $d(n)$ and $p(n)$, respectively. The pilot tones are multiplexed with the data signal in the frequency domain and assume that combined signal in time-domain is given by $x(n)$. Therefore, the received time-domain signal at successive time instants, $y(t, n)$, follows

$$y(t, n) = x(n) \otimes h(t, n) + w(t, n), \qquad (2)$$

where $\otimes$ denotes the convolution operation and w is the discrete AWGN noise. The time-domain pilot signal can be used to correlate with the received time-domain signal in order to obtain an estimation of the channel. Considering channel taps $l = 1, 2, \ldots, L$, the time-domain correlation output between local pilot and mth received frame is given by

$$\widehat{h}(t_m; l) = \sum_{n=1}^{N} y(t_m, mN + n) \odot p\{mN + n\}$$

$$= \sum_{n=1}^{N} [x(mN + l) \otimes h(t_m, l) \qquad (3)$$

$$+ w_m(t_m, mN + l + n)] \odot p\{mN + n\}$$

$$= h(t_m; l) + \widehat{w}(t_m; l),$$

where $\odot$ denotes the correlation operation, $\widehat{h}(t_m; l)$ denotes the estimated time-domain channel in the mth frame and l denotes each path in the CIR, and $\widehat{w}(t_m; l)$ denotes the correlation of the pilot signal with the inherent noise component. Note that the definition of frame depends on how frequent channel needs to be observed. The estimated channel $\widehat{h}(t_1, t_2, \ldots, t_M; l) \equiv h_1, h_2, \ldots, h_M$ at successive time instants, $t_1, t_2, \ldots, t_M$, will be used in the next section to estimate the channel parameters in the current estimation period.

3. Estimation of Channel Parameters

Let us consider that the channel samples observed at time instants $t_1, \ldots, t_M$ are given by $h_1, \ldots, h_M$. Considering a single path at different observation instants, the path amplitude, Doppler frequency, and initial phase can be modeled as

$$h_m[a, \zeta, \gamma] = a \cos\{2\pi\zeta t_m + \gamma\}$$

$$= A_1 \cos 2\pi\zeta t_m + A_2 \sin 2\pi\zeta t_m \qquad (4)$$

$$= h_m[\theta],$$

where A_1 and A_2 are unknown constants, $A_1 = a \sin \gamma$ and $A_2 = a \cos \gamma$, and θ is the set of the channel parameters. The channel observation instant and path numbers are dropped for simplification. For the estimation of channel parameters, here we have followed IEEE Standard 1057 [13]. Note that here in this section we have considered only the real part of the complex channel. Similar analysis holds for imaginary part of the channel estimate as well.

Let us assume that an estimate of Doppler frequency in iteration i, say $\widehat{\zeta}_i$, of ζ is available. Therefore, a Taylor series expansion around the estimate $\widehat{\zeta}_i$ gives [14]

$$\cos t_m = \cos 2\pi\widehat{\zeta}_i t_m - t_m \sin 2\pi\widehat{\zeta}_i t_m \Delta\zeta_i. \qquad (5)$$

Similarly,

$$\sin t_m = \sin 2\pi\widehat{\zeta}_i t_m + t_m \cos 2\pi\widehat{\zeta}_i t_m \Delta\zeta_i, \qquad (6)$$

where $\Delta\zeta_i = \zeta_i - \widehat{\zeta}_i$.

Using the above two approximations in (4), we get that

$$h_m[\theta_i] \approx A_1 \cos 2\pi\widehat{\zeta}_i t_m + A_2 \sin 2\pi\widehat{\zeta}_i t_m$$

$$- A_1 t_m \Delta\zeta_i \sin 2\pi\widehat{\zeta}_i t_m + A_2 t_m \Delta\zeta_i \cos 2\pi\widehat{\zeta}_i t_m, \qquad (7)$$

where θ_i is the parameter vector.

The above equation is nonlinear in the parameters, but it may be linearized using the assumption that Doppler is constant over the observation period; that is, $\Delta\zeta_i = \zeta - \widehat{\zeta} \approx 0$. Putting available estimates of A_1 and A_2 from previous iteration, that is, $A_{1,i-1}$ and $A_{2,i-1}$, the above equation results in an equation linear in the components of θ_i. Therefore, observing the past observations, the following can be written:

$$h = \widehat{Q}_i \theta_i, \qquad (8)$$

where $\widehat{Q}_i$ is the $M \times 3$ matrix, and denoting $2\pi\widehat{\zeta}_i t_i = \widehat{\Omega}_i$, $\widehat{Q}_i$ can be written as

$$\begin{bmatrix} \cos \widehat{\Omega}_1 & \sin \widehat{\Omega}_1 & -\widehat{A}_{1,i-1} t_1 \sin \widehat{\Omega}_1 + A_{2,i-1} t_1 \cos \widehat{\Omega}_1 \\ \cos \widehat{\Omega}_2 & \sin \widehat{\Omega}_2 & -\widehat{A}_{1,i-1} t_2 \sin \widehat{\Omega}_2 + A_{2,i-1} t_2 \cos \widehat{\Omega}_2 \\ \vdots & \vdots & \vdots \\ \cos \widehat{\Omega}_M & \sin \widehat{\Omega}_M & -\widehat{A}_{1,i-1} t_M \sin \widehat{\Omega}_M + A_{2,i-1} t_M \cos \widehat{\Omega}_M \end{bmatrix}. \qquad (9)$$

It is obvious that in order to obtain a unique solution for each of the parameters, at least three observations are necessary. The simple technique is to repeatedly solve the above linear system; that is, at iteration i, $\widehat{Q}_i$ is used to obtain a new set of estimates $\widehat{\theta}_i$ and the solution is given by

$$\theta_i = \left(\widehat{Q}_i^T \widehat{Q}_i\right)^{-1} \widehat{Q}_i^T h. \tag{10}$$

The algorithm follows an iterative process to find the parameters that minimize the sum of squared differences as the following:

$$e = \sum_{m=1}^{M} (h_m - h_m [\theta])^2 \tag{11}$$

$$\text{S.T. } e < e_{\text{thr}},$$

where e_{thr} is the estimation error threshold set to achieve a certain estimation performance.

Once A_1 and A_2 are known, initial phase (γ) and the amplitude (a) from the ith iteration can be obtained as follows:

$$\gamma_i = \tan^{-1} \frac{A_{1,i}}{A_{2,i}}, \tag{12}$$

$$a_i = \frac{A_{1,i}}{\sin \gamma_i} = \frac{A_{2,i}}{\cos \gamma_i}.$$

4. Initializations of the Parameters

4.1. Initial Guess for the Amplitude, a. The initial guess for a in the minimization problem plays an important role in the convergence of the algorithm. In order to have a faster convergence, the initial guess can be obtained by averaging the amplitudes of the past CIRs. Due to the averaging, the impact of noise will be reduced as well [15]. The initial guess for channel magnitude, a_i, can be derived as

$$\tilde{a}_i (t, l) = \frac{1}{M} \sum_{m=1}^{M} \{a_m (t, l)\}, \tag{13}$$

where M indicates the number of CIRs considered from the past observations.

4.2. Initial Guess for the Doppler Frequency, ζ. For a moving vehicle, there could be a constrained set on the Doppler frequency based on the mobility pattern of the vehicle. Basically, it could be determined by the maximum acceleration or deceleration of the vehicle. Therefore, the Doppler shift can be initialized based on the study of mobility pattern in a particular scenario, that is, highway. The constrained can be formulated as

$$\zeta - \Delta\zeta_{\max} < \zeta_i < \zeta + \Delta\zeta_{\max},$$

$$\zeta - \frac{\cos \gamma_l}{\lambda} \left\{ \frac{-\Delta v(t)}{\Delta t} \right\} < \zeta_i < \zeta + \frac{\cos \gamma_l}{\lambda} \left\{ \frac{\Delta v(t)}{\Delta t} \right\}, \tag{14}$$

where $\Delta\zeta_{\max}$ is the maximum change in the Doppler shift because of the change in velocity or angle of arrival. The initial value of the Doppler frequency can be chosen within this range for faster and more accurate convergence.

4.3. Initial Guess for the Initial Phase, γ. If the Doppler phase remains constant over the observation period, taking the difference between the phase of the successive channel estimate would leave the initial phase. For a faster convergence, guess for initial phase can be computed from the differential phase rotations obtained from the past CIRs as follows:

$$\overline{\gamma_1} = \left[\overline{\phi}_1 - \overline{\phi}_2\right]^T, \tag{15}$$

where ϕ is the combined phase shift. It includes the Doppler-induced phase shift and initial phase and is given by $\phi_{t_1,l} = 2\pi\zeta_{t_1}, t_1$. Similarly, for the next two observations, we can write

$$\overline{\gamma_2} = \left[\overline{\phi}_2 - \overline{\phi}_3\right]^T. \tag{16}$$

In general, these successive differences for M-observed channels can be averaged to obtain a guess of the initial phase for the minimization problem as follows:

$$\widetilde{\gamma_i} = \frac{1}{M-1} \sum_{m=1}^{M-1} \overline{\gamma}_m. \tag{17}$$

It should be noted that the convergence would be much faster than the conventional joint estimation technique, as in the conventional technique only one estimation is used to obtain the initial guesses of the channel parameters.

4.4. Determination of Number of Required Observations, M. The performance of channel estimation following this technique will be determined by the number of observations obtained as well as the estimation error introduced in the past observations. Let us consider that past observed CIR has an error e, which is a zero-mean Gaussian noise with variance σ^2. Therefore, according to IEEE STD 1057, the lower bound of the performance for this estimator can be found by Cramer-Rao bound (CRB) as follows:

$$\text{CRB}\left(\widehat{\zeta}\right) = \left(\frac{n}{2\pi t_n}\right)^2 \frac{12}{\text{SNR}\,(M^2 - 1)\,M}, \tag{18}$$

where $\text{SNR} = (A_1^2 + A_2^2)/2\sigma^2$, n is the observation instant, t_n is the observation interval, and M is the total number of observations. It is seen that the CRB is inversely proportional to the total number of observations.

It is important that we study the number of observations needed to obtain certain performance threshold of the estimator. For a certain CRB, the above equation can be rewritten as

$$M^3 - M - C = 0, \tag{19}$$

where $C = (n/2\pi t_n)^2(12/\text{SNR} \times \text{CRB}(\widehat{\zeta}))$. Solving this equation analytically, we get the following three solutions for the number of the observations, M,

$$M = \begin{cases} \frac{1}{6}S + \frac{2}{S}, \\ -\frac{1}{12}S - \frac{3}{S} + \frac{\sqrt{-j}}{12}S, \\ -\frac{1}{12}S - \frac{3}{S} - \frac{\sqrt{-j}}{12}S, \end{cases} \qquad (20)$$

where $S = [108 \times C + 12 \times \sqrt{(-12 + 81C^2)}]^{1/3}$. As the number of the observations will be a real positive number and assuming the condition $81C^2 - 12 \geq 0$, we have the following three real solutions:

$$M = \begin{cases} \frac{S}{6} + \frac{2}{S}, \\ -\frac{S}{12} - \frac{3}{S}, \\ -\frac{S}{12} - \frac{3}{S}. \end{cases} \qquad (21)$$

The above solutions provide the bounds on the number of the observations required in order to maintain the CRB for a given SNR level and sampling frequency.

5. Estimator Performance with Varying Doppler

In this section, we will study the change in Doppler frequency over the observation period and its impact on the performance of the estimator. If the Doppler frequency does not change appreciably over the observation period, the previous technique can be used to estimate the channel parameters. However, the Doppler frequency may change over the observation period, depending on the mobility pattern of the users and angle of arrivals for each path.

It is obvious that the speed of a vehicle is a time-varying function and so is the Doppler frequency for each path. The relationship between the change of Doppler frequency and the speed of a vehicle can be derived as follows [16]:

$$\frac{d\zeta_{t,l}}{dt} = \frac{\cos \gamma_{t,l}}{\lambda} \frac{dv(t)}{dt} + \frac{v(t)}{\lambda} \frac{d}{dt} \{\cos \gamma_{t,l}\}, \qquad (22)$$

where $\Delta t = t_2 - t_1$ is the time difference between two successive channel sampling instants. Therefore, it can be seen that change in Doppler frequency depends on the change in the speed or change in the angle of arrival. Even if there is no change in speed within this short-time period Δt, it is possible that multipath will have different angles of arrivals. However, with no significant change in speed within the short interval of time Δt, the above equation can be written as

$$\delta \zeta_{t_m,l} = \frac{v(t)}{\lambda} \frac{d}{dt} \{\cos \gamma_{t,l}\}. \qquad (23)$$

The above equation would determine the variation of Doppler frequency for each multipath in a CIR over the time difference between successive CIRs. Under high mobility application, in fact, a short interval of time may introduce a significant variation of Doppler shifts for the successive CIRs depending on the duration of the data frame structure; however, change in Doppler frequency will be dominated by the change in the angle of arrivals for each multipath.

Therefore, considering the variation in Doppler, (4) can be written as

$$\begin{aligned} h_m[a, \zeta, \gamma] &= a \cos \left\{ 2\pi \zeta'_{t_m, l} t_m + \gamma_{t_m, l} \right\} \\ &= A_1 \cos 2\pi \zeta'_{t_m, l} t_m + A_2 \sin 2\pi \zeta'_{t_m, l} t_m \qquad (24) \\ &= h_m \left[\theta' \right], \end{aligned}$$

where $\zeta'_{t_m, l} = \zeta \pm \delta \zeta_{t_m, l} = \rho \zeta$. Here, ρ denotes the correlation in Doppler frequency over the past observation instants and $\delta \zeta$ is the amount of variation. When $\rho = 1$, it denotes the constant Doppler case. In order for this problem to be linearized, the following condition has to be satisfied:

$$\zeta_m - \widehat{\zeta}_m = \Delta \zeta_i \approx 0, \qquad (25)$$

where ζ_m and $\widehat{\zeta}_m$ denote the original and estimated Doppler frequencies. It also means that $\rho \approx 1$. Once these conditions are satisfied, a linear equation similar to (7) is formulated and can be solved to obtain the channel parameters. However, if all of the above conditions are not satisfied, the estimation of channel parameters remains a nonlinear problem. Therefore, the estimated parameter will have an error, which can be modeled as follows:

$$h'_m[\theta] = h_m[\theta] + e', \qquad (26)$$

where $h'_m[\theta]$ denotes the estimated set of parameters when the Doppler frequency is slightly different over the past observations and e' is corresponding error when compared to the case of no Doppler variation. In this case, this error can be amalgamated with the error threshold (e_{thr}) in the minimization process. The new threshold can be written as

$$e'_{\text{thr}} = e_{\text{thr}} + e'. \qquad (27)$$

The small variation in Doppler shift over the observation period can be considered as a noise in the estimation process. Therefore, the noise introduced in the estimation process will be higher than the stable Doppler case. Let us consider that the variance of the noise in this case is given by $\sigma'^2 = \sigma^2 + \Delta \sigma^2$, where $\Delta \sigma^2$ is the noise introduced in the past observation of the channel due to varying Doppler and can be considered as zero-mean Gaussian variable when there are sufficient numbers of observations. Therefore, CRB for this estimator under varying Doppler can be found as follows:

$$\text{CRB}\left(\widehat{\zeta}\right) = \left(\frac{n}{2\pi t_n}\right)^2 \frac{12}{\text{SNR}'(M^2 - 1)M}, \qquad (28)$$

where $\text{SNR}' = (A_1^2 + A_2^2)/2\sigma'^2$.

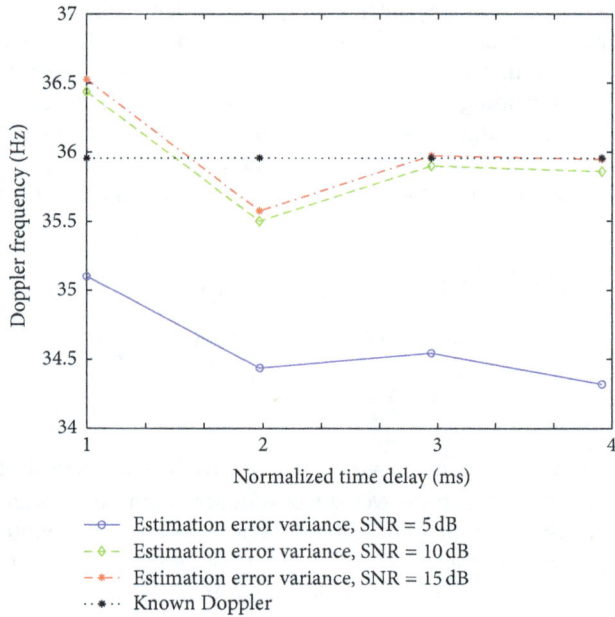

FIGURE 3: Comparison of estimated Doppler frequency with true Doppler in presence of different estimation errors (constant Doppler case).

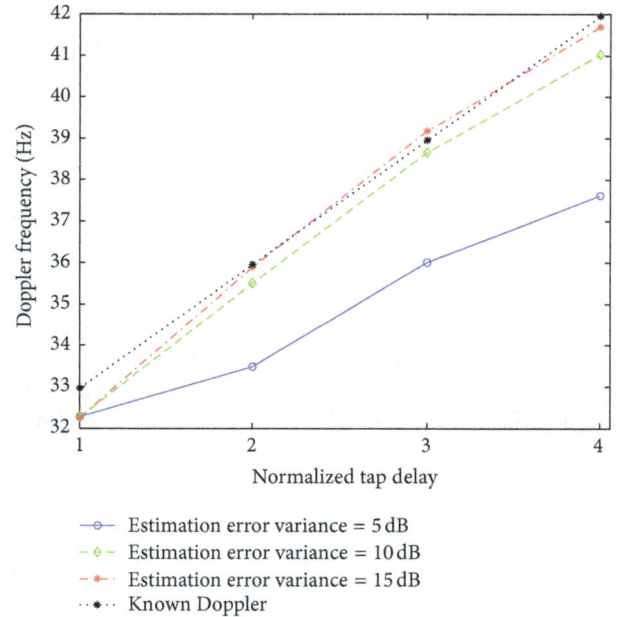

FIGURE 4: Comparison of estimated Doppler frequency with true Doppler in presence of different estimation errors (varying Doppler case).

6. Simulation Results

Digital Video Broadcasting-Terrestrial (DVB-T) system is used to evaluate the performance of the proposed technique. It is assumed in the simulation that the past channel responses may experience a constant or varying Doppler under a receiver velocity from 50 km/h to 100 km/h. DVB-T standard transmission for 2 K mode with 8 MHz bandwidth and 800 MHz carrier frequency is used for the simulation of the proposed technique. The sampling period for the chosen data transmission is $7/64\,\mu s$. In this paper, we have considered an arbitrary statistical channel as well as widely known International Telecommunication Union (ITU) Vehicular Channel Model B to evaluate the system performance [17]. It is assumed that the arbitrary channel is normalized and follows uniform power delay profile.

Figure 3 shows the comparison of estimated and known Doppler frequency with estimation error introduced by the past observations of the CIRs. In this channel model, we assumed that each path experiences same Doppler frequency. It is important to note that as the estimation error variance decreases, the estimated Doppler frequency approaches the true Doppler frequency.

Figure 4 shows the comparison of estimated and known Doppler frequency with estimation error introduced by the past observed CIRs with the Doppler frequency set to be different for different channel taps. Similar to the Figure 3, as the estimation error variance decreases, the estimated Doppler frequency approaches the true Doppler frequency. Therefore, it is expected that the system performance will improve after compensating the Doppler impairment as shown in the next few simulations.

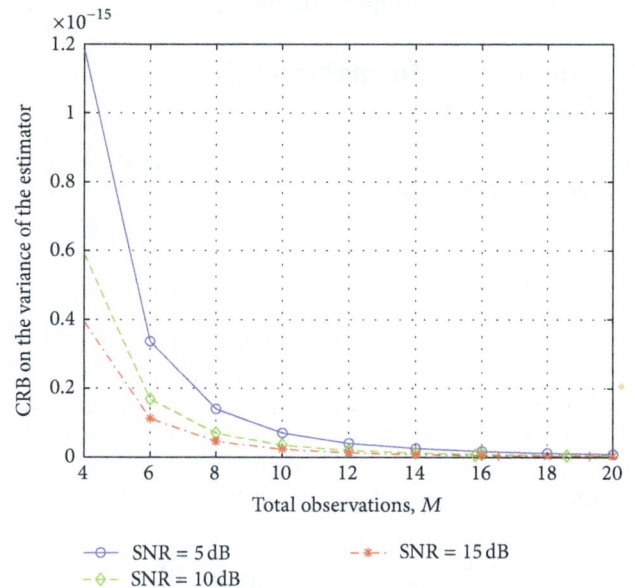

FIGURE 5: Cramer-Rao bound on the variance of the proposed channel estimator.

Figure 5 shows the CRB of the estimator versus the number of past observations under different SNR. It is seen that with an increase in the number of past observations, the CRB decreases. In addition, as the SNR increases, the CRB also decreases, indicating the improvement of the estimation performance.

Figure 6 presents the performance of the DVB-T system transmitting in 2k mode at a carrier frequency of 800 MHz. We have first adopted the widely known ITU Vehicular

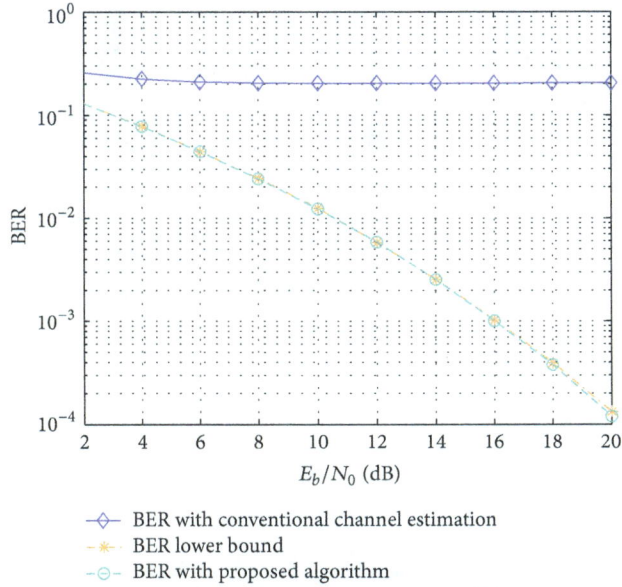

FIGURE 6: Improvement of DVB-T system performance in terms of BER with the proposed algorithm in ITU Vehicular Channel Model B.

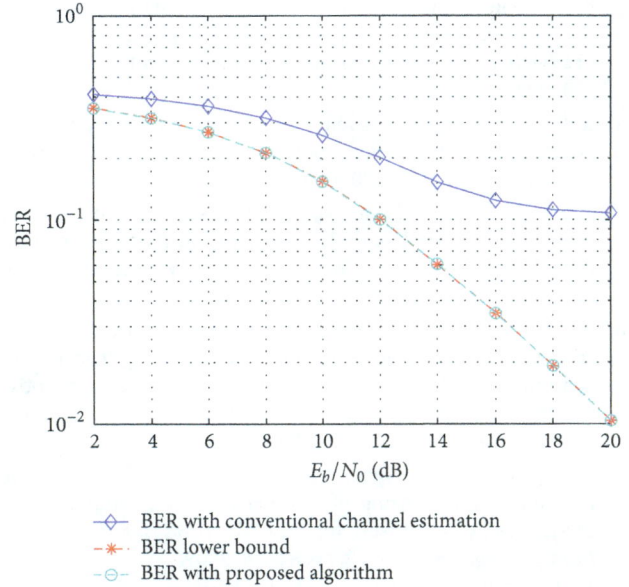

FIGURE 7: Improvement of DVB-T system performance in terms of BER with the proposed algorithm for an arbitrary channel, under a receiver mobility of 100 km/h.

Channel Model B. It is seen that the BER performance is significantly improved once the Doppler impairment is compensated on the estimated channel. The conventional channel estimation shows a large BER due to the nature of the ITU channel model. In conventional scheme, Doppler induced impairment is not compensated after the channel estimation, and therefore the signal constellation is rotated which caused error on the data demodulation. On the other hand, we have used the fact that estimated Doppler can be used to compensate the impairment caused by Doppler frequency itself which eventually contributes to the improvement of BER. As seen from this result, the BER performance with the proposed algorithm is very close to the lower bound (Doppler impairment can be perfectly compensated), indicating the robust performance of the proposed Doppler estimation algorithm.

Figure 7 presents the performance of the DVB-T system under the same system parameters as above, transmitting in 2 k mode at 800 MHz. However, this time we have simulated a worse channel with 200 taps following normalized uniform power delay profile. It is seen that the proposed algorithm performs better even under this worse channel condition. Similarly, as shown in the above figure, the BER performance with the proposed algorithm is also very close to the lower bound, indicating the robust performance of the proposed algorithm. The performance improvement is about 4 dB in the low SNR region and that is more than 6 dB in the high SNR region.

7. Conclusions

In this paper, a robust Doppler frequency and channel estimation technique is presented that is, suitable for mobile communications. The technique exploits past observed CIRs

in order to jointly estimate channel parameters. It is found that the estimated Doppler frequency is very close to the known Doppler frequency as observed in the simulation results. Further, the Doppler frequency estimation algorithm has been evaluated with estimation error variance introduced by the past CIRs. As expected, it is found that with the decrease of estimation error, the Doppler frequency estimation becomes more accurate. CRB of the proposed channel estimator is also evaluated with the number of past observations under different SNR values. It is observed that even in highly mobile channels, with the compensation of Doppler frequency-induced impairment, the system BER performance can be significantly improved. In addition, because of the joint estimation of the channel parameters, the estimation is more efficient and much simpler to implement. The proposed technique is suitable for emerging V2V or V2I communication systems, where delay requirement and high reliability are essential.

Acknowledgment

This work was supported in part by a Canada Graduate Scholarship—Doctoral (CGSD) from the Natural Sciences and Engineering Research Council (NSERC), Canada.

References

[1] N. Goertz and J. Gonter, "Limits on information transmission in vehicle-to-vehicle communication," in *Proceedings of the IEEE Vehicular Technology Conference (VTC '11)*, pp. 1–5, Spring, 2011.

[2] H. Trivedi, P. Veeraraghavan, S. Loke, H. P. Le, and J. Singh, "A survey of lower technologies for vehicle-to-vehicle communication," in *Proceedings of the IEEE Malaysia International Conference on Communicaitons*, pp. 441–446, 2009.

[3] P. A. Bello, "Some techniques for the instantaneous real-time measurement of multipath and Doppler spread," *IEEE Transactions on Communication Technology*, vol. 3, no. 3, pp. 285–292, 1965.

[4] R. Bamler, "Doppler frequency estimation and the Cramer-Rao bound," *IEEE Transactions on Geoscinece and Remote Sensing*, vol. 29, no. 3, pp. 385–390, 1991.

[5] T. Yücek, R. M. A. Tannious, and H. Arslan, "Doppler spread estimation for wireless OFDM systems," in *Proceedings of the IEEE/Sarnoff Symposium on Advances in Wired and Wireless Communication*, pp. 233–236, April 2005.

[6] C. Tepedelenlioğlu, A. Abdi, G. B. Giannakis, and M. Kaveh, "Estimation of Doppler spread and signal strength in mobile communications with applications to handoff and adaptive transmission," *Wireless Communications and Mobile Computing*, vol. 1, no. 2, pp. 221–242, 2001.

[7] H. Nguyen, T. Le-Ngoc, and C. C. Ko, "RLS-based joint estimation and tracking of channel response, sampling, and carrier frequency offsets for OFDM," *IEEE Transactions on Broadcasting*, vol. 55, no. 2, pp. 84–94, 2009.

[8] X. Zhang, F. Xie, W. Wang, and M. Chatterjee, "TCP throughput for vehicle-to-vehicle communications," in *Proceedings of the IEEE International Conference on Communications and Netwroking (ChinaCom '06)*, pp. 1–5, 2006.

[9] M. M. Freda, J. F. Weng, and T. Le-Ngoc, "Joint channel estimation and synchronization for OFDM systems," in *Proceedings of the IEEE Vehicular Technology Conference (VTC '04)*, pp. 1673–1677, 2004.

[10] C. Xiao, Y. R. Zheng, N. C. Beaulieu et al., "Second-order statistical properties of the WSS Jakes' fading channe simulator," *IEEE Transactions on Communications*, vol. 50, no. 6, pp. 888–891, 2002.

[11] M. J. Rahman, X. Wang, S. I. Park, and H. M. Kim, "Robust synchronization technique for mobile DTV broadcasting system," in *Proceedings of the IEEE Conference on Computer and Information Technology (ICCIT '10)*, pp. 216–220, 2010.

[12] H. Minn and V. K. Bhargava, "An investigation into time-domain approach for OFDM channel estimation," *IEEE Transactions on Broadcasting*, vol. 46, no. 4, pp. 240–248, 2000.

[13] P. Handel, "Properties of the IEEE-STD-1057 four-parameter sine wave fit algorithm," *IEEE Transactions on Instrumentation and Measurement*, vol. 49, no. 6, pp. 1189–1193, 2000.

[14] E. Kreyszig, H. Kreyszig, and E. J. Norminton, *Advanced Engineering Mathematics*, John Wiley & Sons, 2010.

[15] C. W. Therrien and M. Tummala, *Probability for Elctrical and Computer Engineers*, CRC Press, 2004.

[16] S. Haykin, *Communication Sytems*, John Wiley & Sons, 2001.

[17] International Telecommunication Union—Recommendations (ITU-R) M.1225: Guidelines for evaluation of radio transmission technologies for IMT-2000.

11

A Semi-Deterministic Channel Model for VANETs Simulations

Jonathan Ledy,[1] **Hervé Boeglen,**[2] **Anne-Marie Poussard,**[1]
Benoît Hilt,[2] **and Rodolphe Vauzelle**[1]

[1] *Laboratoire XLIM-SIC, UMR CNRS 6172, Université de Poitiers, 86034 Poitiers, France*
[2] *Laboratoire MIPS-GRTC, Université de Haute Alsace, 68000 Colmar, France*

Correspondence should be addressed to Jonathan Ledy, ledy@sic.univ-poitiers.fr

Academic Editor: Athanasios Panagopoulos

Today's advanced simulators facilitate thorough studies on Vehicular Ad hoc NETworks (VANETs). However the choice of the physical layer model in such simulators is a crucial issue that impacts the results. A solution to this challenge might be found with a hybrid model. In this paper, we propose a semi-deterministic channel propagation model for VANETs called UM-CRT. It is based on CRT (Communication Ray Tracer) and SCME—UM (Spatial Channel Model Extended—Urban Micro) which are, respectively, a deterministic channel simulator and a statistical channel model. It uses a process which adjusts the statistical model using relevant parameters obtained from the deterministic simulator. To evaluate realistic VANET transmissions, we have integrated our hybrid model in fully compliant 802.11 p and 802.11 n physical layers. This framework is then used with the NS-2 network simulator. Our simulation results show that UM-CRT is adapted for VANETs simulations in urban areas as it gives a good approximation of realistic channel propagation mechanisms while improving significantly simulation time.

1. Introduction

Vehicular Ad hoc NETworks (VANETs) are a very promising research area interesting the scientific community, car manufacturers, and mobile telephony operators. Vehicular applications should be thoroughly tested before they are deployed in the real world. Because the setup of experimental VANETs would imply huge investments, computer simulations are generally preferred.

One of the major issues when using simulators for VANETs concerns the vehicular environment and therefore the realistic modeling of the wireless propagation channel. Indeed, there are still several problems linked to the impact of the mobility and the traffic density on channel statistics yet to solve, for example, packets loss, rate of flow, frequency correlation, and amplitude distribution.

Many research and development works relating to routing [1], communication robustness [2], and information dissemination in VANETs [3] show results obtained with simulations involving very basic radio propagation models available in simulation tools (Friis and two-ray ground models, e.g.). The consequence of the mobility on the

physical layer is most of the time treated in a simplistic and consequently not quite realistic manner. This can lead to erroneous results [4]. Moreover, one finds very few effective and robust channel models which take into account the mobility and especially the transmission environment.

From this, one can understand that the radio propagation model used by the network simulation tool is a key factor in MANETs (Mobile Ad Hoc NETworks) and particularly in the VANETs subclass. Developing a radio channel model, which would describe the realistic radio channel conditions as accurately as possible, has been a continuous challenge. This is precisely what this work addresses.

There already exist reliable channel models which are customizable according to the environment [5], but most of them are dedicated to mobile telephony. In parallel, one finds research works presenting deterministic channel models [6, 7] which are based on ray-tracing or ray-launching methods which allow a realistic modeling of the channel. Unfortunately, these models require very high processing times.

As far as VANETs are concerned, deterministic channel models are not suitable because of the high mobility, the diversity of the environment encountered, and the high number of communicating nodes. The study of the higher layers of the OSI model (in particular the Network and Application layers) requires a low simulation time (i.e., a couple of minutes) in order to allow statistical analyses on large simulation series. To answer the challenge of channel modeling in VANETs, several works propose various methods which can be classified in two categories according to the research domain of their authors.

In the network community, Dhoutaut et al. [8] propose a propagation model based on Markov chains elements and real world experiments which is able to generate packet losses in a very realistic way. Later on, Han and Abu-Ghazaleh proposed another method based on Finite State Markov model [9]. These models are half-way between very detailed models using ray-tracing with computationally intensive algorithms and models using theoretical analysis where physical phenomena are only handled in an aggregate manner. But according to the authors, their models are not yet able to make a clear relation with a real environment.

In the physical channel modeling community, one can find different statistical channel models which have been derived from intensive measurement campaigns. The stochastic parameters of these models are extracted from the measurement data. It has been shown that the measured amplitude samples follow Rice, Rayleigh, or Weibull distributions [10]. In the case of vehicular channels there exist such type of channel models which have been designed by Acosta-Marum and Ingram for the validation of the 802.11p standard [11, 12]. It is a classical tapped delay line Wide Sense Stationary Uncorrelated Scattering (WSSUS) channel model. In [10], Sen and Matolak propose channel models which take into account the sudden appearance/disappearance of scatterers (moving obstacles) by modeling them as first-order, two-state Markov chains. In [13] Keredal et al. propose a geometry-based stochastic MIMO model for Vehicle-to-Vehicle (V2V) communications. However, as stated by Molisch in [14], there is only a small amount of V2V channel measurements available which "does not allow the derivation of statistically significant statements about real-world V2V channels". In conclusion, extreme care has to be taken when choosing and parameterizing available statistical channel models for VANETs.

Another fundamental topic in VANETs simulation concerns the mobility model used for simulations. Many works, like the working by Marfia et al. [15], show the importance of realistic mobility models. Indeed, the use of nonspecific mobility models employed in VANETs simulations may provide bad results, because they ignore the typical behavior of the nodes in this kind of network. In conclusion, a combination of realistic radio wave propagation models and realistic mobility models is a large step towards more realistic simulation environments as shown by Günes et al. [16]. In our work, the selected simulation tools are the VANET-specific mobility generator VanetMobiSim [17] and the generic network simulation tool NS-2 [18].

The rest of this paper is organized as follows. In Section 2, we present the components of UM-CRT that are, respectively, the statistical SCME-UM model and the realistic CRT simulator. In Section 3, we give a detailed presentation of our semi-deterministic model, UM-CRT. Our framework including implementation and the simulation of the 802.11 standard is also presented in this section. Section 4 is dedicated to the evaluation of our model. Finally, Section 5 concludes this paper and deals with future works.

2. Towards a Semi-Deterministic Model

Statistical and deterministic channel models are the two common ways to describe the radio channel behavior in VANETs simulations. In this section, we describe an example of each of these approaches: for the statistical one, the Spatial Channel Model Extended in its *Urban Microenvironment* (SCME-UM) is described and, for the deterministic one, the Communication Ray Tracer (CRT) simulator is presented. We then analyze their main characteristics in order to propose a new semi-deterministic solution which benefits from the advantages of these two classical approaches. This proposition is called UM-CRT because it is based on respectively, SCME-UM and CRT.

2.1. The Spatial Channel Model Extended (SCME). The Spatial Channel Model Extended (SCME) statistical channel model is an evolution of the 3GPP Spatial Channel Model (SCM) [19]. It has been developed within the European WINNER project [20] for the simulation of B3G systems. The SCM model is limited to the simulation of systems at 2 GHz for a maximum transmission bandwidth of 5 MHz, whereas its extension, SCME for SCM Extended, allows for the simulation of systems at 2 and 5 GHz for a maximum transmission bandwidth of 100 MHz [21].

SCM and SCME are so-called geometric models for which scatterers are placed stochastically in the simulation scene. SCME considers clusters of scatterers. Each cluster corresponds to a resolvable path. Each path is made up of several nonresolvable subpaths. Figure 1 is a typical example showing the main geometrical parameters used by the model where a Base Station (BS) antenna array communicates with a Mobile Station (MS) antenna array. This example shows only one cluster labeled n. A subpath is also shown (labeled m).

SCME is a natively Multiple Input Multiple Output (MIMO) model. It allows for the simulation of three types of environments: Urban Macrocell, Suburban Microcell (distance between MS and BS 3 km maximum), and Urban Microcell (distance between MS and BS of 1 km maximum). In the context of urban VANETs, because of intervehicular distances less than one kilometer we have chosen the Urban Microcell (UM) environment. The authors of [21] provide a Matlab implementation of the SCME model [22] which we have used in our framework. This piece of software generates Channel Impulse Responses (CIR) which can then be used in a digital communication chain.

2.2. The Communication Ray Tracer (CRT) Simulator. Let us now describe the Communication Ray Tracer (CRT) software. It is a deterministic propagation simulator developed by the Xlim-SIC laboratory from the University of

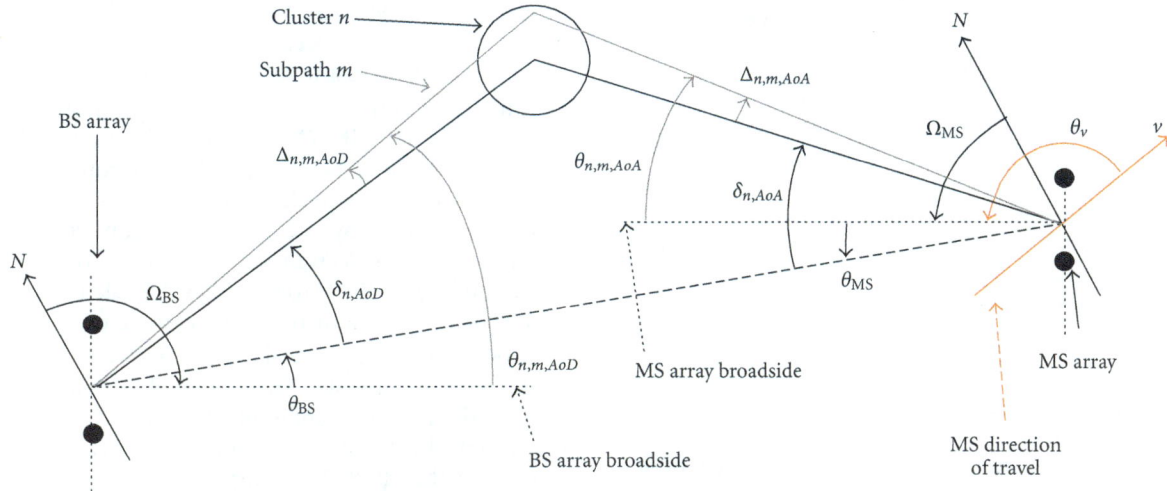

FIGURE 1: Geometric parameters of the SCME model.

Poitiers (France) [23]. CRT is based on an optimized 3D deterministic ray-tracing method to model the propagation of radio waves in real environments (outdoor and indoor). Thus, CRT takes into account all the characteristics (geometric and electric) of the environment and provides the information about the multipath phenomenon, where each path is characterized by its attenuation, delay, angular directions, phase shift, and polarization. In this way, we obtain a complete characterization of the narrow-band and wide-band channel. This provides a realistic approach of the multi-path propagation mechanisms. CRT has been validated by several measurement campaigns [23] and will therefore be considered as our reference.

Figure 2 shows an example of a multi-path propagation simulation in a 3D urban environment using CRT. Each white line is one path followed by the radio waves between the transmitter and the receiver defined by a black circle in Figure 2. The CRT simulation procedure is as follows. Firstly, the user chooses an environment and places the communicating nodes into that scene. In a second stage, the nodes are associated with a moving trajectory. Then, CRT calculates the CIRs for all the communicating nodes in the simulation scenario.

In order to reduce the computation time in a context of mobility, an optimization based on the stationarity property of the channel [24] has been studied. It has been shown that a stationarity area of 8 meters can be considered.

2.3. Basis of the Proposed Semi-Deterministic Model for VANETs.
The presentation of the two previous models shows several key elements.

The SCME-UM statistical model is very efficient to produce CIRs in environments modeled by several clusters placed randomly in the scene. However, it does not take into account the geometrical specificities of real environments. On the contrary, CRT is able to provide CIRs directly connected with the environment by a complete modeling of the interactions between radio waves and building. However,

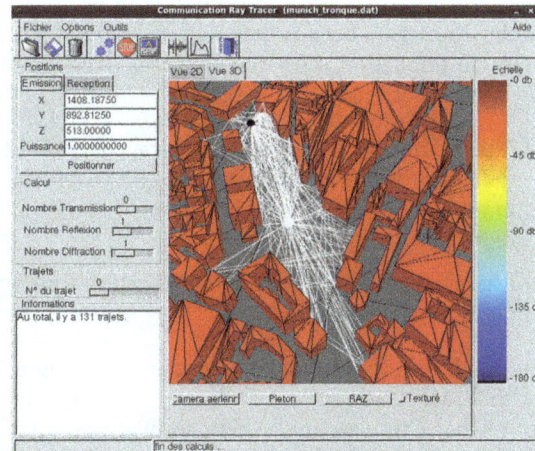

FIGURE 2: A CRT simulation of multipath propagation in a realistic 3D urban environment.

it necessitates an important computation time for each CIR calculation. Therefore, for VANETs applications, which introduce the mobility for each node and a significant number of possible radio links between the nodes in the network, a deterministic solution leads to a huge computation time.

To address this problem, we propose in this paper a hybridisation of the two models presented previously named UM-CRT. Indeed, we will combine at the same time the major wave propagation phenomena (existence or not of the direct path) with its deterministic component and the low computation time allowed by its statistical one.

3. The Semi-Deterministic Channel Model

3.1. The UM-CRT Semi-Deterministic Channel Model.
According to the principle presented in the previous section, UM-CRT is created from the association of two models. Figure 3 shows the relationship between UM-CRT and both CRT and SCME-UM, respectively, in the left and right parts

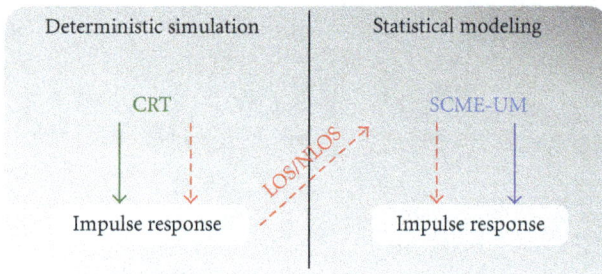

FIGURE 3: Principle of the UM-CRT model.

with fine arrows. UM-CRT is depicted with bold dashed arrows.

Classically, for all radio links existing between nodes, on the one hand, CRT computes CIRs according to the simulation of all the received multi-paths according to an environment modeled in 3D. On the other hand, SCME-UM provides statistically generated CIRs.

Here, we propose to limit the search path by ray-tracing only to the direct path because it is well known that this path has the main impact on the received signal. This limitation has two advantages.

(i) It takes into account the geometrical characteristic of the considered environment: we can determine the Line of Sight (LOS) and Nonline of Sight (NLOS) radio links.

(ii) It considerably reduces the computation time of the deterministic simulation.

On this basis, it is possible to generate representative CIRs with SCME-UM.

Notice that complete multi-path simulations with CRT are possible in order to exploit other information included in the deterministic CIRs such as delay spread. But it will be very time consuming.

Moreover, in order to reduce again the computation time, we propose to consider the stationarity property of the channel introduced previously. We assume that a CIR of a radio link remains constant when the move of its extremities is less than 8 meters in relation to a reference position. So we do not compute the CIRs at each time but only for a finite number of distances between the transmitter and the receiver. In practice, as the SCME-UM is limited to 600 meters between the transmitter and the receiver, we consider a set of 90 distances to calculate the CIRs. These CIRs can be precalculated in order to accelerate the computation time for statistical studies of VANETS performance.

To summarize, for a VANETs scenario based on a set of vehicles, UM-CRT allows computing a statistical CIR of each radio link between two mobile nodes, at each time, according to a LOS/NLOS deterministic analysis.

3.2. The UM-CRT Framework. In order to evaluate the performance of a VANETs scenario in a realistic environment, it is necessary to introduce real transmission conditions in a network simulator. We consider here the NS2 platform.

These transmission conditions are based on the channel model and on the specific digital communication chain considered. This constitutes a realistic physical layer.

Firstly, we explain the considered physical layer. Secondly, we introduce the UM-CRT framework in an NS2 context. With this framework, it is possible to compute the performance analysis according to some QoS metrics (Packet Delivery Ratio, Delay, etc.) in several configurations, as it will be shown in Section 4. Finally, we show the validity (ou accuracy) of Bit Error Rate (BER) calculated with our framework in the Munich city center as an exemple.

Concerning the physical layer in VANETs context, there exists the 802.11p standard adopted at the the end of 2010 [25]. It is an adaptation of the 802.11a standard for Dedicated Short Range Communication (DSRC) between vehicles. Although 802.11p improves the robustness against channel frequency selectivity, doubling the OFDM symbol time degrades the robustness of the system against time selectivity which depends on the Doppler frequency shift directly related to the vehicles speed.

The 802.11p standard does not account for the Multiple-Input Multiple-Output technology although it is known to improve significantly the reliability and the throughput of data transmission [26–28]. We believe that this issue is going to be a natural evolution of the standard. From the time being, and in order to assess the communication robustness performance of MIMO systems in VANET scenarios, we have decided to use the 802.11n standard in our simulations where we set up a 2×2 antenna system. Last but not least, the 802.11n standard allows transmissions in the 5 GHz frequency band. The 802.11p and 802.11n implementations have been written in C++ using the IT++ library [29].

In [30] Hamidouche et al. study the impact of a realistic physical layer on the H.264/AVC video transmission over ad hoc networks in an urban environment. They also propose an error model which is based on a Bit Error Rate (BER) computation. We will make use of this error model into our model-building process.

From the impulse response calculated by SCME-UM, we use the 802.11 physical layer described previously to calculate a BER. Each BER value gives an accurate information about the quality of the communication between two nodes.

In order to evaluate the accuracy of our realistic BER computation approach, we consider the 802.11p standard in the environment in which we will run our simulations, that is to say a VANETs scenario with 40 vehicles moving in the Munich city center. Figure 4 shows the BER evolutions according to SNR observed for all radio links associated to the scenario. The red curve is obtained with the UM-CRT model and the blue one with the CRT simulation. The last one is considered as the reference because the channel impulse responses are computed deterministically.

We can observe that the results are very close and consequently the approach proposed in our framework is valid in terms of BER.

Finally, this realistic physical layer based on UM-CRT and called UM-CRT framework is introduced in the NS2

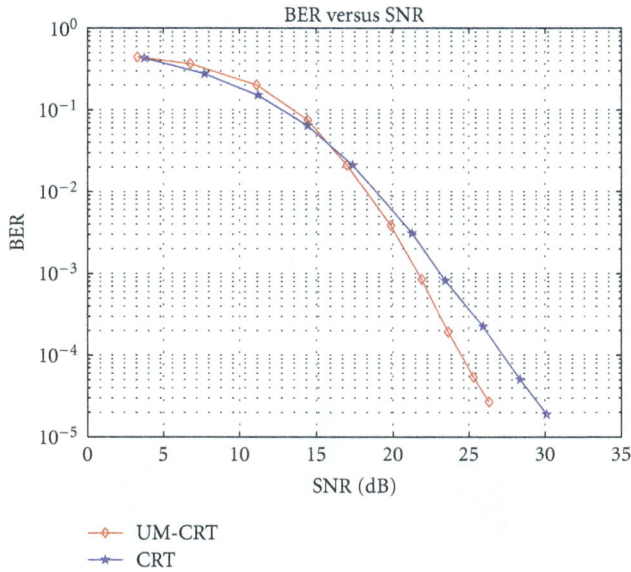

FIGURE 4: BER evolutions according to SNR obtained by UM-CRT and CRT.

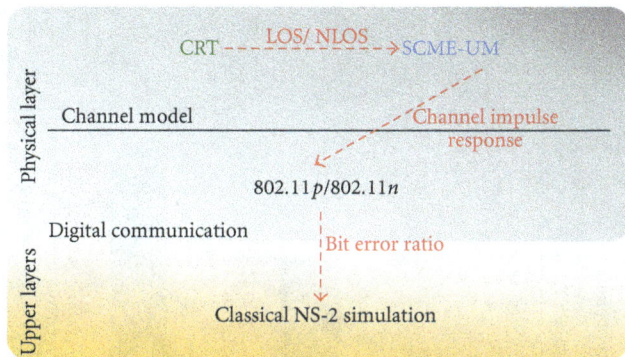

FIGURE 5: The UM-CRT framework.

platform. It communicates with the upper layers. All these steps are summarized in Figure 5.

Figure 6 shows an example of an application of the UM-CRT framework in our simulation environment with buildings shown in red. This figure shows an instantaneous representation of a 40-vehicle scenario with several communications. The colored lines represent the BER values between each transmitter-receiver couple of vehicles, characterizing the radio link quality. The darker the color, the higher the BER. This tool is useful to determine easily which route between nodes is bad or good.

To conclude, the NS2 platform modified with the UM-CRT framework constitutes a VANETs simulator which allows to evaluate QoS performance with realistic transmission conditions in a specific environment.

4. UM-CRT Evaluation

To evaluate the UM-CRT propagation model, we compare it with the CRT simulator using NS-2 simulations in a typical urban environment: that is, the center of Munich City

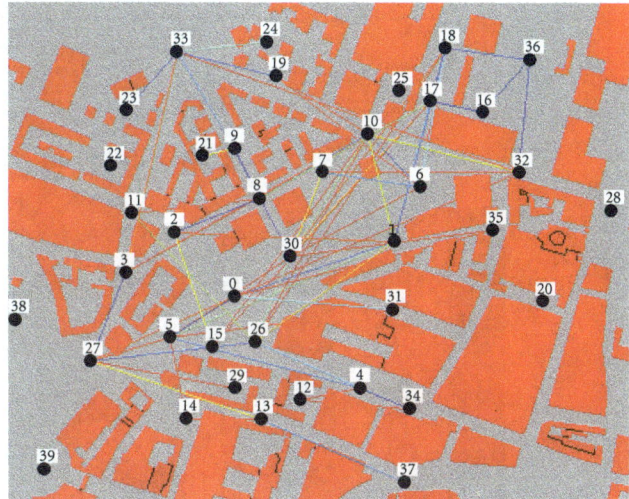

FIGURE 6: Instantaneous representation of the radio link quality in terms of BER between vehicles in the Munich city center.

(cf. Figure 6). As mentioned above, CRT will be our reference model and can be considered realistic enough to match real world implementation results.

4.1. Accuracy Evaluation. Please note that in this evaluation, the results of the statistical SCME-UM model used alone are not presented because they do not take into account a real propagation environment. LOS and NLOS results will always either be nearly perfect (~100% of packets reach their destination) or bad (~0% of packets reach their destination).

The simulations were run in SISO and MIMO modes in the 5 GHz band. In order to have enough different cases for comparing UM-CRT to CRT, we ran simulations with different VANET realistic mobility scenarios generated by VanetMobiSim. Each of them is defined with different starting points, traffic lights configurations, and mobilities. The mobility is random and the nodes' speed is variable with time and limited by 3 maximum allowed speeds (0, 8, and 15 m/s). The routing protocol we used is AODV in its basic setup. The traffic generated for inter-node communications is UDP based. The simulations were performed on a Linux Mandriva system with an updated NS-2.29 simulator version.

We first compare UM-CRT to CRT in the SISO mode (802.11p). The comparison is done in terms of packet delivery ratio (PDR) between a transmitter and a receiver. This is a common criterion used in VANETs performance evaluation. We also compare them in terms of number of hops and end-to-end delay. We finally evaluate how MIMO impacts the PDR (802.11n case). Note that in these evaluations we present averaged results for every simulation over 5 different 40-second simulation time.

Figure 7 shows that the PDR varies in an important manner depending to the maximum allowed speed. Remember that our main goal is to compare the proximity of the results of UM-CRT and CRT in every situation. We can observe that the higher the speed is, the closer the PDR results are. We can also see that in the static situation (0 m/s), UM-CRT results

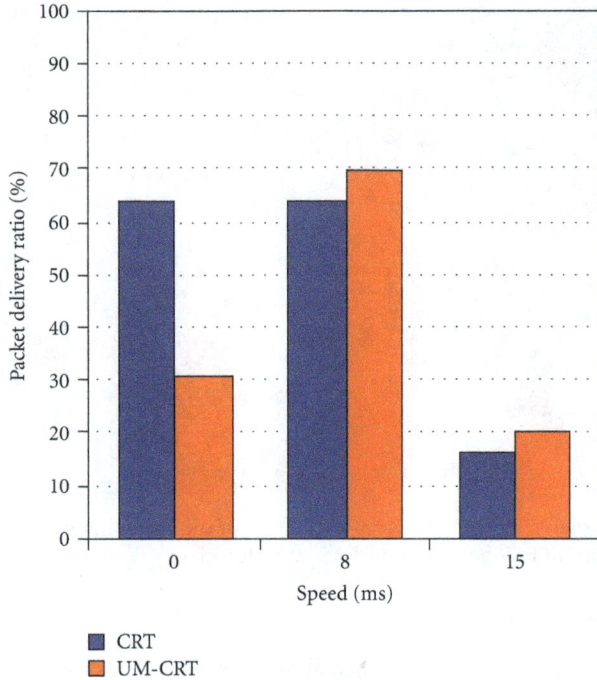

FIGURE 7: UM-CRT PDR evaluation.

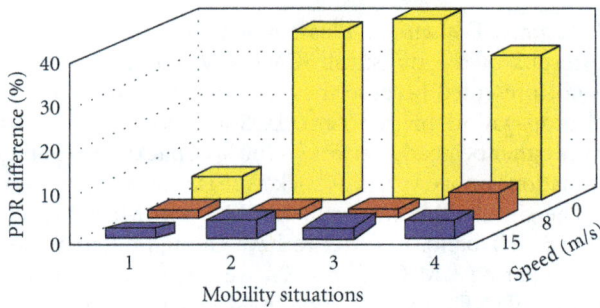

FIGURE 8: UM-CRT versus CRT PDR in four mobility situations.

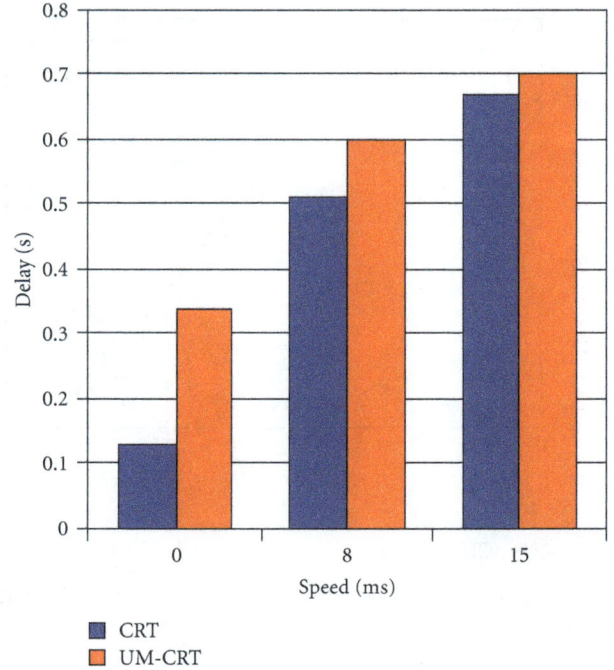

FIGURE 9: Delay evaluation.

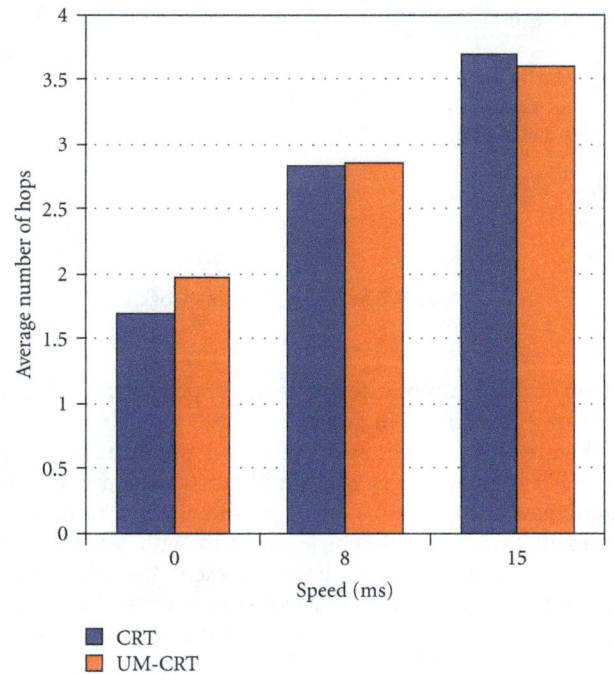

FIGURE 10: Average number of hops evaluation.

do not match the CRT reference results. Additionally, the known negative influence of the speed on the PDR can be observed in the 15 m/s part of the figure.

In Figure 8 we have a global view of the PDR difference between UM-CRT and CRT in several mobility situations. This confirms the similarity between UM-CRT and CRT and therefore the realism of our semi-deterministic model when we are in nonstatic situations. In the cases with mobility, the difference between the two models is less than 10%.

Figures 9 and 10 confirm that concerning the end-to-end delay and the average number of hops UM-CRT matches better CRT; that is, it becomes more realistic, when the speed is higher. If we look further, the similarity between the two models can also be observed through an increasing delay and average number of hops that both increase when the speed increases. For these two parameters, the difference between the two models decreases above 0,1 seconds for the delay and can be considered as very similar for the number of hops, when we are not in a static situation.

Results have the same trend in the case of a static situation in the simulations. This is a current limitation of our model which is not suitable for null speed. In this case, the LOS-NLOS criterion is not suitable alone to produce results that match the deterministic model.

To summarize, when we are not in a static situation, the UM-CRT model gives results quite similar to CRT in all

TABLE 1: Simulation time comparison.

	CRT				UM-CRT				SCME-UM			
Scenario	1	2	3	4	1	2	3	4	1	2	3	4
Preprocessing Full CRT (h)	18	18	17	18								
Preprocessing LOS/NLOS (h)					1	1	1	1	1	1	1	1
Simulation (h)	57	65	55	63	0.5	0.5	0.5	0.5	0.5	0.5	0.5	0.5
TOTAL Time (h)	75	83	72	81	1.5	1.5	1.5	1.5	1.5	1.5	1.5	1.5

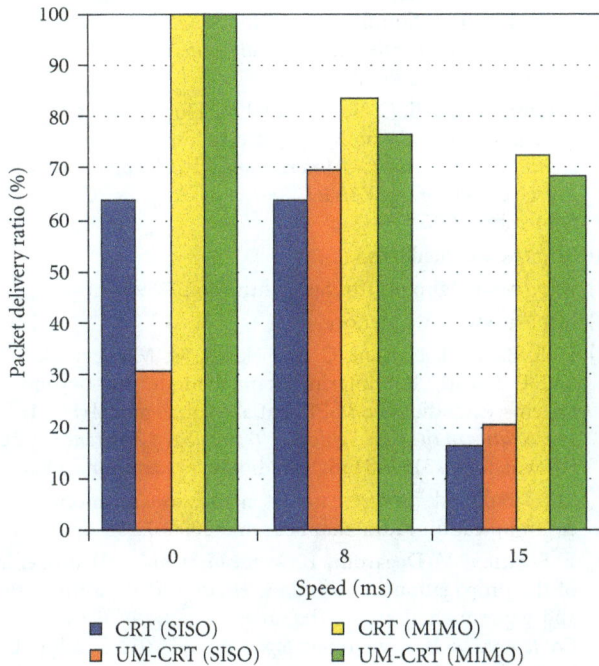

FIGURE 11: MIMO Impact.

Indeed, results between CRT and UM-CRT are close. As expected, we can also notice that a MIMO channel is more robust than a SISO one. The received packets rates are better in MIMO cases, no matter the nodes speed. Approximately 70% of the packets are received with a 15 m/s speed in this scenario for the MIMO mode, whereas it is less than 50% for the SISO mode.

From these results, one can conclude that the MIMO mode improves the transmission's robustness in VANETs. So MIMO technology allows reducing transmission power with a packet delivery ratio equal to the SISO mode in order to limit perturbations, or it can help to improve the packet delivery ratio. In both cases, MIMO technology seems to be very interesting to answer VANETs' challenges.

4.2. Computation Time Evaluation. As shown in Section 4.1, our semi-deterministic propagation model is not only very realistic (except for static nodes situations) but it also decreases the simulation time.

In Table 1, one can observe that each CRT simulation lasts at least three days (equivalent to 40-second simulated time in NS-2) whereas in the case of SCME-UM this reduces to only 1.5 hours. Section 4.1 has shown the realism of UM-CRT. Table 1 showed that the computation time for UM-CRT has been reduced significantly. This can be explained by the principle used in the model (see Section 3) which consists in reading a cache containing CRT and SCME-UM impulse responses which are then used for BER computations.

5. Conclusion and Future Works

In this paper we have presented UM-CRT, a semi-deterministic channel propagation model for VANETs. UM-CRT, which was integrated into the NS-2 network simulator, is based on the stochastic SCME-UM model and the deterministic CRT channel simulator.

The implementation of this new model allows to run network simulations in a very fast way. Indeed the computation time is reduced from more than 70 hours to less than 2 hours. This makes UM-CRT quite suitable for VANETs simulations having a large number of high-mobility nodes. Moreover, our results showed UM-CRT to be appropriate for mobility scenarios and realistic vehicular networks simulations, typically urban scenarios.

Furthermore, results show that UM-CRT is also adapted for the MIMO technology. As expected, simulations involving this configuration have clearly showed the robustness of the MIMO scheme compared to SISO one.

situations. This can be explained by the number of statistical outcomes that increase because of the mobility. So we can conclude that the UM-CRT model gives results quite close to the deterministic model.

Furthermore, as the speed gets higher, one can see that the channel deteriorates (the received packets rate decreases) and that it is more difficult to achieve a reliable communication (delay and average number of hops increase). We can see that, for a maximum speed of 15 m/s, the received packet rate does not exceed 50%. This is a second expression of the determinism that our statistically based model produces.

We will now see the impact of the use of UM-CRT in a MIMO mode instead of a SISO mode.

The impact of the MIMO mode on the packet delivery ratio is presented in Figure 11. It confirms the tendencies observed in the SISO case. The UM-CRT model gives quite similar results to CRT for the MIMO mode too. However, there is an exception in the static case for MIMO: CRT and UM-CRT results match perfectly at 0 millisecond whereas they do not at 8 milliseconds or 15 milliseconds. This can be explained by the robustness of the channel which has improved compared to the SISO case. It is therefore possible to obtain 100% of received packets even with fixed nodes.

We currently focus our work on the selection of new relevant criteria extracted from the CIR such as the RMS delay spread or the link capacity.

References

[1] M. Boban, G. Misek, and O. K. Tonguz, "What is the best achievable QoS for unicast routing in VANET?" in *Proceedings of the IEEE Globecom Workshops*, pp. 1–10, New Oreleans, La, USA, December 2008.

[2] S. Y. Wang, "The effects of wireless transmission range on path lifetime in vehicle-formed mobile ad hoc networks on highways," in *Proceedings of the IEEE International Conference on Communications*, vol. 5, pp. 3177–3181, Seoul, Korea, May 2005.

[3] S. Yousefi, S. Bastani, and M. Fathy, "On the performance of safety message dissemination in vehicular ad hoc networks," in *Proceedings of the 4th European Conference on Universal Multiservice Networks*, pp. 377–387, Toulouse, France, February 2007.

[4] F. J. Martinez, C. K. Toh, J. C. Cano, C. T. Calafate, and P. Manzoni, "Realistic radio propagation models (RPMs) for VANET simulations," in *Proceedings of the IEEE Wireless Communications and Networking Conference*, Budapest, Hungary, April 2009.

[5] 3GPP, "Spatial channel model for MIMO simulations," TR 25.996 V9.0.0 (2009-12), http://www.3gpp.org/ftp/Specs/archive/25_series/25.996/25996-900.zip.

[6] I. Stepanov and K. Rothermel, "On the impact of a more realistic physical layer on MANET simulations results," *Elsevier Ad Hoc Networks Journal*, vol. 6, no. 1, pp. 61–78, 2006.

[7] R. Delahaye, A.-M. Poussard, Y. Pousset, and R. Vauzelle, "Propagation models and physical layer quality criteria influence on ad hoc networks routing," in *Proceedings of the 7th International Conference on Intelligent Transport Systems Telecommunications*, pp. 433–437, Sophia Antipolis, France, June 2007.

[8] D. Dhoutaut, A. Régis, and F. Spies, "Impact of radio propagation models in vehicular ad hoc networks simulations," in *Proceedings of the 3rd ACM International Workshop on Vehicular Ad Hoc Networks (VANET '06)*, pp. 40–49, Los Angeles, Calif, USA, September 2006.

[9] S. Han and N. B. Abu-Ghazaleh, "Estimated measurement-based Markov models: Towards flexible and accurate modeling of wireless channels," in *Proceedings of the 5th IEEE International Conference on Wireless and Mobile Computing Networking and Communication*, pp. 331–337, Marrakech, Morocco, October 2009.

[10] I. Sen and D. W. Matolak, "Vehicle-vehicle channel models for the 5-GHz band," *IEEE Transactions on Intelligent Transportation Systems*, vol. 9, no. 2, Article ID 4517519, pp. 235–245, 2008.

[11] G. Acosta-Marum and M. A. Ingram, "A BER-based partitioned model for a 2.4GHz vehicle-to-vehicle expressway channel," *Wireless Personal Communications*, vol. 37, no. 3-4, pp. 421–443, 2006.

[12] G. Acosta-Marum and M. A. Ingram, "Six time- and frequency- selective empirical channel models for vehicular wireless LANs," *IEEE Vehicular Technology Magazine*, vol. 2, no. 4, Article ID 4498409, pp. 4–11, 2007.

[13] J. Karedal, F. Tufvesson, N. Czink et al., "A geometry-based stochastic MIMO model for vehicle-to-vehicle communications," *IEEE Transactions on Wireless Communications*, vol. 8, no. 7, pp. 3646–3657, 2009.

[14] A. F. Molisch, F. Tufvesson, J. Karedal, and C. F. Mecklenbrauker, "A survey on vehicle-to-vehicle propagation channels," *IEEE Wireless Communications*, vol. 16, no. 6, pp. 12–22, 2009.

[15] G. Marfia, G. Pau, E. De Sena, E. Giordano, and M. Gerla, "Evaluating vehicle network strategies for downtown Portland: opportunistic infrastructure and the importance of realistic mobility models," in *Proceedings of the 5th International Conference on Mobile Systems, Applications and Services*, pp. 47–51, San Juan, Puerto Rico, USA, June 2007.

[16] M. Günes, M. Wenig, and A. Zimmermann, "Realistic mobility and propagation framework for MANET simulations," in *Proceedings of the 6th International Conference on Networking*, Atlanta, Ga, USA, 2007.

[17] J. Haerri, F. Filali, C. Bonnet, and M. Fiore, "VanetMobiSim: generating realistic mobility patterns for VANETs," in *Proceedings of the 3rd ACM International Workshop on Vehicular Ad Hoc Networks (VANET '06)*, Los Angeles, Calif, USA, September 2006.

[18] http://www.isi.edu/nsnam/ns.

[19] http://www.3gpp.org/ftp/Specs/html-info/25996.htm.

[20] http://www.ist-winner.org.

[21] D. S. Baum, J. Hansen, G. Del Galdo, M. Milojevic, J. Salo, and P. Kyösti, "An interim channel model for beyond-3G systems: extending the 3GPP spatial channel model (SCM)," in *Proceedings of the 61st Vehicular Technology Conference (VTC '05)*, vol. 5, pp. 3132–3136, Stockholm, Sweden, June 2005.

[22] http://radio.tkk.fi/en/research/rf_applications_in_mobile_communication/radio_channel/scme-2006-08-30.zip.

[23] F. Escarieu, V. Degardin, L. Aveneau et al., "3D modelling of the propagation in an indoor environment : a theoretical and experimental approach," in *Proceedings of the European Conference on Wireless Technologies (ECWT '01)*, London, UK, September 2001.

[24] R. Delahaye, *Simulation réaliste et efficace de la couche physique pour l'aide au routage des réseaux ad hoc*, Ph.D. thesis, University of Poitiers, France, 2007.

[25] IEEE Standard for information technology—Telecommunications and information exchange between systems—Local and metropolitan area networks—Specific requirements, Part 11: Wireless LAN Medium Acces Control (MAC) and Physical layer (PHY) specifications, Amendment 6: Wireless Access in Vehicular Environments, IEEE std 802.11p, 2010.

[26] G. J. Foschini, "Layered space-time architecture for wireless communication in a fading environment when using multi-element antennas," *Bell Labs Technical Journal*, vol. 1, no. 2, pp. 41–59, 1996.

[27] E. Telatar, "Capacity of multi-antenna Gaussian channels," *European Transactions on Telecommunications*, vol. 10, no. 6, pp. 585–595, 1999.

[28] C. Oestges and B. Clerckx, *MIMO Wireless Communications*, Elsevier, 2007.

[29] http://sourceforge.net/apps/wordpress/itpp.

[30] W. Hamidouche, R. Vauzelle, C. Olivier, Y. Pousset, and C. Perrine, "Impact of realistic MIMO physical layer on video transmission over mobile ad hoc network," in *Proceedings of the IEEE 20th Personal, Indoor and Mobile Radio Communications Symposium (PIMRC '09)*, Tokyo, Japan, September 2009.

12

A Methodology to Estimate Capacity Impact due to Connected Vehicle Technology

Daiheng Ni, Jia Li, Steven Andrews, and Haizhong Wang

Department of Civil and Environmental Engineering, University of Massachusetts Amherst, Amherst, MA 01003, USA

Correspondence should be addressed to Daiheng Ni, ni@ecs.umass.edu

Academic Editor: Nandana Rajatheva

Recent development in connected vehicle technology or equivalently vehicular ad hoc networks (VANET) has stimulated tremendous interests among decision makers, practitioners, and researchers due to the potential safety and mobility benefits provided by these technologies. A primary concern regarding the deployment of connected vehicle technology is the degree of market penetration required for effectiveness. This paper proposes a methodology to analyze the benefit of highway capacity gained from connected vehicle technology. To fulfill this purpose, a model incorporating the effects of connected vehicle technology on car following is formulated, building on which a rough estimate of the resulting capacity gain is derived. A simulation study is conducted to verify the model, and an illustrative example is provided to show the application of the methodology. This work provides decision makers and practitioners with a basic tool to understand the mobility benefit obtained from connected vehicle technology and how such benefit varies as market penetration changes.

1. Introduction

Recent development in connected vehicle technology (CVT), formally known as IntelliDrive or vehicle infrastructure integration (VII) in transportation community and as ad hoc networks (VANET) in wireless network community, has stimulated tremendous interests among decision makers, practitioners, and researchers due to the potential safety and mobility benefits provided by these technologies. Supported by the dedicated short range communication (DSRC) standard, connected vehicle technology will enable road vehicles to communicate with each other as well as to roadside infrastructure in the future; see an illustration in Figure 1. Thus, highways and streets will become an environment that encompasses ubiquitous computing and communication. Consequently, a new class of applications can be developed to dramatically increase safety, throughput, and energy efficiency. For example, CVT may serve as an ever-vigilant copilot to watch for potential hazards such as abrupt braking by a leading vehicle, a side vehicle in blind spot, and even a collision from behind. In addition, CVT supports various functionalities to relieve congestion such as notifying downstream congestion, alternative routes, and even parking information.

Moreover, CVT opens the door of enroute entertainment such as downloading music and video, checking e-mails, and maintaining social connections like Facebook. All of these possibilities depend on large-scale deployment of connected vehicle technology. However, a deployment decision has to take many factors into consideration. Among others, primary factors are the infrastructure needed for success and the degree of market penetration (i.e., percent of vehicles equipped with connected vehicle technology) required for effectiveness.

The above question is very difficult to answer because of the following: field experiments require a large-scale connected vehicle technology testbed which has yet to be deployed; simulation is unavailable since existing traffic simulation packages are not designed to model traffic enabled by connected vehicle technology; analytical modeling is prohibitive because of the complexity and interdependency involved in connected vehicle systems. To bypass these difficulties, this paper carries a humble goal by following a simplified modeling approach which is complemented by Monte Carlo simulation. In addition, the focus is to explore a feasible approach to conduct preliminary estimation of the mobility

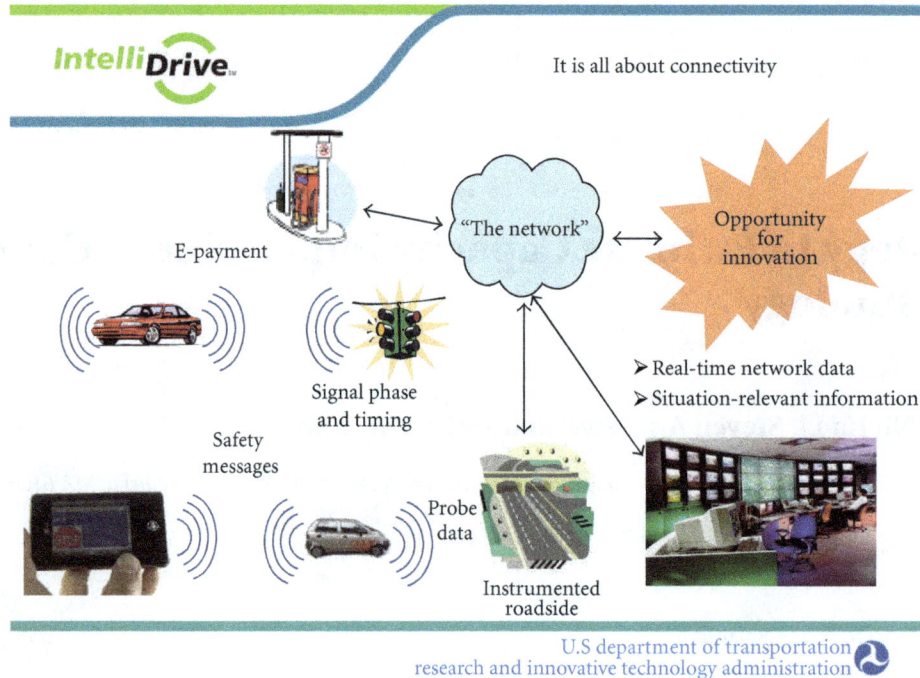

FIGURE 1: Connected vehicle technology.

benefit of connected vehicle technology, that is, the increase of highway capacity brought about by connected vehicle technology and how the result changes as market penetration varies. There are two building blocks in this approach. The first is to incorporate the effects of connected vehicle technology into driving behavior modeling. For such a purpose, a car-following model was derived based on the classical Gipps' model [1] by attributing the effects of connected vehicle technology to the change in the distribution of drivers' perception-reaction time. Recognizing that connected vehicle technology may bring other profound changes in traffic operations than merely perception-reaction time, the proposed model has to be kept as tractable as possible to make the analysis feasible yet capturing the major effect of connected vehicle technology. For this reason, some aspects of traffic flow, such as lane changing and hysteresis, are not modeled. Based on the proposed model, the second building block is a probabilistic analysis to provide an estimate of highway capacity. In this part, the major tools utilized are Wald's formula in probability theory and a theorem regarding the product moment of stopping time. An analytical approximate formula for capacity is obtained therein. A Monte Carlo simulation study is conducted to provide an alternative to verify this estimate since field tests are not possible at this time. The result obtained in this paper provides decision makers and practitioners with a basic tool to understand mobility benefit resulted from connected vehicle technology and how such a benefit varies as connected vehicle technology market penetration increases. In addition, using the methodology proposed in this paper, researchers can fine tune the assumption about the effects of connected vehicle technology to further investigate its benefits.

The remainder of this paper is organized as follows. In Section 2, relevant studies on this subject are briefly reviewed to provide a context in which the current paper fits. Next, in Section 3, the effects of connected vehicle technology are incorporated into the modeling of driving behavior by rectifying the Gipps' model. Following that, Section 4 is the probabilistic analysis and simulation verification. Section 5 provides an illustrative example to show the application of the methodology. Finally, the findings and results are summarized at the end.

2. Existing Studies

The idea of studying traffic flow benefits due to advanced technologies such as adaptive cruise control (ACC) systems and automated highway systems (AHS) has been addressed in the past. A great deal of studies have been identified which provided insights into highway capacity and traffic stability. A good survey of these studies can be found in [2]. A few additional references that present the necessary context for this study are added here. In their early studies on flow benefits of AHS, authors in [3, 4] investigated how ACC affected traffic flow and found that the improvement in capacity is small. Also focused on ACC, authors in [5] studied the impact of ACC on traffic flow stability and found that car following based on a constant time headway is essentially unstable.

While traffic operation in a separate lane hosting only ACC vehicles represents an ideal condition, analyzing mixed traffic flow consisting of ACC-automated vehicles and manually operated vehicles poses more challenges. Reference [6] presented such a study. Their simulation results related the capacity trend to mixed ratios of ACC-equipped cars and their market penetration. They found that the capacity benefit became significant when ACC-equipped cars exceeded 50% market penetration. When all cars were equipped with the technology, they found a 33% increase in capacity.

In addition to considering mixed traffic, incorporating inter vehicular communication such as cooperative ACC (CACC) represents a more realistic scenario. Refernce [7] used Monte Carlo simulation to estimate lane capacity under varying proportions of autonomous ACC (AACC) and CACC. They concluded that AACC could only have a small impact on highway capacity (at most a 7% increase), while significant capacity gain could be expected with increased CACC market penetration (potentially more than doubling the capacity). Authors in [8] studied similar subject matter with a focus on the impacts of CACC on a highway-merging scenario. Based on the traffic flow simulation model MIXIC, they found improved traffic stability and a slightly increased capacity compared to the non-AAC-equipped scenario.

In European Union, a simulation study was conducted on cooperative systems deployment impact assessment (CODIA). This study reported reduced average speed and hence increased journey times due to vehicle-infrastructure cooperation, and such an increase exhibited a quadratic "line of best fit" as market penetration varies from 0 to 100 percent [9].

Inspired by these original studies, our work considers a more general scenario which incorporates three types of driving modes enabled by connected vehicle technology (denoted as CVT thereafter), namely non-CVT, CVT assisted, and CVT automated. In the *non-CVT* mode, drivers operate their vehicles without any assistance from connected vehicle technology, just as what a regular driver does. In the *CVT-assisted* mode, drivers receive connected vehicle technology assistances such as driver advisories (e.g., downstream congestion) and safety warnings (e.g., emergency brake), but these drivers still assume full control of their vehicles. The *CVT-automated* mode means that a vehicle is operated by CVT-enabled automatic driving features; however, the driver may break the loop and take over at any time as the need arises. In relation to these modes, existing studies emphasized the CVT-automated mode since ACC, AACC, and CACC can be considered as special cases of this mode. This paper broadens the perspective by also considering the effect due to CVT-enabled assistance to drivers (such as driver advisories and warnings). Moreover, this research takes a probabilistic approach and analytically relates the capacity benefit to the attributes of these driving modes and their varying market penetration rates. It is noted that CVT may result in increased throughput due to reduced accidents and suppressed congestion, and benefit of this nature is typically scenario dependent. As a generic approach, this paper explores the upper bound of such benefit, that is, the increased capacity, given that accidents and congestion have been prevented.

3. Incorporating Connected Vehicle Technology Effects

3.1. Assumptions and Simplifications. Connected vehicle technology can bring about many fundamental changes to transportation systems such as ubiquitous situational awareness, more efficient system control, more advanced safety features and. However, one thing remains the same: drivers will still have full control even though it may be delegated to CVT-enabled systems. Hence, it is reasonable to begin with driver modeling in order to predict the operations of CVT-enabled transportation systems. Among others, the major effects of connected vehicle technology on drivers are changes in the way that information is acquired, processed, and applied. For example, on-board radar can tell the subject driver exactly how far the leading and/or trailing vehicle is and how fast the gap closes, and wireless communication can warn the subject driver of an abrupt braking by the leading vehicle or the approaching of a fast vehicle behind. Given the mix of CVT-enabled and regular vehicles in the traffic, it is likely that rear ends might be resulted due to sudden and unexpected "automated" braking. Hopefully, CVT is able to monitor such hazard and warn the subject driver in advance. On-board computer can synthesize these sources of information and present the subject driver with driver advisories which allow CVT-assisted drivers to have a better understanding of their local and global contexts than drivers without such assistance. As such, assisted drivers may need less time to look out for information (e.g., accident ahead) and could plan accordingly in advance. Thus, they could focus more on understanding the information (e.g., expect emergency brake) and make control decisions. In addition, if a vehicle is so equipped, the information can also be processed before the result is delivered (e.g., a warning to slow down). These assistances may significantly reduce drivers' perception time, and they only need to concentrate on executing decisions which is related to reaction time. Moreover, the reaction time needed to execute decisions can be further shortened if a vehicle is running in the CVT-automated mode.

Therefore, central to driver modeling in CVT-enabled transportation systems is the modeling of driver perception-reaction time. This modeling strategy is further supported by the following two considerations. First, the perception-reaction time is a very, if not the most, significant parameter governing drivers' car-following behavior which directly affects traffic density and highway capacity. Such a parameter is very sensitive to stimuli from drivers' local context (such as in-vehicle assistance systems). This is also evident in various microscopic traffic models, in particular, the Gipps-type model [1] which follows a "safe-distance" argument. Other aspects of driving, such as vehicle handling, are intrinsic characteristics of drivers and less influenced by external information brought by connected vehicle technology. Secondly, connected vehicle technology provides real-time information to drivers. Though field experiments have yet to be conducted to verify this postulation, evidence in psychology literature such as in [10] indicates that perception-reaction time strongly depends on the type and intensity of stimulus. Since CVT-enabled systems constitute a new type of stimulus with high intensity, such systems would shorten drivers' perception-reaction time, according to the literature.

Hence, it seems plausible to attribute the effects of connected vehicle technology to the distribution of drivers' perception-reaction time. For example, in the non-CVT mode, a driver typically needs to go through the full perception-reaction process and thus may necessitate a relatively long perception-reaction time (perhaps, a few seconds) on average. In addition, drivers without any assistance have less situational awareness which results in more uncertainty in their responses. This may give rise to a larger variance in their perception-reaction time. In the CVT-automated mode, the perception process is taken care of by connected vehicle technology, and the reaction process is handled by the automatic driving system. Thus, the resulting perception-reaction time can be minimal. Also, human drivers are not involved in the driving loop; therefore, the variance of perception-reaction time may be close to zero. In the CVT assisted mode, a wide range of possibilities may occur to the distribution. On one hand, it seems intuitive that CVT assistances such as advisories and warnings can greatly reduce drivers' perception time. On the other hand, such a new service may demand more attention to understand and familiarize and thus require a longer perception time, which is particularly true during confidence-building process. Before experimental data become available, the above discussion on perception-reaction time and their distributions remain open to debate. Nevertheless, it is reasonable to assume that the perception-reaction time of non-CVT, CVT-assisted, and CVT-automated drivers follow different distributions. Figure 2 is provided only for illustration purpose. Note that no assumption is made here regarding the actual distributions nor their relation. In particular, one should not take the figure as an implication that CVT-assisted and CVT-automated drivers have shorter perception-reaction times than non-CVT drivers. This treatment keeps the subsequent formulation generic and flexible for analysts to customize their models. For example, analysts can plug in suitable perception-reaction time distributions based on their own understanding or experiments in the field or on driving simulators. Remarkably, systems such INRETS MSIS [11], which integrated a driving simulator and a behavioral traffic simulation, can be a reasonable surrogate to field experiments in order to characterize the impacts of connected vehicle technology on drivers before its large-scale deployment. It is important to clarify at the beginning that the purpose of this paper is to formulate a methodology in generic terms to estimate the capacity benefit attainable from wide deployment of CVT, so that researchers, planners, and decision makers have a tool that can be tailored to their specific applications. It is recognized that the distributions of perception-reaction time under different driving modes are important inputs to the methodology. However, it is not the main focus of this paper to quantify these distributions which will be kept generic in subsequent discussion.

In addition, it is assumed that CVT-automated and CVT-assisted modes are always able to reap the benefits of VANET, that is, such vehicles are always assumed to be in a vehicular

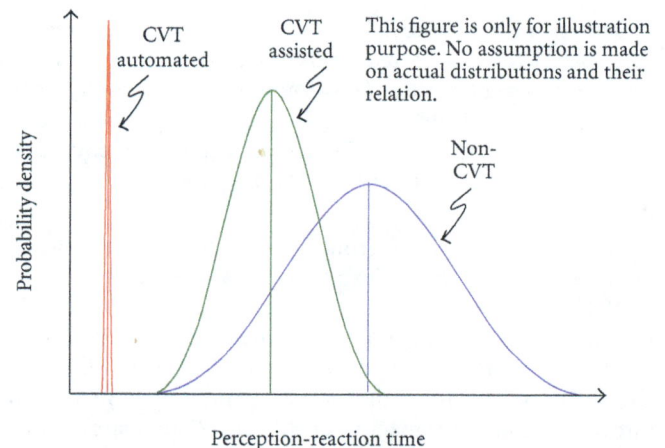

FIGURE 2: Perception-reaction time under different driving modes.

ad hoc network. It is recognized that such an assumption is not very true, especially under low CVT market penetration rates. Fortunately, this assumption is acceptable considering the following. First, it tends to overestimate highway capacity when there are not many CVT-equipped vehicles. Though not desirable, such an estimate does provide an upper bound of the capacity gained by connected vehicle technology. Second and perhaps more importantly, the validity of such an assumption increases when the deployment of connected vehicle technology is relatively significant, a scenario at which connected vehicle technology aims and under which connected vehicle technology makes the most sense. In order to fully account for this limitation, one must consider the dynamics of and interdependence between vehicular ad hoc networks and vehicular traffic. If this complication was to be taken into account, an analytical approach would no longer be adequate. Therefore, the goal of this research is to conduct a preliminary estimation of capacity benefit. Considering that field data are rare and the actual effects of connected vehicle technology are still subject to discussion, an easily understood and tractable approach seems more desirable to fulfill the purpose.

It is further assumed that CVT market penetration rate (i.e., the percent of total vehicles operating in each of the three modes) is known. With the above setup, it is straightforward to derive a car-following model with perception-reaction time as a parameter. Compared with the original [1] model, the new model rectifies the perception-reaction time which considers CVT-enabled driving modes and incorporates their market penetration rates. This model is then used to derive an equilibrium flow-density relationship, from which maximum flow rate (i.e., the capacity) can be derived. Considering the random nature of the perception-reaction time, a probabilistic analysis is performed to investigate the properties of the capacity, and a Monte Carlo simulation is used to verify the results obtained above.

3.2. Model Formulation. Perception-reaction time plays a significant role in car following and traffic operations. Gipps' model [1] stipulates that, at any moment, a vehicle should

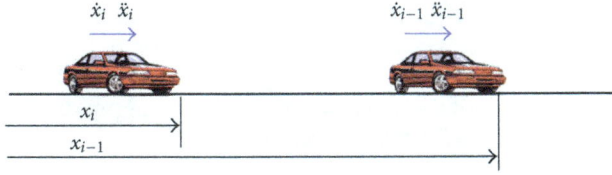

FIGURE 3: Vehicles in car following.

leave enough room in front of it in order to be able to stop safely behind its leading vehicle in the event that the leading vehicle applies emergency brake. Figure 3 shows two vehicles in car following. The leading vehicle with ID $i - 1$ is at position x_{i-1} with speed $\dot{x}_{i-1}$ and acceleration $\ddot{x}_{i-1}$. The subject vehicle i is at position x_i with speed $\dot{x}_i$ and acceleration $\ddot{x}_i$. The minimum safe distance should allow the subject vehicle to stop behind the leading vehicle after a perception-reaction time τ_i and a deceleration process at a comfortable level $b_i = \ddot{x}_i < 0$. To be specific, the stopped position of the leading vehicle if it applies an emergency brake at time t is

$$x_{i-1}^* = x_{i-1}(t) - \frac{\dot{x}_{i-1}^2(t)}{2B_{i-1}}, \tag{1}$$

where B_{i-1} is the maximum deceleration ($\ddot{x}_{i-1}^{\max} < 0$) applied by the leading vehicle. The stopped position of the subject vehicle is

$$x_i^* = x_i(t) + \dot{x}_i\tau_i - \frac{\dot{x}_i^2(t)}{2b_i}. \tag{2}$$

To ensure safety, the following relationship must hold:

$$x_{i-1}^* - l_{i-1} \geq x_i^*, \tag{3}$$

where l_{i-1} is the length of the leading vehicle. Putting the above together, the safe distance for the subject vehicle can be expressed as

$$D_i(t) = x_{i-1}(t) - l_{i-1} - x_i(t) \geq \dot{x}_i\tau_i - \frac{\dot{x}_i^2(t)}{2b_i} + \frac{\dot{x}_{i-1}^2(t)}{2B_{i-1}}, \tag{4}$$

which corresponds to a spacing of

$$S_i(t) = x_{i-1}(t) - x_i(t) \geq l_{i-1} + \dot{x}_i\tau_i - \frac{\dot{x}_i^2(t)}{2b_i} + \frac{\dot{x}_{i-1}^2(t)}{2B_{i-1}}. \tag{5}$$

Under equilibrium conditions, vehicles tend to behave uniformly and thus lose their identities. After suppressing time t, the spacing becomes

$$S \geq l + \dot{x}\tau - \frac{\dot{x}^2}{2b} + \frac{\dot{x}^2}{2B} = \left(\frac{1}{2B} - \frac{1}{2b}\right)\dot{x}^2 + \tau\dot{x} + l. \tag{6}$$

Following Gipps' argument, an additional delay θ to τ offers extra protection for the subject vehicle, so the above inequality is turned into

$$S = l + \dot{x}(\tau + \theta) - \frac{\dot{x}^2}{2b} + \frac{\dot{x}^2}{2B} = \left(\frac{1}{2B} - \frac{1}{2b}\right)\dot{x}^2 + (\tau + \theta)\dot{x} + l. \tag{7}$$

Thus, the safe car-following distance, or equivalently the safe spacing, is explicitly expressed as a function of speed $v \equiv \dot{x}$ (under equilibrium conditions, it is also the traffic speed) with parameters τ, θ, B, b, and l. Among all the parameters, τ and θ characterize the behavior of drivers and are independent of speed v and spacing $S \cdot B$, b, and l are vehicle properties and can be assumed as constants. Since we also know that density k is related to spacing S,

$$k = \frac{1}{S}, \tag{8}$$

the flow q is obtained by substituting k and v into the fundamental relation

$$q = kv = \frac{v}{Gv^2 + \tau'v + l}, \tag{9}$$

where $\tau' = \tau + \theta$ and $G = (1/2B) - (1/2b)$. In this relation, v can be viewed as the primary input. v and τ' are independent variables. The maximum attainable q is of interest. To find the maximum q (denoted q_m), we solve the equation

$$\left.\frac{dq}{dv}\right|_{v_m} = -\left.\frac{G - (l/v^2)}{(Gv + \tau' + (l/v))^2}\right|_{v_m} = 0. \tag{10}$$

We get the root

$$v_m = \sqrt{\frac{l}{G}}, \tag{11}$$

and correspondingly,

$$q_m = \frac{1}{2\sqrt{Gl} + \tau'}. \tag{12}$$

To verify that q_m is indeed a maximum as v varies, one may simply check the second derivative of q at v_m. It turns out that this is true.

4. Probabilistic Analysis

4.1. The Stopping Time Formulation with Random τ. Note that the above discussion does not incorporate the random nature of perception-reaction time τ nor its distributions under different driving modes. To begin with, the above equation allows the calculation of capacity q_m given perception-reaction time τ if it is uniform across the entire driver population. Such a case is simple but unrealistic. A step forward would be to assume uniform perception-reaction time under each driving mode. Combined with market penetration (analogous to the probability of each driving mode), these perception-reaction times can be used to estimate the average (i.e., mathematical expectation) capacity as the result of the entire driver population. This subsection deals with a more realistic scenario which assumes different distributions of perception-reaction time under different driving modes, and the estimation of capacity is to compute the mathematical expectation of q_m based on these underlying distributions and their market penetration rates.

Denote $f_{no}(t)$ the probability density of perception-reaction time of drivers under the non-CVT with mean τ_{no}

and variance $\text{Var}(\tau_{no})$. Similarly, the probability density of the CVT-assisted mode is $f_{as}(t)$ with mean τ_{as} and variance $\text{Var}(\tau_{as})$; the probability density of the CVT-automated mode is $f_{au}(t)$ with mean τ_{au} and variance $\text{Var}(\tau_{au})$. In addition, market penetration rates of road vehicles operating in non-CVT, CVT-assisted, and CVT-automated modes are denoted as p_{no}, p_{as}, and p_{au} respectively. They satisfy the following relationships: $0 \leq p_{no}, p_{as}, p_{au} \leq 1$, and $p_{no} + p_{as} + p_{au} = 1$. Therefore, the perception-reaction time of an individual driver i, τ_i, is a random variable which can be modeled by drawing first from the percent/probability of market penetration to determine which driving mode this driver uses and then from the distribution of perception-reaction time of that particular mode.

Henceforth, we will investigate the properties of q and q_m as τ takes on random values. Usually, the first-order second moment analysis (FOSM) is sufficient to fulfill this purpose. However, since FOSM is based on the Taylor expansion of functions, the accuracy of approximation relies heavily on the convergence rate of the Taylor series in the neighborhood of the expansion. For the higher-order moment, this is especially true. In this situation, it is unfortunate that the expression of q_m is ill posed to adopt the FOSM. This is because q_m, written in the form of $f(x) = 1/(a + bx)$, corresponds to a slowly converging expansion series when $|a + bx| \sim 0$, a result of comparable values of a and b.

Thus, we tackle the problem in a different way. In particular, we introduce the stopping time concept such that the expansion-based analysis like FOSM is avoided. The procedure is as follows. First, we redefine the flow as

$$q = kv = \frac{N}{L}v, \tag{13}$$

where v is the traffic speed, L is the length of a segment of highway in consideration, and N is the number of vehicles within the length. Flow q can be written as $N/(L/v)$ and interpreted as the number of vehicles occupying a certain length of road divided by the time they take to traverse the road. Under equilibrium conditions, this definition is equivalent to the original definition. Then we can adopt the concept of random walk. It is easy to see that N is actually the stopping time (stopping time, a standard concept in probability theory, can be roughly regarded as a "random time" whose value depends on current and historical values of a stochastic process). A rigorous definition is found in [12] where a random walk $\sum_{i=1}^{n} S_i$ with positive drift $E(S_i)$ has

$$N = \inf_n \left\{ n : \sum_{i=1}^{n} S_i > L \right\}, \tag{14}$$

where inf indicates the infinium of a set,

$$S_i = Gv^2 + \tau'v + l, \tag{15}$$

then

$$\mu_q \equiv E(q) = \frac{v}{L}E(N). \tag{16}$$

Moreover,

$$E(N) = \frac{L}{E(S_i)} = \frac{L}{Gv^2 + \mu_{\tau'}v + l}, \tag{17}$$

where the first equality is due to Wald's equation, with its form and derivation given in [12]. Application of this equation requires $E(S_i) < \infty$, which is obviously true from a realistic point of view. Thus, we obtain the approximation of expected capacity when speed is v,

$$\mu_q \equiv E(q) \sim \frac{v}{Gv^2 + \mu_{\tau'}v + l}, \tag{18}$$

where $\mu_{\tau'} = E(\tau')$. Plugging in the optimal speed $v = v_m = \sqrt{l/G}$, we obtain the maximum value of approximation of expected capacity,

$$\mu_{q,m} \equiv E(q_m) \sim \frac{1}{2\sqrt{Gl} + \mu_{\tau'}}. \tag{19}$$

To obtain the variance of q and q_m, we need to utilize a formula regarding the variance of stopping time given in [13]. In the current scenario, it is

$$\text{Var}(N) = \mu^{-3}\sigma^2 L + \mu^{-2}K + o(1), \tag{20}$$

where K is a constant independent of L,

$$\mu = E(S_i) = Gv^2 + \mu_{\tau'}v + l,$$
$$\sigma^2 = \text{Var}(S_i) = \sigma_{\tau'}^2 v^2. \tag{21}$$

By substituting them into the definition of q, by letting L be large enough, and by only keeping the dominating term, we get the variance of flow in the general case

$$\sigma_q^2 \equiv \text{Var}(q) \sim \frac{\sigma_{\tau'}^2 v^4}{(Gv^2 + \mu_{\tau'}v + l)^3 L}. \tag{22}$$

Plugging in the optimal speed $v = v_m = \sqrt{l/G}$, we obtain the approximate variance of the maximum flow, that is, capacity,

$$\begin{aligned}
\sigma_{q,m}^2 \equiv \text{Var}(q_m) &\sim \frac{\sigma_{\tau'}^2 l^2}{\left(2l + \mu_{\tau'}\sqrt{l/G}\right)^3 G^2 L} \\
&= \frac{\sigma_{\tau'}^2 Gl}{\left(2Gl + \mu_{\tau'}\sqrt{Gl}\right)^3}\frac{l}{L}.
\end{aligned} \tag{23}$$

We see that two quantities, $\sqrt{Gl}$ and l/L, together with characteristics of perception-reaction time τ' determine the variance. It is notable that the involved quantities are all easily measured, indicating the advantage of our estimate formula in terms of calibration.

4.2. Simulation Verification. To verify the creditability of the above approximation formulas, it is necessary to device an independent approach, preferably simulation since large-scale field experiments are infeasible. Considering that a traffic simulation package that is capable of capturing the effects of connected vehicle technology is not available, this study employs a Monte Carlo simulation which repeatedly samples a traffic system by drawing from underlying distributions.

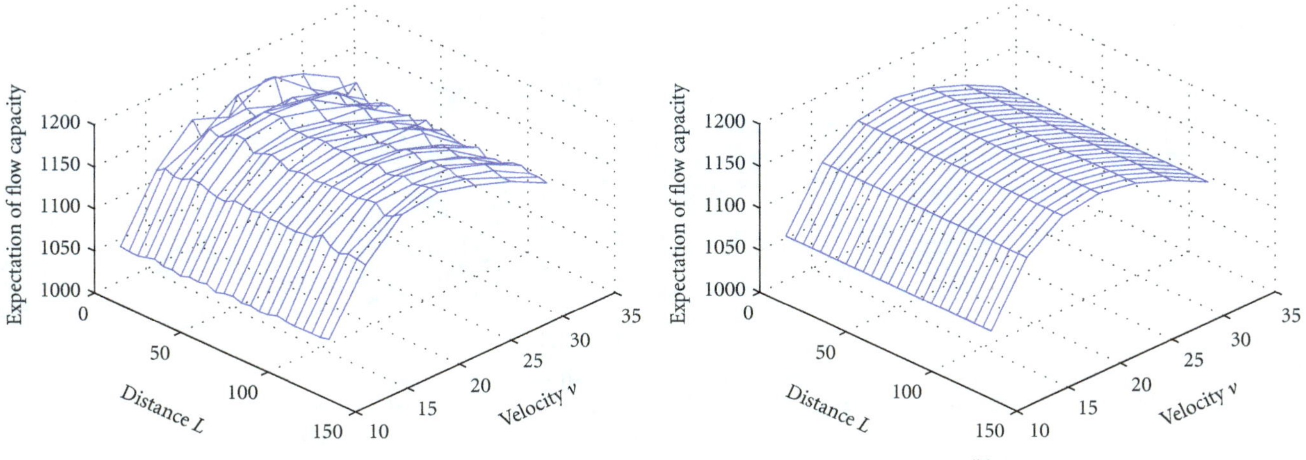

FIGURE 4: The comparison of simulated and approximate expectation of flow μ_q (in passenger cars per hour per lane or pcphpl) ((a) simulated value; (b) approximate value).

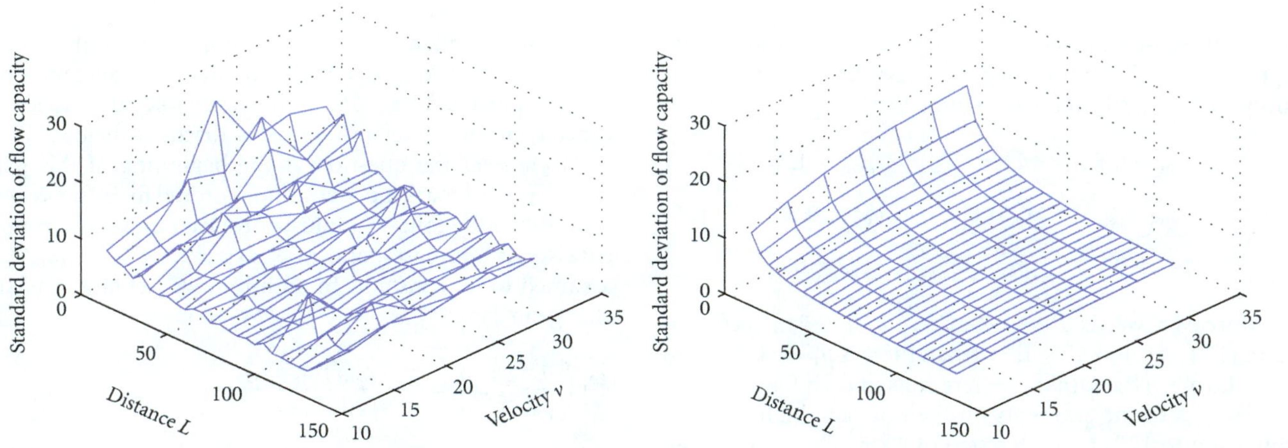

FIGURE 5: The comparison of simulated and approximate standard deviation of flow σ_q (in pcphpl) ((a) simulated value; (b) approximate value).

The number of trials depends on the precision required with one trial as follows:

(1) select road length in consideration, denoted L, and initialize $L_0 = 0$,

(2) randomly sample the perception-reaction time τ and calculate the cumulative length, which is defined as $L_{j+1} = L_j + S$,

(3) if $L_j > L$, denote $N = \max j$, calculate $q = vN/L$, and else return to (2).

Each trial in the simulation mimics the instantaneous vehicle spatial distribution on the road, such that the count of vehicles N at each moment is obtained. In the simulation, we assume that the random perception-reaction time has a density of the following form:

$$f_\tau(t) = p_{au} f_{au}(t) + p_{as} f_{as}(t) + p_{no} f_{no}(t), \tag{24}$$

where $\sum_i p_i = 1$ and $i \in \{au, as, no\}$. Here, $f_i(\cdot)$'s represents the density of the perception-reaction time distribution

of the ith group, and p_i's are interpreted as the market penetration of the corresponding groups. For the purpose of illustration, we consider an ideal and simplified case. We assume that the f_i's are the density of uniform random variables. The function f_i is of the form

$$f_i(t) = \frac{1}{u_i - l_i} I(t \in [l_i, u_i]). \tag{25}$$

It is easy to see that a random variable with the above density f_i has an expectation $\mu_i = (u_i + l_i)/2$ and a variance $\sigma_i = (u_i - l_i)^2/12$. We then have

$$\mu_\tau = \sum_{i=1}^{3} p_i \mu_i,$$

$$\sigma_\tau = \sqrt{\sum_{i=1}^{3} p_i \mu_{i,2} - \mu_T^2} = \sqrt{\sum_{i=1}^{3} p_i (\sigma_i^2 + \mu_i^2) - \mu_T^2}. \tag{26}$$

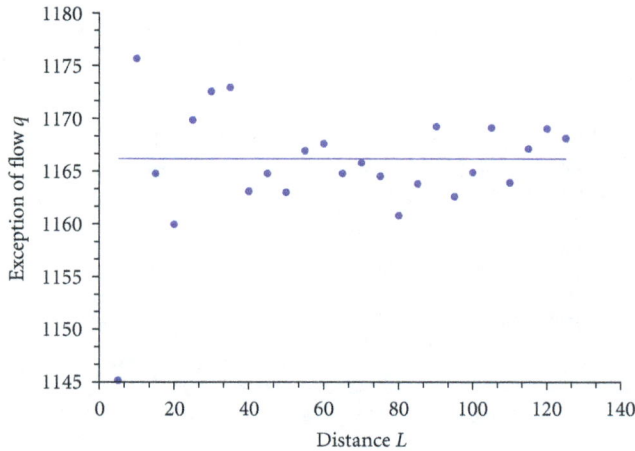

FIGURE 6: The comparison of simulated and approximate expectation of maximum flow $\mu_{q,m}$ (distance L, m; flow q, pcphpl; dots: simulated values; solid line: approximate values).

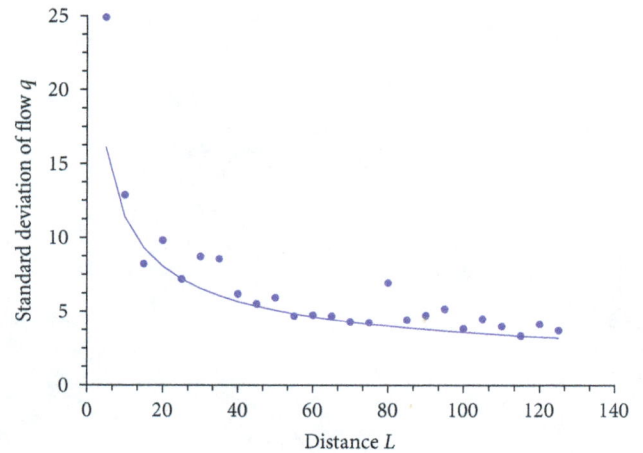

FIGURE 7: The comparison of simulated and approximate standard deviation of maximum flow $\sigma_{q,m}$ (distance L, m; flow q, pcphpl; dots: simulated values; solid line: approximate values).

Let the first, second, and third terms in expression of f_T represent the CVT-automated, CVT-assisted, and non-CVT groups, respectively, then we assume that

$$\mu_{au} = 0.5, \quad \mu_{as} = 1.0, \quad \mu_{no} = 1.5,$$
$$\sigma_{au} = 0, \quad \sigma_{as} = 0.2, \quad \sigma_{no} = 0.5, \quad (27)$$
$$p_{au} = 0.2, \quad p_{as} = 0.5, \quad p_{no} = 0.3.$$

Moreover, we fix $B = -4\,\text{m/sec}^2, b = -2\,\text{m/sec}^2, l = 4.5\,\text{m}$ (15 feet), and $\tau' = 1.5\,\tau$ (then there is $\mu_{\tau'} = 1.5\,\mu_\tau$ and $\sigma_{\tau'} = 1.5\,\sigma_\tau$). The number of iterations in each loop is 1000. The lengths of the segments on the one-lane highway vary from 5 km to 125 km, with a step of 5 km. The traffic speed varies from $v_{opt} - 10.5$ km/hr to $v_{opt} + 10.5$ km/hr, with a step of 3.5 km/hr. The v_m is the optimal speed defined above.

Observation of the simulation results and the conclusions drawn are as follows. First, in the case under consideration, the approximate expectation μ_q and standard deviation σ_q of the flow are close to the simulation results, as shown in Figures 4 and 5. Second, in particular, the comparison of $\mu_{q,m}$ and $\sigma_{q,m}$ with the simulation is shown in Figures 6 and 7. The relative error of $\mu_{q,m}$ is very small, around 1%. For the standard deviation $\sigma_{q,m}$, we see the fit of approximation to simulation is also near perfect, especially as the distance L gets larger. To summarize, the simulation of the case when τ takes a specific distribution numerically justifies the approximations obtained by probabilistic analysis, and it intuitively illustrates the quality of these approximations.

5. An Illustrative Example

The simulation study in the above section is able to provide an estimate of the capacity gain in absolute terms. Such a result, however, is lower than the typical capacity under ideal conditions, that is, 2400 pcphpl for a basic freeway section. This is due to the conservative nature of the Gipps model, which is less capable of capturing the close-following

behavior in reality. Developing a more realistic model may resolve the problem, but the mathematical tractability may be lost as well. Therefore, it is reasonable to describe the capacity benefit in relative terms, as discussed below.

To answer the question at the beginning of this paper (i.e., degree of market penetration required for effectiveness), we provide the following illustrative example. This example consists of four cases, and in each case, the ratio p_{au}/p_{as} is assumed to be constant. In addition, we define the relative change in capacity as

$$r\left(\frac{p_{au}}{p_{as}}, p_{no}\right) = \frac{q_m(p_{au}, p_{as}, p_{no})}{q_m(0,0,1)}$$
$$= \frac{q_m((1 - p_{no})((p_{au}/p_{as})/(p_{au}/p_{as} + 1)))}{q_m(0,0,1)},$$
$$\frac{(1 - p_{no})(1/(p_{au}/p_{as} + 1)), p_{no})}{q_m(0,0,1)}, \quad (28)$$

where $q_m(\cdot,\cdot,\cdot)$ is the capacity corresponding to market penetration (p_{au}, p_{as}, p_{no}), and the second equality is $p_{au} + p_{as} + p_{no} = 1$. This formula could be interpreted as the ratio of increased capacity over the original capacity (i.e., $p_{no} = 100\%$). By employing this definition, we will hopefully overcome the lower estimate by the Gipps' model. We obtain the values of r in four cases, that is, when $p_{au}/p_{as} = 0.1, 1, 10, 100$. The results are as shown in Figure 8. It is found that the increase of capacity ranges between 20% and 50% when connected vehicle technology is fully deployed (i.e., $p_{no} = 0$), with the former case corresponding to $p_{au}/p_{as} = 0.1$ and the latter case $p_{au}/p_{as} = 100$. The result seems to suggest that the change of the p_{au}/p_{as} ratio from 1 to 10 has much stronger effect than that from 10 to 100. A plausible interpretation is that, with high percentage of CVT-assisted vehicles in the traffic, drivers have more chances to negotiate and hence more room of improvement. As the traffic is dominated by CVT-automated vehicles, they

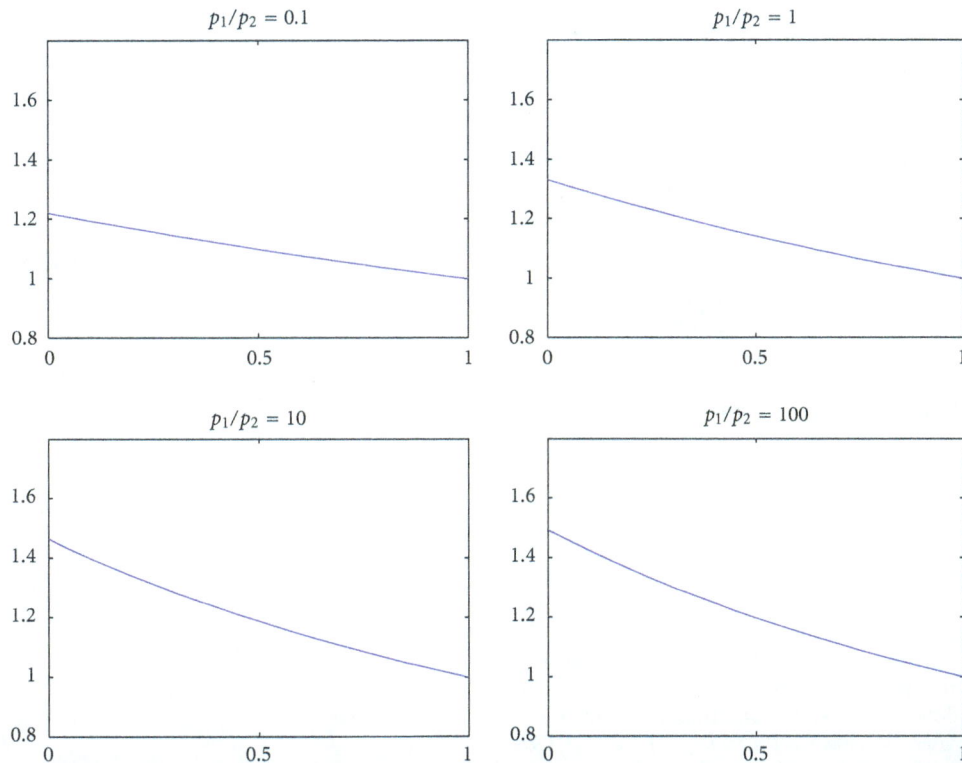

FIGURE 8: The relative benefits of CVT as function of market penetration of CVT in different cases (x-axis: market penetration of nonassisted vehicles; y-axis: ratio of increased capacity to the original capacity).

move like a train whose already optimized performance allows little room for improvement. Note that the above example is only a rough estimate under some assumptions and simplifications. Nevertheless, the example does indicate that the benefit from employing connected vehicle technology could be quite significant even when the market penetration of CVT-automated vehicles is small (given full CVT deployment). As more accurate information regarding the involved parameters becomes available, the estimate can be fine-tuned, and more accurate results are expected. The outcomes can be used to help make the decision on connected vehicle technology deployment in future.

6. Concluding Remarks

The purpose of this study was to provide a methodology to estimate the capacity benefit obtained from CVT deployment. To fulfill this purpose, the classical Gipps' model was modified to incorporate the effects of CVT, and a probabilistic approach was presented to analyze the impact of such effects on highway capacity. In particular, approximate formulas of expectation and variance of the capacity were derived in a random setting. A simulation study was conducted to numerically verify the creditability of these formulas. Also included was an illustrative example that applied the proposed methodology. Under some assumptions and simplifications, the example showed that the increase of capacity ranges between 20% and 50% when CVT is fully deployed. Note that the lower bound, which corresponds

to a relatively small market penetration of CVT-automated vehicles, may still yield quite significant capacity gain. Findings of this study, in general, agree with previous studies [6–8] on related subject, but this paper provides more in-depth information such as the curve of capacity gain as market penetration changes and the methodology to estimate such curve.

Though the proposed methodology was motivated by capturing the effects of CVT (i.e., its impacts on three driving modes), such a methodology is readily applicable to a traffic system involving multiple classes of drivers (e.g., teenagers, middle-aged, and senior drivers whose perception-reaction times may follow different distributions). Meanwhile, the study may be further extended to more complicated and realistic scenarios (e.g., nonequilibrium flow and nonhomogeneous types of vehicles) where more involved simulation is expected before field experiments in a large-scale testbed become feasible. Since the distributions of perception-reaction time under different driving modes are critical to estimation results, further studies particularly those of human factors are necessary to quantify these distributions.

Acknowledgments

This research is partly supported by U.S. Department of Transportation University Transportation Center Program. The authors would like to thank the anonymous reviewers for their valuable comments and suggestions to improve the quality of the paper.

References

[1] P. G. Gipps, "A behavioural car-following model for computer simulation," *Transportation Research Part B*, vol. 15, no. 2, pp. 105–111, 1981.

[2] J. VanderWerf, S. Shladover, N. Kourjanskaia, M. Miller, and H. Krishnan, "Modeling effects of driver control assistance systems on traffic," *Transportation Research Record*, no. 1748, pp. 167–174, 2001.

[3] B. Rao and P. Varaiya, "Flow benefits of autonomous intelligent cruise control in mixed manual and automated traffic," *Transportation Research Record*, vol. 1408, pp. 36–43, 1993.

[4] B. Rao, P. Varaiya, and F. Eskafi, "Investigations into achievable capacities and stream stability with coordinated intelligent vehicles," *Transportation Research Record*, vol. 1408, pp. 27–35, 1993.

[5] S. Darbha and K. R. Rajagopal, "Intelligent cruise control systems and traffic flow stability," *Transportation Research Part C*, vol. 7, no. 6, pp. 329–352, 1999.

[6] T. H. Chang and I. S. Lai, "Analysis of characteristics of mixed traffic flow of autopilot vehicles and manual vehicles," *Transportation Research Part C*, vol. 5, no. 6, pp. 333–348, 1997.

[7] J. VanderWerf, S. Shladover, M. Miller, and N. Kourjanskaia, "Evaluation of the effects of adaptive cruise control systems on highway traffic flow capacity and implications for deployment of future automated systems," in *Proceedings of the 81st TRB Annual Meeting*, 2001, Pre-Print CD-ROM.

[8] B. van Arem, C. J. G. van Driel, and R. Visser, "The impact of cooperative adaptive cruise control on traffic-flow characteristics," *IEEE Transactions on Intelligent Transportation Systems*, vol. 7, no. 4, pp. 429–436, 2006.

[9] K. Kulmala, P. Levikangas, N. Sihvola et al., "Co-operative systems deployment impact assessment (CODIA) deliverable 5: final study report," Tech. Rep., VTT Technical Research Centre of Finland, 2008.

[10] R. D. Luce, *Response Times: Their Role in Inferring Elementary Mental Organization*, Oxford University Press, New York, NY, USA, 1986.

[11] S. Espi and J.-M. Auberlet, "ARCHISIM: a behavioural multi-actors traffic simulation model for the study of a traffic system including ITS aspects," *International Journal of ITS Research*, vol. 5, no. 1, pp. 7–16, 2007.

[12] R. Durrett, *Probability: Theory and Examples*, Duxbury Press, 1996.

[13] T. Lai and D. Siegmund, "A nonlinear renewal theory with applications to sequential analysis. II," *Annuals of Statistics*, vol. 7, no. 1, pp. 60–76, 1979.

13

A Real-Time Embedded Blind Spot Safety Assistance System

Bing-Fei Wu, Chih-Chung Kao, Ying-Feng Li, and Min-Yu Tsai

Institute of Electrical and Control Engineering, National Chiao Tung University, Hsinchu 300, Taiwan

Correspondence should be addressed to Chih-Chung Kao, kevinkao@cssp.cn.nctu.edu.tw

Academic Editor: David Fernández Llorca

This paper presents an effective vehicle and motorcycle detection system in the blind spot area in the daytime and nighttime scenes. The proposed method identifies vehicle and motorcycle by detecting the shadow and the edge features in the daytime, and the vehicle and motorcycle could be detected through locating the headlights at nighttime. First, shadow segmentation is performed to briefly locate the position of the vehicle. Then, the vertical and horizontal edges are utilized to verify the existence of the vehicle. After that, tracking procedure is operated to track the same vehicle in the consecutive frames. Finally, the driving behavior is judged by the trajectory. Second, the lamps in the nighttime are extracted based on automatic histogram thresholding, and are verified by spatial and temporal features to against the reflection of the pavement. The proposed real-time vision-based Blind Spot Safety-Assistance System has implemented and evaluated on a TI DM6437 platform to perform the vehicle detection on real highway, expressways, and urban roadways, and works well on sunny, cloudy, and rainy conditions in daytime and night time. Experimental results demonstrate that the proposed vehicle detection approach is effective and feasible in various environments.

1. Introduction

In recent years, the driving safety has become the most important issue in Taiwan. Numbers of car accidents and casualties increase year by year. According to the accident data by Taiwan Area National Freeway Bureau, the main reason of occurring accidents is human negligence. Therefore, collision-forewarning technologies get great attention increasingly, and several kinds of driving safety assisted products are promoted, including lane departure warning systems (LDWSs), blind spot information systems (BLISs), and so forth. These products could provide more information about the vehicle surroundings with the driver, so that the driver could make the correct decision when driving on the road. BLIS could monitor whether the vehicles appear in the side of host car or not and inform the driver when the driver intends to change lanes. Radar is another solution for BLIS. However, the cost is much higher than the camera. Consequently, vision-based blind spot detection becomes popular in this field.

There are many vision-based obstacle detection systems proposed in the literature. Most of them focus on detecting lane marking, the obstacles in front of the host car for lane departure warning [1–3], and collision avoidance applications [4–7]. Lane detection is exploited for driving safety assistant in early periods. Moreover, a complete survey was addressed in [1]. Besides, the front obstacle detection was discussed enthusiastically in the past decade. Online boosting algorithm is proposed to detect the vehicle in front of host car [2]. The online learning algorithm can conquer the online tuning problem for a practical system. O'Malley et al. [3] presented a rear-lamp vehicle detection and tracking for night condition. The rear-lamp pairs are used to recognize the front vehicle and track lamp pairs by Kalman filter. Liu and Fujimura [4] proposed a pedestrian detection system by stereo night vision. Human became hot spot in night vision and would be tracked by blob matching. Labayrade et al. [5] integrated 3D cameras and laser scanner to detect multiobstacles in front of the vehicle. The width, height, and depth of the obstacle were estimated by stereo vision and the precise obstacle position can be provided by laser scanner. This cooperative fusion approach achieved an accurate and robust detection.

Although most attention was fascinated with the front-view obstacle and lane detection, some researchers attended to solve the problems in blind spot obstacle detection.

Wong and Qidwai [6] installed six ultrasonic sensors and three image sensors in the car. They applied fuzzy inference in algorithm to forewarn the driver and reduce the car accident possibility. Achler and Trivedi [7] mounted an omnidirectional camera to monitor the area surrounding by the host vehicle. Hence, the obstacles in both sides of blind spot could be detected at the same time. The wheel was filtered with Type Filter and then the system could determine whether the vehicle exists or not. Ruder et al. [8] designed a lane change assistance system with far range radar, side radar, and stereo vision. The sensor fusion and Kalman filter were used to track the vehicle stably. Díaz et al. [9] applied optical flow algorithm to segment the vehicle in the blind spot, and several scale templates were established for tracking. Batavia et al. [10] also monitored the vehicle in rear image with optical flow algorithm and edge feature. Stuckman et al. [11] used infrared sensor to get the information of the blind spot area. This method was implemented on digital signal processor (DSP) successfully. Adaptive template matching (AdTM) algorithm [12] was proposed to detect the vehicle which was entering the blind spot area, and the algorithm defined levels to determine the behavior of the tracked vehicle. If this vehicle was approaching, the level would increase, otherwise the level would decrease. Multi-line-CCD was equipped by Yoshioka et al. [13] to monitor the blind spot area. This sensor could obtain the height of a pixel in the image because of the parallax between two lenses. Thus, this method could obtain the height of the vehicle. Techmer [14] utilized inverse perspective mapping (IPM) and edge extraction algorithm to match the pattern and to determine whether a vehicle exists in the blind spot or not. Furukawa et al. [15] applied three cameras to monitor front area, left behind area and right behind area. Horizontal segment by edge was used and the template matching was operated by orientation code matching, which is one of the robust matching techniques. Most importantly, these algorithms required less resources for operation and were implemented in one embedded system. Jeong et al. [16] separated the input image into several segmentations and determined these segmentations belonging to the foreground or background by gray level. Afterwards, scale invariant feature transform (SIFT) was implemented to generate robust features to check whether the vehicle exists or not. Finally, modified mean-shift was used to track the detected vehicle. C. T. Chen and Y. S. Chen [17] estimated the image entropy of the road scene in the near lane. The obstacle could be detected and located by analyzing the lane information. Although they could track the obstacles in real-time, they only judged whether the tracked vehicle was approaching or not by considering the location in the previous frame and current frame. Consequently, the false alarm would be easily triggered. Four prespecified regions were defined to identify the dangerous level in [18]. Sobel edge was extracted and morphological operation was applied to generate clearer edge image. However, if only considering the edge information, the system would easily alarm falsely by shadow and safety island.

The blind spot detection (BSD) for vehicles and motorcycles in daytime and nighttime is proposed in this paper.

Moreover, one of the most important issues in this field is the execution efficiency of the system. If the efficiency is not high enough, this system could not achieve it to real-time results, and it could not remind the driver immediately. There would be no use value in the system which is described above. There are a lot of methods which had been proposed in recent years to prevent collision in blind spot zone, but most of those methods are implemented on PC, which are not suitable as an automobile electronics. Although there are some methods which were implemented on DSP platform, low frame rate and robustness became the serious problems to them. In this paper, edge, shadow, and lamps in spatial domain are applied to increase execution efficiency. Therefore, the performance of vehicle detection in this system is the main topic here, especially overcoming the complex problems in harsh environments, such as driving on urban roads. Using the general features in spatial domain and keeping high performance has been implemented through the method introduced in this paper. Developing on DSP platform, the frame rate of this system could still achieve 59 fps at most. For CIF images, this efficiency is high enough that the system can provide real-time information with the drivers, so that drivers can make the most correct decisions in time. The system has high frame rate on TI DM6437 platform, and through the long verification with on-road field tests on highways, expressways, and urban roadways, and works well on sunny, cloudy, and rainy conditions in the daytime and nighttime. This shows that the system has very well robustness, so that it can work anytime at everywhere and provide warning function with the drivers. The warning functions which could be a buzzer or LED light would be triggered to alarm to the driver.

Section 2 briefly introduces the working flow of the presented method in this paper. The algorithms of vehicle detection in the daytime and nighttime are introduced in Sections 3 and 4, respectively. The experiment results and comparisons would be shown in Section 5. Finally, the conclusions would be addressed in Section 6.

2. System Overview

Since the features for vehicle detection in the daytime are obviously different from the features in the nighttime, the utilized features and verification procedures would be distinct. Moreover, considering the practical application, BSD should work day and night. Because it is very difficult to distinguish between the daytime and the nighttime, both of vehicle detection algorithms in the daytime and in the nighttime should be processed in each frame. Daytime and nighttime algorithms which detect and track different features with the same workflow in Figure 1 have been implemented into our system and make this system more practical and robustness. The algorithm for the nighttime follows the algorithm for the daytime in our system, so that there is no need to determine what time it is now.

There are three main detecting modes in the algorithm of vehicle detection. Those are the full searching mode, the tracking mode, and the partial searching mode, respectively.

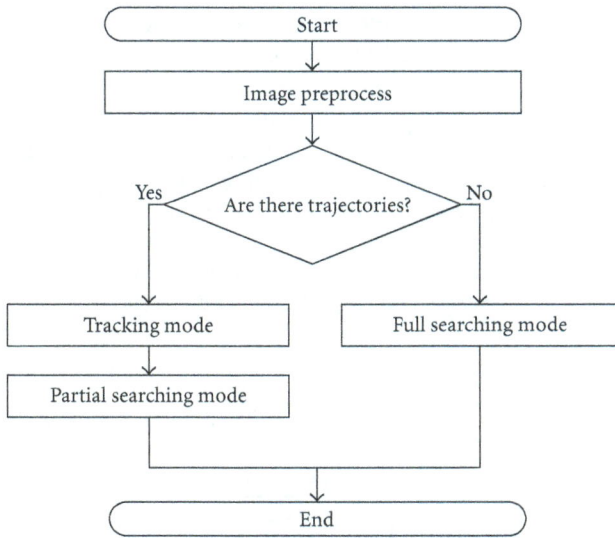

FIGURE 1: The flow chart of the proposed system.

Image preprocessing is performed to extract the edge, shadow, lamp features for the vehicle detection. First, there is no vehicle which is detected and tracked as a trajectory. Therefore, the system would search the possible vehicle candidates in the whole region of the interest (ROI) of the image in full search mode. If the vehicle is detected and tracked successfully in the successive video frames, the vehicle trajectory is generated, and the system would process the tracking mode in the next frame. Because of the data saved from the full searching mode, we already know where the vehicles exist; thus, there is no need to search the whole ROI again. We can only search the region where the vehicles exist and determine their behavior in the tracking mode. According to the locations of vehicles which had been saved in the last frame, the searching region would be set adaptively. After detecting, candidate matching and the vehicle behavior would be judged. In the end, the system triggers the warning signal to remind the driver. Partial searching mode always follows the tracking mode to search if there is any other vehicle or motorcycle in ROI. However, the partial searching mode would not search the zone in where there is already a vehicle appearing and has been detected by the tracking mode.

3. Vehicle Detection Algorithm in the Daytime

There are seven topics presented in the following subsections. First, we introduce the definition of our ROI and give comparison with ISO specification in Subsection 3.1. The follows in Subsections 3.2 and 3.3 are edge detection and the shadow searching. After detecting a vehicle, we search the correct boundary of a vehicle in the Subsection 3.4. Then, the candidates are verified in Subsection 3.5. Matching the vehicle appearing in the consecutive frames to generate the vehicle trajectory would be discussed in Subsection 3.6. Finally, the vehicle behavior judgment is performed in the Subsection 3.7.

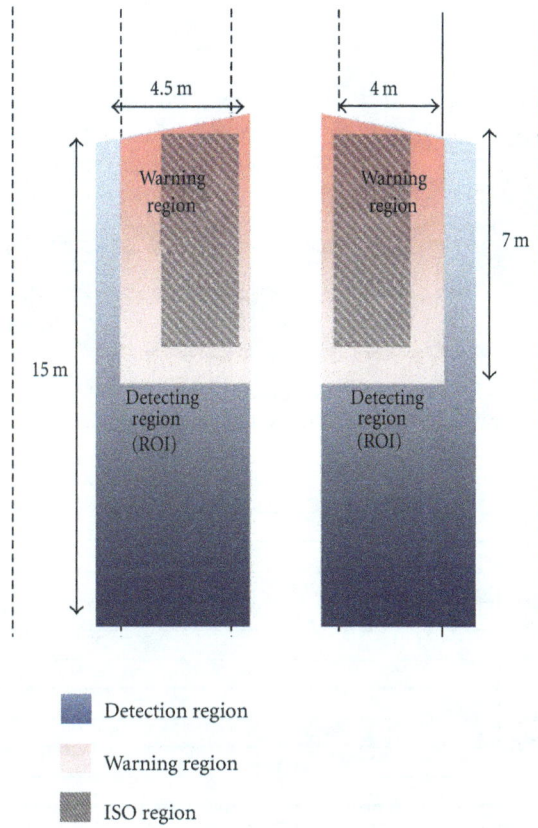

FIGURE 2: Definitions of 3 regions.

3.1. Define the ROI in the Daytime.
We have to define ROI in this system clearly at first. Referring to the definition of lane change decision aid systems (LCDASs) in ISO document, we obtain the definition of blind spot area and delimit ROI in our system. As shown in Figure 2, the ISO definition of blind spot area is 3 meters to the side of the car and 3 meters behind the car. The ROI in our system is larger than the ISO definition and completely covers it.

The specification of detecting region in this system is 4.5 meters to the side of the car and 15 meters behind the car. The specification of the warning region is 4 meters to side of the car and 7 meters behind of the car. When a vehicle is approaching to the warning region, the system would send a warning signal to the driver. The detecting region and the warning region are drawn in blue and red in Figure 3, respectively.

3.2. Image Preprocessing for Vehicle Detection in the Daytime.
Shadow and edge features are chosen for vehicle detection in the proposed system in the daytime. Shadow under the car could illustrate the location of the vehicle. Therefore, extracting the shadow region is the first step to detect the vehicle. Wheels always provide great amount of vertical edge, and there are a lot of horizontal edges on the air dam of most vehicles. The information would be fairly useful when detecting vehicles.

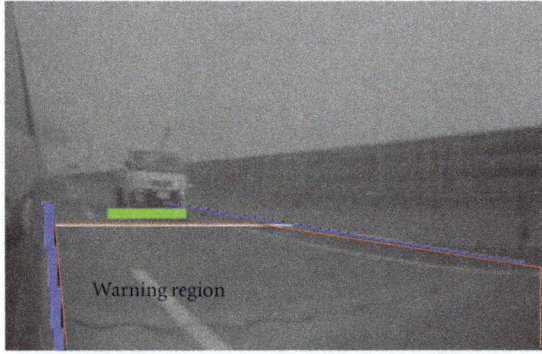

FIGURE 3: The area surrounded by the blue line is detecting region, and the area surrounded by the red line is warning region.

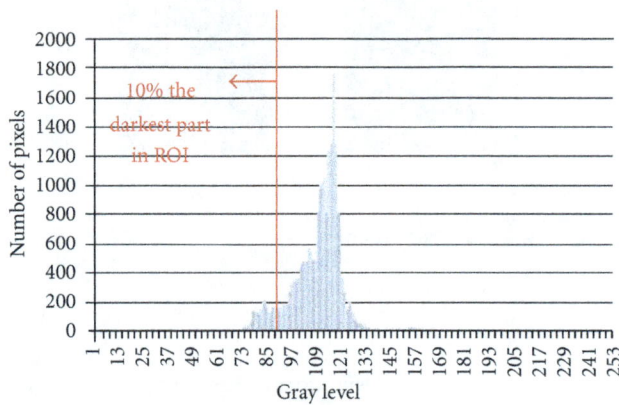

FIGURE 4: The gray level histogram with the pixels in ROI. The darkest part of this histogram might be shadow.

Several weather conditions may happen when driving on the road, so a fixed threshold to extract shadow region may fail in outdoor scene. However, no matter what the weather is, it is supposed that the color of shadow must be the darkest part in ROI. Therefore, we establish a gray level histogram of the pixels in ROI for adaptive shadow threshold. As shown in Figure 4, we assume that 10% of the darkest part of this histogram might be shadow, thus adaptive threshold g^* for shadow detection in this frame is calculated by (1). N is the number of total pixels in ROI, $h(g)$ is the number of pixels at gray level g, and 0.1 is chosen for δ here. According to the states of the road surface, the threshold for shadow extraction could be set dynamically to 92 in the sunny day, 78 in the rainy day, and 56 under the bridge by this method:

$$g^* = \arg\left(\sum_{g=0}^{255} h(g) > \delta N\right). \tag{1}$$

In addition, edge is a useful feature in spatial domain for vehicle detection, and Sobel mask is used to extract the edge feature here. Therefore, there are three kinds of features that could be used to recognize vehicles.

As in Figure 5(a), shadow pixels are drawn in white, and the horizontal edge pixels are drawn in gray. Otherwise, the pixels are set in black. The extracted vertical edge pixels

would be set to 0 in another plane, and the nonvertical edge pixels are set in 255 in Figure 5(b).

3.3. Shadow Searching. The first step of vehicle detection is shadow searching. Every pixel between point S_L and point S_R in each row is checked. Although every pixel is checked, the computation loading would not increase very much. The definitions of points are calculated by (2) and illustrated in Figure 6, where L_P and R_P are the positions of the left and right boundaries of ROI in each row, and v is the row index:

$$S_M(v) = \frac{(L_P(v) + R_P(v))}{2},$$
$$S_L(v) = \frac{(L_P(v) + S_M(v))}{2},$$
$$S_R(v) = \frac{(R_P(v) + S_M(v))}{2}, \tag{2}$$
$$S_{LRM}(v) = \frac{(S_L(v) + S_M(v))}{2}.$$

When one of these pixels is the shadow pixel, vertical projection would be processed in this row to find whether there are continuous shadow pixels in this row, as shown in Figure 7. The length of shadow pixels and the length of ROI in this row are denoted by λ_S and λ_L, respectively. If λ_S is larger than $1/8\,\lambda_L$, it would be considered as shadow under a vehicle and keep searching the rest upper part of the ROI. After that, there might be several shadow candidates that have to be confirmed later.

3.4. Correct Boundaries of the Vehicle. Although the location of shadow under the vehicle is found, several severe conditions would lead to incorrect detection. Early in the morning or in the evening, the sun would irradiate with a certain angle instead of direct sunlight. This would cause the shadow under a vehicle to become elongate. In the rainy days, severe road surface reflection occurs and causes the same situation as described above, as exhibited in Figure 8. Therefore, the boundaries of vehicle should be confirmed again using average intensity and vertical edges.

In Figure 9, the length of shadow pixels λ_S is found in Subsection 3.3. The row of shadow found in Subsection 3.3 is not the real bottom of the car. Therefore, we have to correct the location from Subsection 3.3. The searching zone is extended $1/2\,\lambda_S$ upward from the row RS, and the real bottom is searched in this zone L'R'RL. Shadow should be the darkest intensity because of the shadow property. In this zone, the average gray level value in each row is calculated, and the row with the minimum gray level would be considered as new bottom of the vehicle. The new bottom of the vehicle is obtained in (3), where average gray level value in the jth row is denoted as $I(j)$. Hence, we could update the bottom location of the vehicle to v^* where is much closer to the bottom of the car:

$$v^* = \underset{j=1\sim(1/2)\lambda_s}{\arg\min}\ \overline{I}(j). \tag{3}$$

After obtaining the bottom position of the vehicle, the left boundary would be searched through the vertical

(a)

(b)

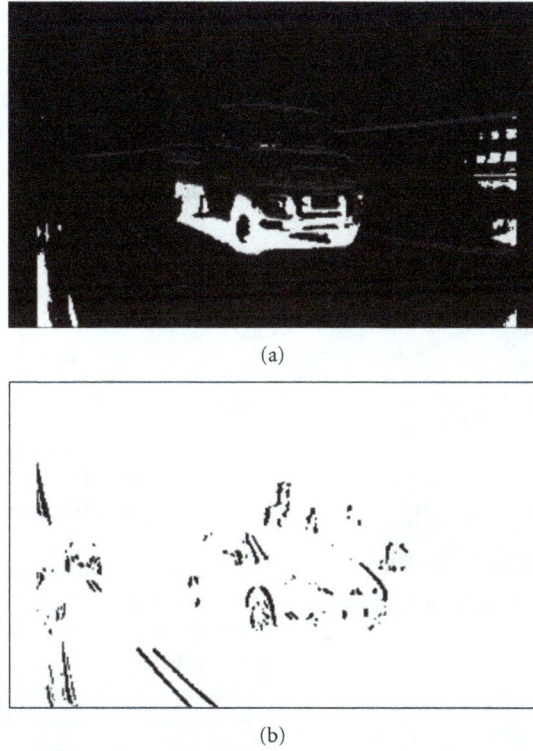

FIGURE 5: (a) The feature plane with horizontal edge and shadow. (b) The vertical edge feature plane.

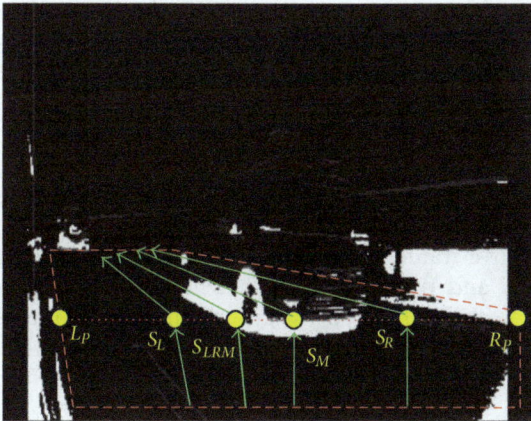

FIGURE 6: The illustration of four shadow searching points.

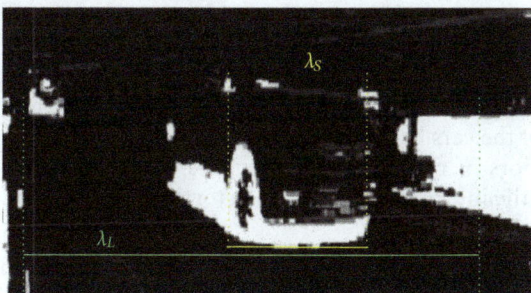

FIGURE 7: Finding the bottom of the vehicle by vertical projection.

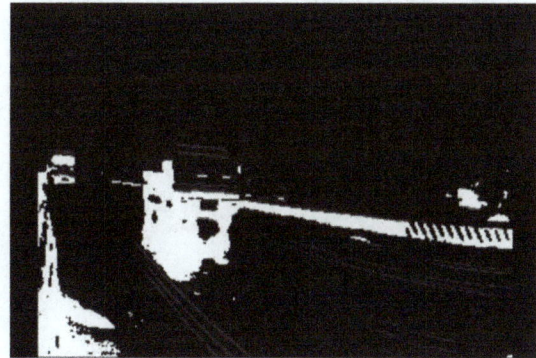

FIGURE 8: The shadow under a vehicle in feature plane is elongated in rainy days.

edges. If the continuous vertical edges exist, it would be considered as a vehicle candidate. As seen in Figure 10, the horizontal projection would be performed to check if there are continuous vertical edges of wheels in this region.

The default height of vehicle, λ_H, which is 3/4 λ_S, that is, the width of the shadow in Subsection 3.3. The position of the left boundary is computed in (4):

$$u_L = \underset{u=1\sim\lambda_s}{\mathrm{argmax}}\, h(u). \qquad (4)$$

The amount of vertical edge after horizontal projection is $h(u)$. The column of the maximum amount of vertical edge is u_L. If $h(u_L)$ is larger than 1/4 λ_H, it would be considered as

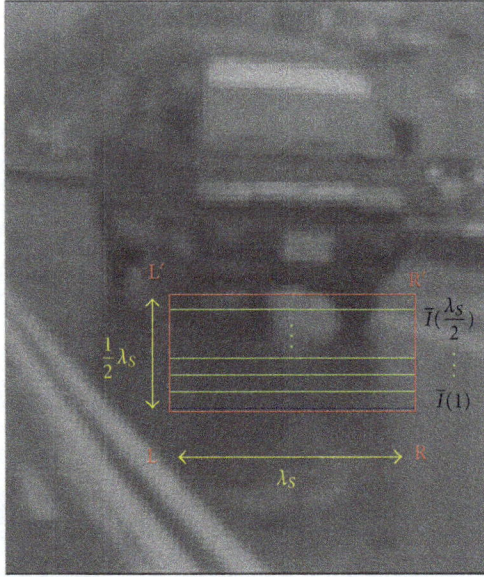

FIGURE 9: The shadow under a vehicle in feature plane is elongated in rainy days.

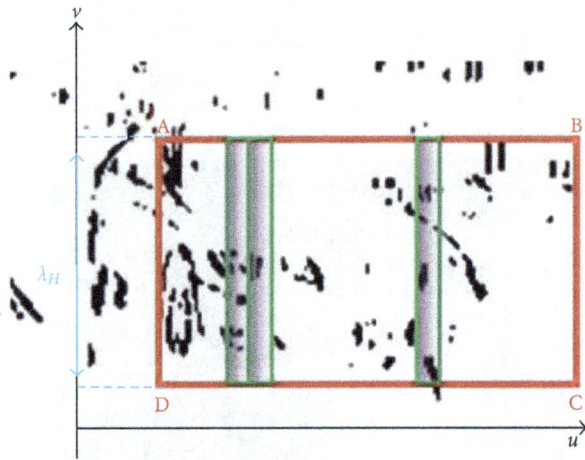

FIGURE 10: The horizontal projection of the vertical edge.

the vertical edge of the wheel. The left side of the boundary of the vehicle would be updated to the column u_L. The vehicle wheel in right side which is always cloaked from the air dim is hard to find through the edges. Therefore, the right side boundary of the vehicle uses the default value from λ_S. When finish checking up all the shadow candidates with vertical edge, there are some vehicle candidates that have been filtered out here. Next step is confirming all vehicle candidates with horizontal edge characteristic.

3.5. Vehicle Verification. Although the shadow under the vehicle and vertical edges is used to detect the vehicle, the verification procedure should be performed to confirm the detection. The horizontal edge is a good feature for

vehicle verification. There are a lot of horizontal edges in the air dam of the most vehicles, so we search the horizontal edge for vehicle verification. As shown in Figure 11, we extend a region for searching continuous horizontal edges. Vertical projection would be processed to check if there are continuous horizontal edges of air dim in this region. The vehicle is verified by (5) and (6):

$$v^* = \underset{v=1 \sim \lambda_H}{\mathrm{argmax}}\ h(v), \tag{5}$$

$$\alpha = \begin{cases} \text{true,} & \text{if } h(v^*) > \dfrac{\lambda_S}{4}, \\ \text{false,} & \text{otherwise.} \end{cases} \tag{6}$$

The amount of horizontal edge after vertical projection is $h(v)$. The row index is denoted as v, λ_H is the height of the detected vehicle, λ_S is the width of the detected vehicle, and the row of the maximum amount of horizontal edge is v^*. Vehicle identification is denoted as α. If the condition is met in (6), this candidate would be considered as a real car and would be tracked in the next frame. The verification of motorcycle is the same as the vehicle verification. The criterion in (6) is important so that most false alarms could be avoided. However, the tradeoff is that some motorcycles with less strong horizontal edges would be deleted.

3.6. Candidate Matching. So far, the real cars have been retained. When the full searching mode is processed, there is no idea about the correlation of the same car in consecutive frames. In order to track the detected car in successive frames, the candidate matching function is designed to conquer this problem. Matching candidate function is always executed not only to generate the new trajectory of the detected vehicle or motorcycle in the beginning but also to match the detected vehicle to the tracked vehicle. The first step in this function is finding the closest vehicle in consecutive frames. If the car position is closer in the consecutive frame, some characteristics are verified in (7), where H and W represent the height and the width of the image, respectively. β_C, β_L, γ_C, γ_L, ρ_C, and ρ_L, are the height, center column location, and center row location of the same vehicle in the current frame and last frame, respectively:

$$\begin{aligned} |\beta_C - \beta_L| &\leq \frac{H}{16}, \\ |\gamma_C - \gamma_L| &\leq \frac{W}{8}, \\ |\rho_C - \rho_L| &\leq \frac{3*W}{32}. \end{aligned} \tag{7}$$

In the end, the trajectory is produced, and the information of the cars has been inherited, updated, and stored. The behaviors of the cars or motorcycles could be determined using the information of the trajectory in tracking mode.

3.7. Behavior Judgment. There are three definitions for the behaviors of cars in this system: relative approaching, relative backing, and relative static, respectively. These definitions are

(a)   (b)

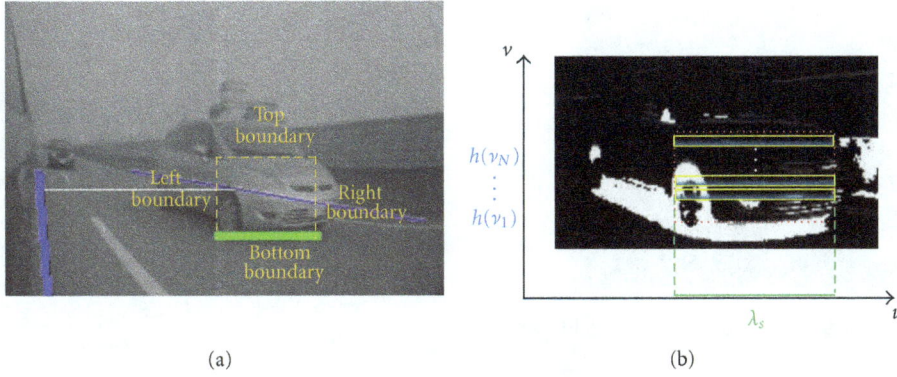

FIGURE 11: (a) Boundary setting for vertical projection. (b) The vertical projection of the horizontal edge.

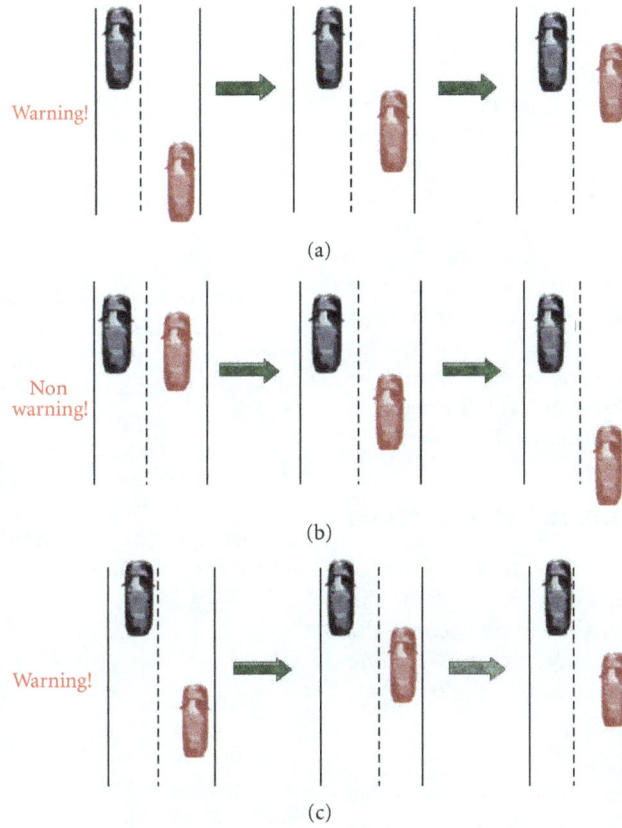

(a)

(b)

(c)

FIGURE 12: (a) Relative approaching. (b) Relative backing. (c) Relative static.

shown in Figure 12. The black car is the host car, and the red car is the tracked car in the blind spot area:

$$\delta_1(t) = v(t) - v(t-1), \tag{8}$$

$$A(t) = \begin{cases} 1, & \text{if } \delta_1(t) > 0, \\ 0, & \text{otherwise}, \end{cases} \tag{9}$$

$$RA = \begin{cases} 1, & \sum_{i=t-M-1}^{i=t} A(i) > 0.6{}^*M, \\ 0, & \text{otherwise}, \end{cases} \tag{10}$$

$$\delta_2 = |v(t) - v(t-1)|, \tag{11}$$

$$S(t) = \begin{cases} 1, & \text{if } \delta_2(t) < 3, \\ 0, & \text{otherwise}, \end{cases} \tag{12}$$

$$RS = \begin{cases} 1, & \sum_{i=t-M-1}^{i=t} S(i) > 0.6{}^*M, \\ 0, & \text{otherwise}. \end{cases} \tag{13}$$

The relative approaching is computed and judged through (8), (9), and (10), where v and t represent the

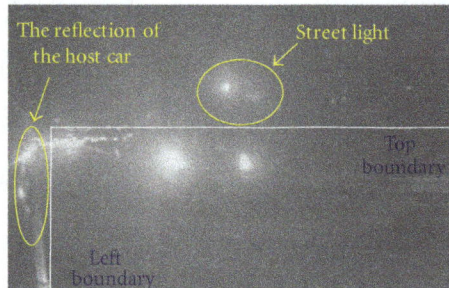

FIGURE 13: The boundary definition of ROI in the nighttime.

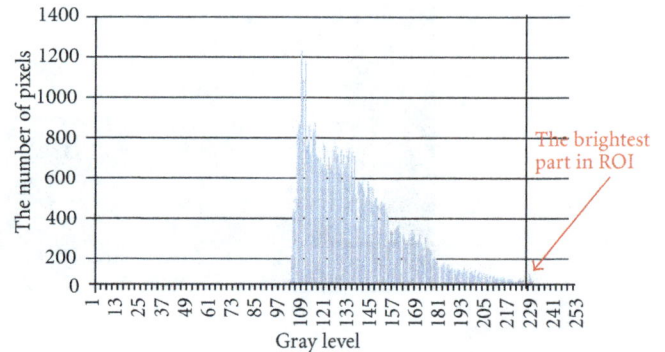

FIGURE 14: The gray level histogram in ROI in the nighttime.

bottom position of the tracked vehicle and time index, respectively. M which is the monitor window is set to 9 here. The relative static is judged by (11), (12), and (13). RA and RS symbolize the relative approach and relative static, respectively. If judgments of both conditions fail, the situation is considered as relative backing. If relative approaching vehicles or relative static vehicles exist, the system should send a warning signal to the driver, otherwise the relative backing vehicles do not have to be warned. Relative approaching and relative backing mean that the speed in the host car is faster or slower than the speed of the tracked vehicle, respectively. Relative static represents that the speed of the host vehicle is similar to that of the tracked vehicle. Because relative static means that there is a tracked vehicle close to the host car, the warning also should be trigger to prevent the collision. Buzzers or LEDs are used for warning signal, which depends on the demands of users.

4. Vehicle Detection Algorithm in the Nighttime

The main feature used in the nighttime is the lamps of vehicles, because most drivers would turn on the lamps in the evening or at night. Therefore, the lamps become the significant feature to identify the vehicle. Before searching the lamps, the ROI for the nighttime has to be determined at first.

4.1. ROI Definition for the Nighttime Detection. Unlike the shadow feature under the vehicle, the lamps are always equipped higher than the air dam. Consequently, the lamps would not be within the day time ROI. Therefore, the ROI for the daytime is not suitable for the nighttime. Figure 13 illustrates the ROI definition for the nighttime vehicle detection. The reflection of the host car and the street lamps could be filtered out.

4.2. Lamp Extraction. As described above, the lamps within the ROI are the targets to be searched and detected. At first, the system should find out where the lamps appear in ROI. It is supposed that the bright objects might be the lamps in ROI, so extracting the bright objects in ROI is the first step in image preprocessing for the nighttime. Calculating the threshold in (14) to extract this feature, we get statistic of the gray level value of pixels in ROI and establish a histogram

for this frame, as shown in Figure 14. The 1% brightest part in ROI is considered as bright objects:

$$g^* = \arg\left(\sum_{g=0}^{255} h(g) > \delta N\right). \tag{14}$$

The amount at gray level g is denoted as $h(g)$, and the calculated threshold in this frame is g^*. N is the number of pixels in the nighttime ROI, and δ is 0.99 here. The sizes of bright objects might be seriously affected if the dynamic threshold changed too fast. In order to avoid the dynamic threshold changing severely, the past thresholds would be referenced in this frame, as in (15). The calculated threshold in this frame and in the last frame are $g^*(t)$ and $g(t-1)$, respectively. The used threshold in the current frame is $g(t)$, where t is time index:

$$g(t) = \frac{7}{8}g^*(t) + \frac{1}{8}g(t-1). \tag{15}$$

The bright pixels would be extracted if their gray level is higher than the threshold. Then, to obtain the clean feature plane, the erosion is performed to remove the noise, and the result is displayed in Figure 15. After that, the connected component labeling is performed to save the rectangle information of the bright objects.

4.3. Lamp Verification. Although the bright objects mentioned above are extracted, some false bright objects caused from the reflection of the pavement and the island would be included. Therefore, we have to filter out the wrong lamp candidates in this flow. First, the intensity variance and size judgments of the bright objects would be performed. The intensity mean and variance are calculated in (16) and (17), where I_i is the gray level value of the ith pixel of this bright object, M, $\bar{I}$, and S are the total number of pixels, intensity mean, and variance of this bright object. After that, the width and height of the bright object should meet the requirements

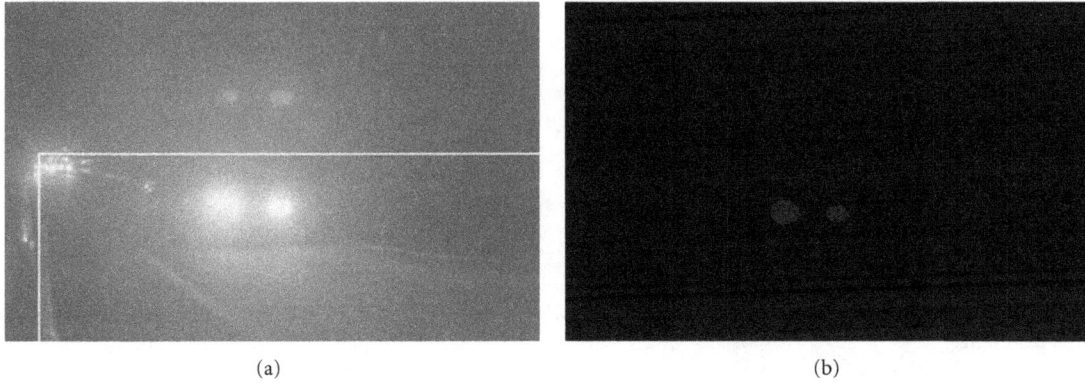

FIGURE 15: (a) The source image in the nighttime. (b) The extracted bright object plane of the left image.

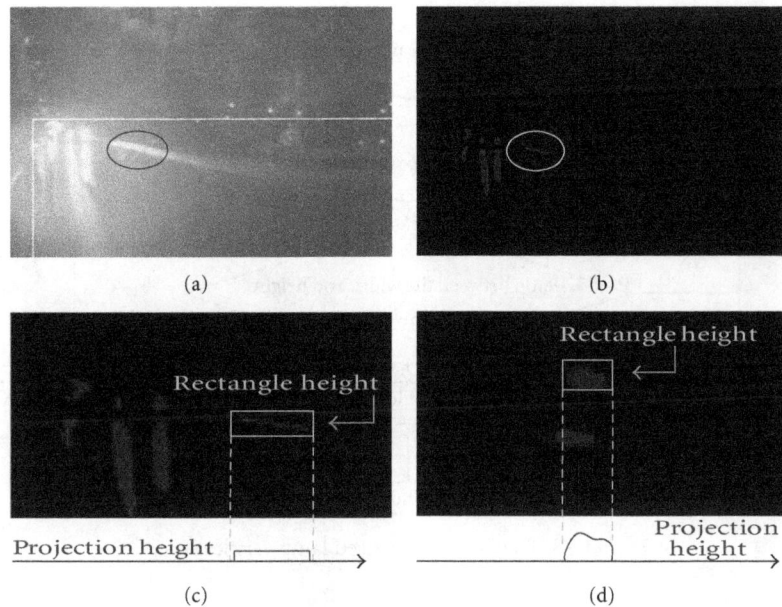

FIGURE 16: (a) The source image with the reflection of the safety island at night. (b) The extracted bright object plane. (c) Vertical projection of the bright object on the safety island. (d) Vertical projection of the bright object of the lamp.

in (18) and (19). H_k, W_k, H, and W are the height and width of the kth bright object in the image, respectively,

$$\bar{I} = \frac{1}{M} \sum_{i=1}^{M} I_i, \tag{16}$$

$$S = \left(\frac{1}{M} \sum_{i=1}^{M} I_i^2 \right) - \bar{I}^2, \tag{17}$$

$$\frac{H}{80} < H_k < \frac{H}{4}, \tag{18}$$

$$\frac{W}{60} < W_k < \frac{W}{4}. \tag{19}$$

The reflection of bright objects on the pavement caused by rainy day could be filtered out directly through the width, height, and area judgments. However, when driving through the urban ways, the bright object caused by safety island appears frequently in the image of the left side of the camera, as shown in Figure 16. Because the rectangle size of safety island bright object is similar to the lamp object, the vertical projection of the bright object is operated to filter out the reflection bright object. The heights of the safety island after vertical projection would be lower than the heights of the general lamps after projection, as shown in Figure 16. So, if the projection height is lower than the half of the rectangle height, this object must be considered as a reflection bright object on the safety island and it would be filtered by this method. Otherwise, this bright object would be retained and considered as a lamp object.

4.4. Filter the Lamps in the Second Next Lane. Since we only focus on the vehicle or motorcycle in the next lane, this subsection is used to judge and filter the lamps appearing in the second next lane. Because of the image captured by the camera lacks the depth information in 2D plane, the

(a)   (b)

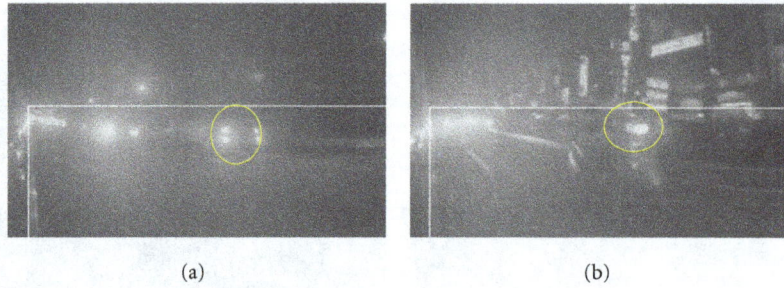

FIGURE 17: (a) The lamps belong to the vehicle in the second lane which are not the targets for this system. (b) The lamp belongs to the locomotive in the next lane.

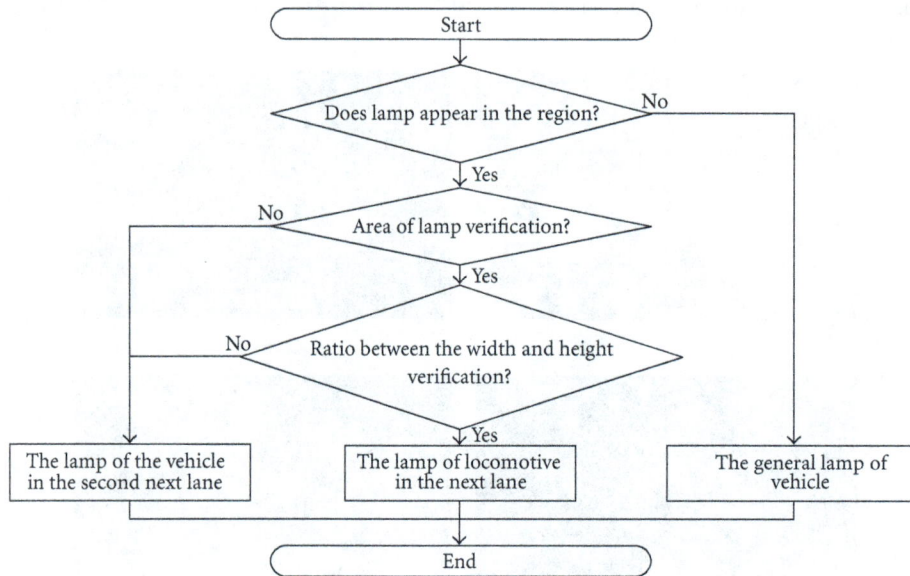

FIGURE 18: The flow of advanced lamp verification.

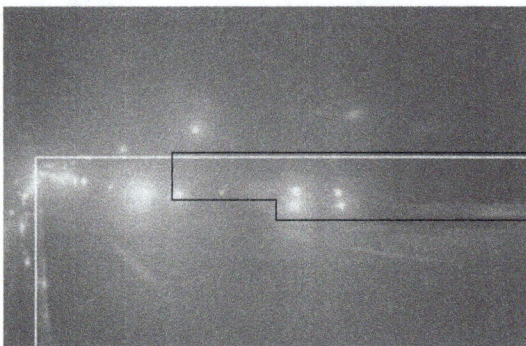

FIGURE 19: The definition of the region which might be in the second next lane.

Figure 18 illustrates the flow of lamps advanced conditions judgment, and the definition of the region of the second next lane is shown in Figure 19. If a bright object appears in this region, it might be the lamp of the vehicle in the second next lane, and the area and ratio verification would be processed in (20). Φ_A, Φ_W, and Φ_H are the thresholds of area, width, and height of the lamp rectangle, respectively. 150, 9, and 7 are chosen here in this system. ζ which is the threshold of the ratio between the width and height is 0.25:

$$A(k) > \Phi_A,$$
$$W(k) > \Phi_W,$$
$$H(k) > \Phi_H, \qquad (20)$$
$$\frac{W(k)}{H(k)} > \zeta.$$

positions of lamps from the vehicle in next two lanes and from the locomotive in next lane would appear in the same region, as shown in Figure 17. If the lamp belongs to the locomotive in the next lane, it needs to be kept tracking. However, if the lamp belongs to the vehicle in the second next lane, it is not the target to be tracked in this system.

4.5. *Lamps Tracking.* This part would only be processed in the tracking mode. It is used to build the relations between lamps in consecutive frames. The concept in this part is identical to match candidates. The first step is checking conditions between lamps, as in (21), (22), and (23). H_C, W_C, H_L, and W_L are the height and width of the same lamp in

(a)

(b)

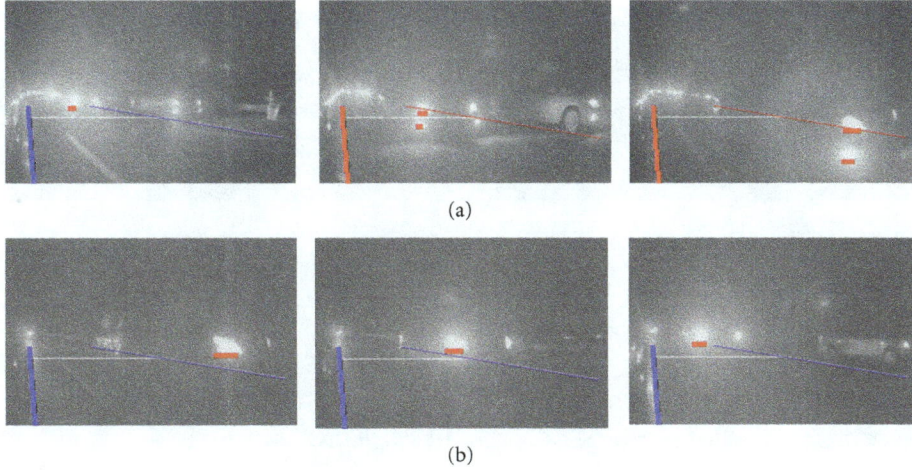

FIGURE 20: (a) Relative approaching. (b) Relative backing.

the current frame and in the last frame, respectively. T_H, T_W, U_R, and L_R are the thresholds for these conditions, and 10, 10, 1.35, and 0.7 are chosen for them here, respectively. If the conditions mentioned above are matched and the computed distance in (24) is minimal, the trajectory of this lamp is generated. The distance between the ith lamp in the current frame and the jth lamp in the last frame is d_{ij}, and the column and row of center coordinates of the lamp rectangle are denoted as x and y:

$$|H_L - H_C| < T_H, \tag{21}$$

$$|W_L - W_C| < T_W, \tag{22}$$

$$L_R \le \frac{H_L/W_L}{H_C/W_C} \le U_R, \tag{23}$$

$$d_{ij} = |x_i - x_j| + |y_i - y_j|. \tag{24}$$

4.6. Behavior Judgment for the Nighttime. The lamps usually move from left to right or from right to left in images in the nighttime. The vertical movements of lamps are not obvious, so column information is more important for behavior judgment here. The judgment is same to the judgment at daytime in Section 3, but the input parameter changes from the row index to the column index of the trajectory. Figure 20 depicts the output image sequence from left to right.

5. Experimental Results

This system has been on-road tested through the long verification. Several challenge video sequences are tested and the results are illustrated in Figure 21. If ROI is drawn in red color, the status represents that the system detected and tracked a vehicle which is relative approaching or relative static in ROI. Otherwise, ROI would be drawn in blue color. The buzzers or LEDs could be the warning signal to the

drivers through the GPIO on DSP, instead of red lines. In general, the proposed system could detect and track the vehicle, bus or motorcycle exactly in the daytime and in the nighttime. Moreover, the system can detect vehicles, even if the reflection is serious at rainy night. For the quantitative evaluation of the vehicle and the motorcycle detection performance, detection ratio, and false alarm ratio commonly used for evaluating performance in information retrieval [19] are adopted in this study. The measures are defined as:

$$\begin{aligned} \text{DR} &= \frac{T_p}{T_p + F_n}, \\ \text{FAR} &= \frac{F_p}{T_p + F_p}, \end{aligned} \tag{25}$$

where T_p (true positives) represents the number of correctly detected vehicles, F_p (false positives) represents the number of falsely detected vehicles, and F_n (false negatives) is the number of the missing vehicles.

Table 1 exhibits the quantitative results of the proposed approach for the vehicle and motorcycle detection and tracking, including the testing time, the detection ratio, and the false alarm ratio on high-speed road (H) and urban road (U) in sunny days and rainy days. The detection ratio achieves to 94%. The false alarm ratio in rainy days are much higher than others because that the camera was set up out of the car. When driving in rainy day, the lens on the camera is easily wet by rainwater. Therefore, this problem would lead to a lot of false alarms. The false alarm ratio is lower than 9.41% except rainy days. The following part evaluates the performance of the proposed system and compares it to the region-based method [17] and the edge-based method [18]. Both of these methods are implemented on DSP platform, so they could be compared here on TI DM6437. Table 2 shows the experiment data of the both methods and our systems. Sobel edge is the only feature used in [18], so that there are no results for the nighttime. Some representative comparative results of vehicle detection for the challenge sequences by

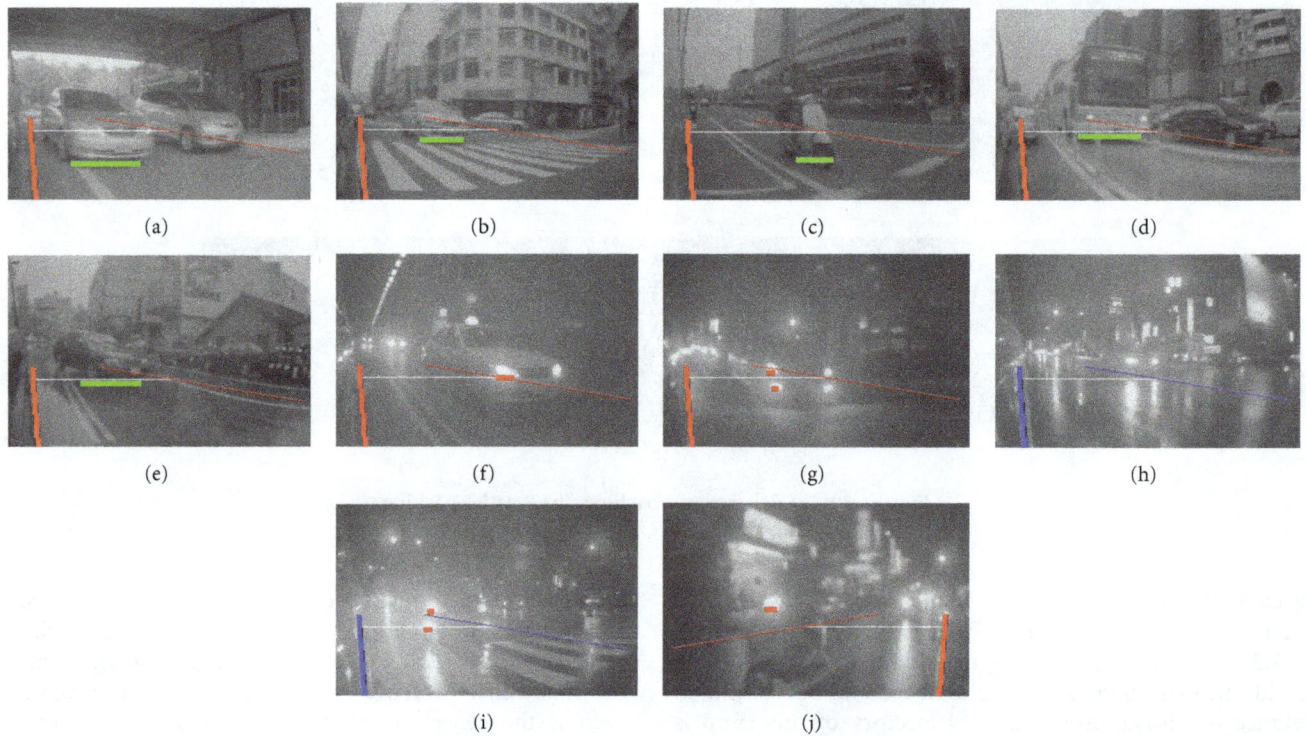

FIGURE 21: Parts of output images to show the experiment results in complex environment. (a) The general cars on complex urban road in the daytime. (b) The tracked vehicle across the zebra crossing. (c) The locomotive on urban road in the daytime. (d) The bus on urban road in the daytime. (e) The car on urban road in rainy day. (f) The car in the tunnel. (g) The general cars in the nighttime. (h) The reflection bright objects on the pavement caused by rainy night. (i) The general car on urban road at rainy night. (j) The locomotive on urban road at rainy night.

TABLE 1: The experimental data of the proposed approach.

Time (minutes)	Daytime				Nighttime			
	Sunny		Rainy		Normal		Rainy	
	H	U	H	U	H	U	H	U
	128	238	105	65	200	145	90	90
T_p	100	548	72	77	95	372	43	131
F_n	0	36	0	4	4	9	0	8
F_p	2	53	1	8	7	36	25	31
DR (%)	100	93.84	100	95.06	95.96	97.64	100	94.24
FAR (%)	1.96	8.82	1.37	9.41	6.86	8.82	36.76	19.14

TABLE 2: The experiment data for comparison.

Methods	C. T. Chen and Y. S. Chen [17]				Çayir and Acarman [18]		Proposed			
Processing time	10 (ms)				16 (ms)		17 (ms)			
Time	Daytime		Nighttime		Daytime		Daytime		Nighttime	
Road	H	U	H	U	H	U	H	U	H	U
Test minutes	30	60	30	28	30	60	30	60	30	28
DR (%)	100	100	100	100	82.35	77.27	100	95.45	100	96.15
FAR (%)	20.93	56	41.54	54.73	22.22	22.73	0	0	0	7.41

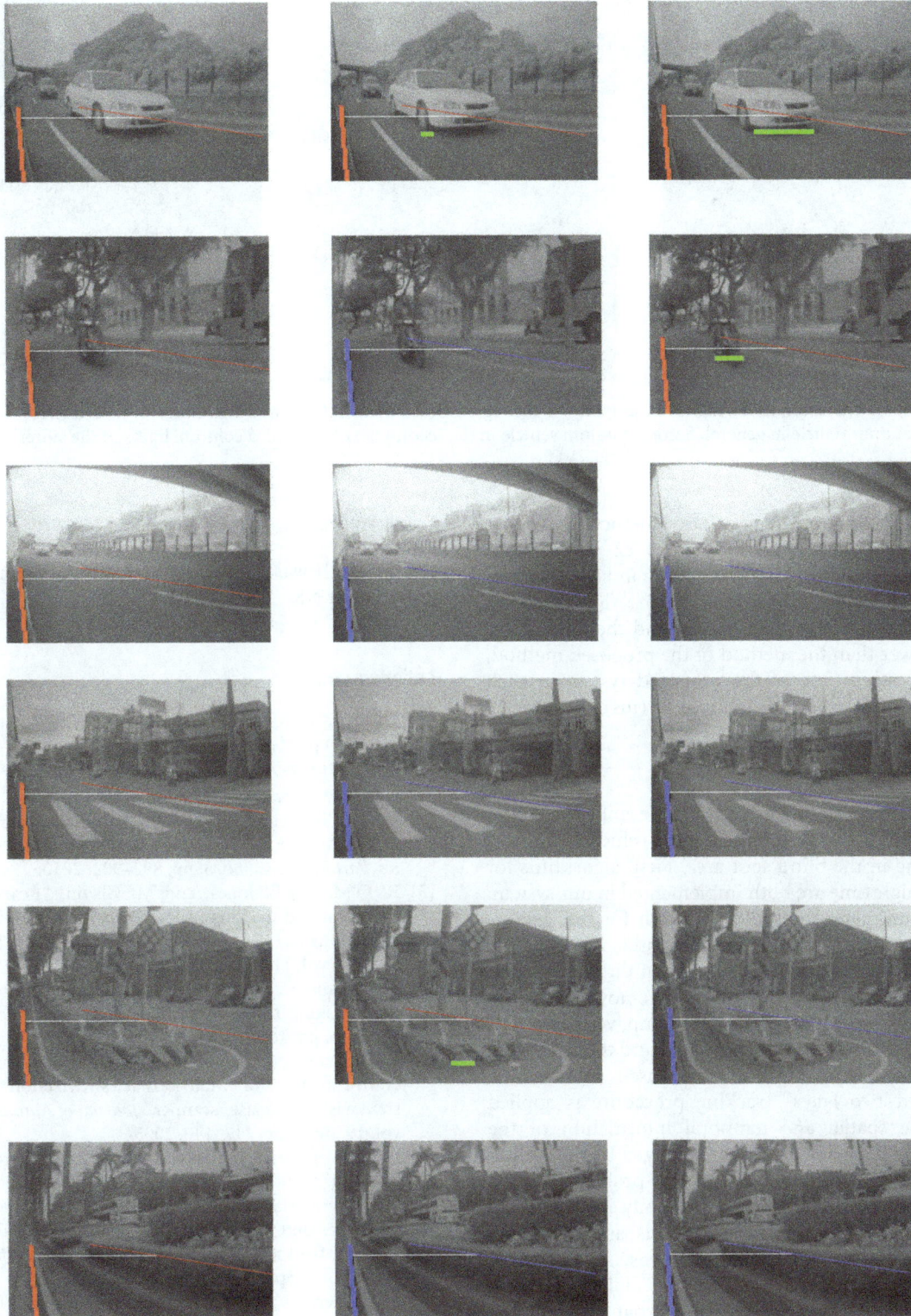

FIGURE 22: Comparative results. First column is the region-based by C. T. Chen and Y. S. Chen [17]. Second column is the edge-based by Çayir and Acarman [18]. Third column is the proposed method. First row: vehicle in general. Second row: motorcycle in general. Third row: shadow under the bridge. Fourth row: zebra crossing marking. Fifth row: safety island. Sixth row: haystack.

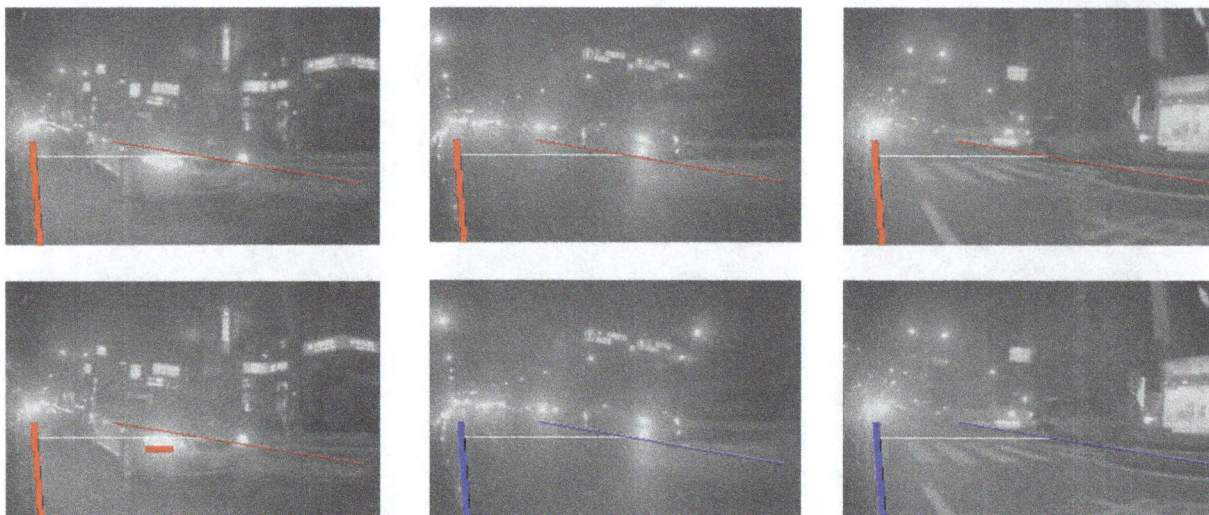

FIGURE 23: Comparative results in the nighttime. First row is the region-based by C. T. Chen and Y. S. Chen [17]. Second row is the proposed method. First column: vehicle in general. Second column: vehicle in the second next lane. Third column: lights of the store.

the proposed approach, the region-based method, and edge-based method are illustrated in Figures 22 and 23 and Table 2. The detection ratio is always 100% in [17], but false alarm ratio is also very high, at least 20.93%. The detection ratio of [18] is not better than others, and the false alarm ratio is not lower than the method of the proposed method, too. The processing time of the proposed system in each frame with the optimized code is about 17 (ms).

6. Conclusions

The proposed approach presents a real-time embedded blind spot safety assistance system to detect the vehicle or motorcycle appearing in the blind spot area. First, algorithms for daytime and nighttime are both implemented in our system. Second, the automatic threshold approach for the shadow and the bright object segmentation is used to overcome the problem of varying light conditions in outdoor scenes. Next, edge feature method is utilized to remove the noises in the complex environments. Then, lamp verification is used to distinguish between the headlight and the reflection, and the advance lamp verification is used to filter the vehicle in next two lanes. Tracking procedure is applied to analyze the spatial and temporal information of the vehicle, and the behavior judgment is judged by the vehicle trajectory. This algorithm solves most of problems caused in various kinds of environments, especially in the urban. It also overcomes the weather conditions and keeps the high performance no matter what time it is. The proposed algorithms were also implemented on a TI DM6437 and tested with the real highway and the urban road in the daytime and nighttime. From the experimental results, it is obvious that the proposed approach not only works well in highway but also has good performance in the urban. With comparing with other solutions, the experimental results also show that the proposed approach has better performance both in detections and false alarms.

Acknowledgment

This research was supported by National Science Council, under Grant NSC 100–2221-E-009-041.

References

[1] J. C. McCall and M. M. Trivedi, "Video-based lane estimation and tracking for driver assistance: survey, system, and evaluation," *IEEE Transactions on Intelligent Transportation Systems*, vol. 7, no. 1, pp. 20–37, 2006.

[2] W. C. Chang and C. W. Cho, "Online boosting for vehicle detection," *IEEE Transactions on Systems, Man, and Cybernetics, Part B*, vol. 40, no. 3, pp. 892–902, 2010.

[3] R. O'Malley, E. Jones, and M. Glavin, "Rear-lamp vehicle detection and tracking in low-exposure color video for night conditions," *IEEE Transactions on Intelligent Transportation Systems*, vol. 11, no. 2, Article ID 5446402, pp. 453–462, 2010.

[4] X. Liu and K. Fujimura, "Pedestrian detection using stereo night vision," *IEEE Transactions on Vehicular Technology*, vol. 53, no. 6, pp. 1657–1665, 2004.

[5] R. Labayrade, C. Royere, D. Gruyer, and D. Aubert, "Cooperative fusion for multi-obstacles detection with use of stereovision and laser scanner," *Journal of Autonomous Robots*, vol. 19, no. 2, pp. 117–140, 2005.

[6] C. Y. Wong and U. Qidwai, "Intelligent surround sensing using fuzzy inference system," in *Proceedings of the 4th IEEE Conference on Sensors 2005*, pp. 1034–1037, November 2005.

[7] O. Achler and M. M. Trivedi, "Vehicle wheel detector using 2D filter banks," in *Proceedings of the IEEE Intelligent Vehicles Symposium*, pp. 25–30, June 2004.

[8] M. Ruder, W. Enkelmann, and R. Garnitz, "Highway lane change assistant," in *Proceedings of the IEEE Intelligent Vehicles Symposium*, vol. 1, pp. 240–244, 2002.

[9] J. Díaz, E. Ros, S. Mota, G. Botella, A. Cañas, and S. Sabatini, "Optical flow for cars overtaking monitor: the rear mirror blind spot problem," in *Proceedings of the 10th International Conference on Vision in Vehicles*, pp. 1–8, Granada, Spain, 2003.

[10] P. H. Batavia, D. A. Pomerleau, and C. E. Thorpe, "Overtaking vehicle detection using implicit optical flow," in *Proceedings of the International IEEE Conference on Intelligent Transportation Systems (ITSC '97)*, pp. 729–734, November 1997.

[11] B. E. Stuckman, G. R. Zimmerman, and C. D. Perttunen, "A solid state infrared device for detecting the presence of car in a driver's blind spot," in *Proceedings of the 32nd Midwest Symposium on Circuits and Systems*, pp. 1185–1188, August 1989.

[12] M. Krips, J. Velten, A. Kummert, and A. Teuner, "AdTM tracking for blind spot collision avoidance," in *Proceedings of the IEEE Intelligent Vehicles Symposium*, pp. 544–548, June 2004.

[13] T. Yoshioka, H. Nakaue, and H. Uemura, "Development of detection algorithm for vehicles using multi-line CCD sensor," in *Proceedings of the International Conference on Image Processing (ICIP '99)*, pp. 21–24, October 1999.

[14] A. Techmer, "Real-time motion analysis for monitoring the rear and lateral road," in *Proceedings of the IEEE Intelligent Vehicles Symposium*, pp. 704–709, June 2004.

[15] K. Furukawa, R. Okada, Y. Taniguchi, and K. Onoguchi, "Onboard surveillance system for automobiles using image processing LSI," in *Proceedings of the IEEE Intelligent Vehicles Symposium*, pp. 555–559, June 2004.

[16] S. Jeong, S. W. Ban, and M. Lee, "Autonomous detector using saliency map model and modified mean-shift tracking for a blind spot monitor in a car," in *Proceedings of the 7th International Conference on Machine Learning and Applications (ICMLA '08)*, pp. 253–258, December 2008.

[17] C. T. Chen and Y. S. Chen, "Real-time approaching vehicle detection in blind-spot area," in *Proceedings of the 12th International IEEE Conference on Intelligent Transportation Systems (ITSC '09)*, pp. 1–6, October 2009.

[18] B. Çayir and T. Acarman, "Low cost driver monitoring and warning system development," in *Proceedings of the IEEE Intelligent Vehicles Symposium*, pp. 94–98, June 2009.

[19] I. Cohen and G. Medioni, "Detecting and tracking moving objects for video surveillance," in *Proceedings of the IEEE Computer Society Conference on Computer Vision and Pattern Recognition (CVPR '99)*, pp. 319–325, June 1999.

14

Automotive Technology and Human Factors Research: Past, Present, and Future

Motoyuki Akamatsu,[1] Paul Green,[2] and Klaus Bengler[3]

[1] *Human Technology Research Institute, AIST, Japan*
[2] *University of Michigan Transportation Research Institute (UMTRI), USA*
[3] *Institute of Ergonomics, Technische Universität München, Germany*

Correspondence should be addressed to Motoyuki Akamatsu; akamatsu-m@aist.go.jp

Academic Editor: Tang-Hsien Chang

This paper reviews the history of automotive technology development and human factors research, largely by decade, since the inception of the automobile. The human factors aspects were classified into primary driving task aspects (controls, displays, and visibility), driver workspace (seating and packaging, vibration, comfort, and climate), driver's condition (fatigue and impairment), crash injury, advanced driver-assistance systems, external communication access, and driving behavior. For each era, the paper describes the SAE and ISO standards developed, the major organizations and conferences established, the major news stories affecting vehicle safety, and the general social context. The paper ends with a discussion of what can be learned from this historical review and the major issues to be addressed. A major contribution of this paper is more than 180 references that represent the foundation of automotive human factors, which should be considered core knowledge and should be familiar to those in the profession.

1. Introduction

In many fields of technology, examinations of the past can provide insights into the future. This paper examines (1) the driver- and passenger-related technology that was developed as a function of time and (2) the research necessary for those developments, as they affected both vehicle design and evaluation. This paper also examines how those developments were influenced by (1) advances in basic technology, (2) requirements from government agencies and international standards, and (3) even the news media. All of this is done roughly chronologically, with developments grouped into three time periods—before World War II, after World War II until 1989, and since 1990.

In the history of research, a research topic becomes popular at some time because of a societal need, researcher interest, technology trends, the introduction of a new method, or a new theory. As a consequence, the number of researchers in the field grows, as does the number of publications, which in turn leads to products, services, and new ideas. These factors have certainly affected the growth of the human factors profession.

The history of automotive technology and human factors research can be viewed similarly. Its history can be divided into three periods. They are (1) the decades before World War II (Section 2), (2) World War II until 1989 (Section 3), and (3) 1990 and beyond (Section 4). This last period is continuing, so it is a bit more difficult to be retrospective in grouping decades. Therefore, Section 4 is divided by research topics, not by decades. For each topic, research activities are described chronologically to help readers to understand how the research has progressed for these 20 years to reach the current status.

2. A Short History of Human Factors Aspects of Automotive Technology before World War II

2.1. Early Stage of Automobiles (1886–1919). Over the course of the first half-century after the invention of the automobile

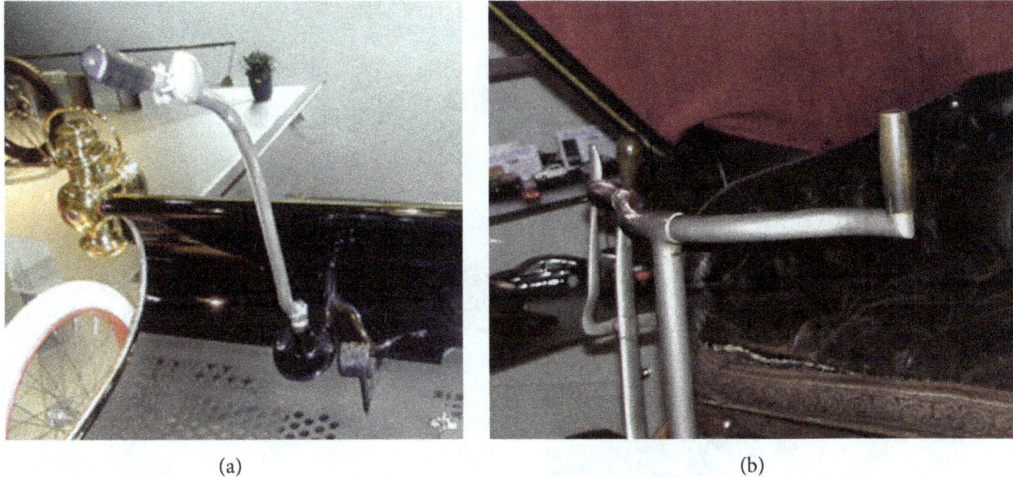

FIGURE 1: Tiller (Oldsmobile 1902 (a)) and bar handle (Peugeot Type 24 1899 (b)) (the author's (MA) photo collection).

by Karl Benz in 1886, various changes were made to self-powered vehicles so they were better suited to human abilities, changes based on experience with animal-drawn vehicles. Interestingly, the seatbelt had been introduced for steam-powered horseless carriages in the 1800s, but its purpose was to keep passengers on their seat, not to keep them safe in the event of a collision [1]. The steering mechanism in very early automobiles was a tiller, a lever arm that connected to the pivot point of the front wheels, a design derived from small boats. Tillers were easy to use for very slow speeds and lightweight vehicles (such as those with three wheels). However, steering a jolting tiller with sheer muscle power was difficult for heavy four-wheel vehicles moving at high speed. A bar handle with grips at both ends to be held with both hands was introduced that could be held more firmly than the tiller. A round steering wheel, able to be turned by muscle power and easier to hold in the hands, was first introduced around 1895 (Figure 1).

The brake system for very early self-powered vehicles consisted of a wooden block pressed against one of the wheels using a hand-operated lever, a technology adapted from horse carriages. A foot pedal to operate the band brake first appeared in Benz Velo in 1894 (Figure 2). The foot-operated pedal could exert greater force than a hand brake and allowed a driver to use both hands to hold a steering wheel. This could be why the steering wheel and the foot pedal appeared in the same period.

Early automobiles were not equipped with any gauges. Oil-pump gauges were the first instruments installed inside vehicles, allowing drivers to confirm the oil flow and to inject additional oil when necessary. Water-pressure gauges were also introduced around 1900. Durability was the biggest issue in the early stage of automobiles. Therefore, general monitoring of the condition of unreliable vehicles by the driver was critical and consumed considerable attention.

The speedometer was introduced after 1900. It was mounted outside of the bulkhead separating the engine and cab, where its cable easily fits. The speedometer was introduced to highlight the vehicle's high-speed capability.

In the USA, the state of Connecticut imposed a speed limit of 8 mph within the city and 12 mph outside of the city in 1901, thus encouraging speedometer installation [2].

The manufacturer Panhard et Levassor first placed a radiator in the front end of the vehicle for effective cooling. A thermometer was installed on top of the radiator in the early 1910s, allowing the driver to read the temperature from the driver's seat. Making sure the instrument was visible to the driver and was easy to install were important design considerations. In many cases, the hood ornament on contemporary vehicles is a remnant of these instruments.

After around 1910, instruments such as tachometers and clocks were installed inside automobiles. These were directly fixed on the surface of the bulkhead, and visibility to the driver was poor (Figure 3(a)). In the late 1910s, instrument panels (or dashboards) were installed separately from the bulkheads (Figure 3(b)). The instrument panel configurations were inconsistent. Some manufacturers concentrated the gauges in the central area of the panel and others distributed them across the panel.

An indication of the importance of the industry was the growth of organizations to support it. In 1901, the later German Verband der Automobilindustrie (VDA) association of automotive industry was founded as Verein Deutscher Motorfahrzeug-Industrieller (VDMI). VDMI was established to promote road transport, defend against "burdensome measures by the authorities" (taxation, liability obligations), support customs protection, and monitor motor shows. In 1923, the VDMI was renamed the Reichsverband der Automobilindustrie (RDA). The present name Verband der Automobilindustrie (VDA) was given to this umbrella organization of the German automotive industry in 1946 (http://www.vda.de/en/verband/historie.html).

To exchange engineering ideas to facilitate the growth of the automotive industry, the Society of Automotive Engineers (SAE) was established in 1905 in the USA. The first SAE meeting was held in 1906, and since then the Transactions of the Society have been published. In the USA, standardization work began in 1910 with the first issue of the *SAE Handbook*

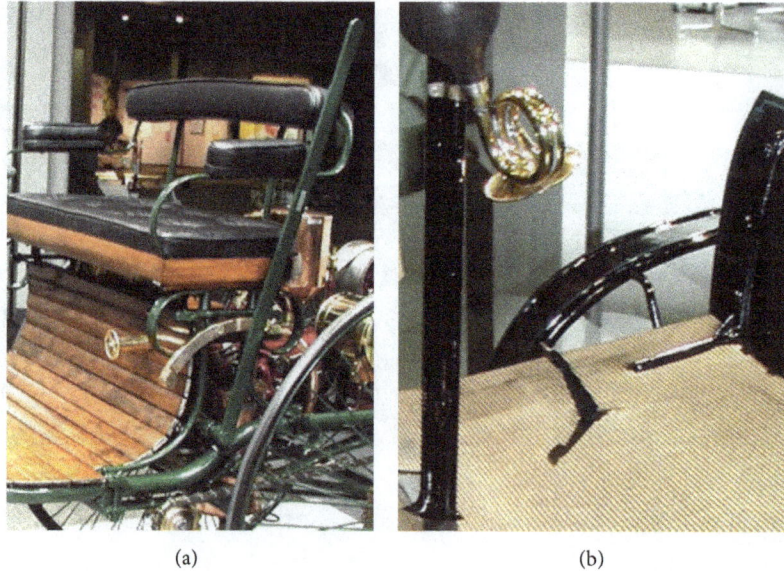

(a)   (b)

FIGURE 2: Hand brake lever (Benz Patent Motor Vehicle 1886 (a)) and foot brake pedal (Benz Velo 1893 (b)) (the author's (MA) photo collection).

(a)   (b)   (c)

FIGURE 3: Meters on bulkhead (Alpha Romeo 24PH 1910 (a)), meters in instrument panel (Dodge Brothers Touring 1915 (b)), and meter cluster (Buick Series 50 1932 (c)) (the author's (MA) photo collection).

of Standards and Recommended Practices. The number of members reached more than 4,300 at the end of the 1910s [3].

In summary, the first human factors development was designing controls for the primary driving task, such as the steering wheel and the brake pedal, which allowed for operation of a heavy self-powered vehicle using only muscle power. The second development was introducing gauges to inform the driver about the mechanical condition of the vehicle and then driving condition (speedometer). In addition, industry associations established in this early stage, such as VDA and SAE, played important roles in the development and dissemination of information related to automotive technology.

2.2. The Dawn of Automotive Human Factors Design (1920–1939). During these two decades, the basic controls and displays of the motor vehicle continued to evolve. An ignition-timing lever had accompanied the steering wheel from early

on. Horn buttons began to be installed in the center of the steering wheel in the late 1920s.

With regard to information presentation, gauge clusters first appeared in 1920s, often on a separate panel. Grouping gauges allowed drivers to read them at a glance. However, most gauge clusters were placed in the center of the instrument panel.

Before the 1920s, switches or knobs typically did not include labels to indicate their function. Drivers had to learn and memorize the function of each. Labels first appeared on controls and on the surface of instrument panels in the 1920s.

In the 1930s, speedometers and other instruments began to be installed directly in front of drivers to improve their visibility (Figure 3(c)), a practice that became common in the 1940s. American and many European luxury automobiles in this period were equipped with a shift lever on the steering column.

In early vehicles, one signaled the intention to turn using a *winker*, a mechanically operated arm or flag that extended

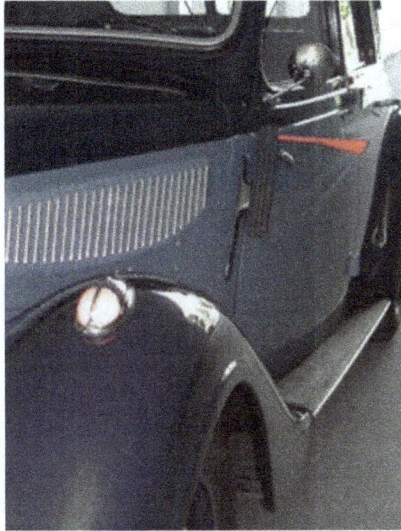

FIGURE 4: Turn indicator (BMW 335 1939, courtesy of H. Bubb).

from the side of the vehicle, first appearing in the 1910s. The exterior signal became a mechanical semaphore in the 1930s (Figure 4) and, finally, an electric light in the 1950s in Germany. A turn-signal switch or turn-signal lever was also being installed in the steering column by the late 1930s (Figure 5).

The seat-sliding mechanism, which adjusts the driving position, appeared in the 1920s. It allowed drivers with different body sizes to find a reasonable distance between the pedals and the seat.

Until the 1930s, the focus of automotive technology was on meeting basic functional requirements, primarily mechanical, to provide a durable vehicle. The shift at that time was toward designing vehicles that could go faster, with the 1934 Chrysler Airflow and its emphasis on aerodynamics as an example. Consequently, cabs shrunk and the car body became more rounded. This, in turn led to efforts to design the cab layout to fit the human body size and provide increased seating comfort while maintaining outward visibility. In an early book on automotive engineering written by Wunibald Kamm, an automobile engineer and an aerodynamicist famous for his Kamm-tail theory, provided an example of desired cabin dimensions (Figure 6) [4].

Thus, basic human factors design features, such as easy-to-operate steering equipment and switches, visible gauges, and a reasonable driving position, were introduced during the 1930s and 1940s. Note that, throughout that period, design decisions to accommodate human operators and passengers were based largely on heuristics from engineers' experience. Also numerous features were designed and implemented to ease the driving task, such as synchronized gears and improved windshield wipers, as well as switchable low and high beams. For additional information on these and prior developments, see [1, 2, 5].

The number of traffic crashes increased after World War I as the production of automobiles increased. In 1920, German psychologist Narziss Ach outlined the importance of psychology and technology in preventing crashes from the perspective of a scientific discipline that he called *psychotechnik*, which is closely related to human factors [6]. At the end of the 1930s, Forbes pointed out that understanding the limitations of driver capabilities such as visual characteristics and reaction time, "human factors" in traffic crashes, was necessary [7]. Both engineers and psychologists were aware of the importance of the human element in vehicle design and traffic safety in this period.

3. Human Factors Activities after World War II until 1989: The Era of Occupant Accommodation and Safety

3.1. Establishment of Human Factors as a Field of Endeavor (1940 to 1949). Although one can identify the roots of human factors being in early work in industrial engineering, such as that of Taylor and Gilbreth, activities at Bell Labs on communication quality, and other examples, human factors as a profession did not take off until WWII [8]. Human factors research was introduced during World War II to adapt military technologies to human operators to make systems more effective and reliable [9–11]. This research field was then expanded to the commercial aviation and automotive industries after World War II.

There was not an immediate transfer of human factors ideas from military to civilian activities. In part, this was because the initial transfer was from military organizations to defense contractors, which took several years, and Europe and Japan were recovering from World War II.

However, this period was not without some progress. Passive-safety technology was introduced at the end of the 1940s. The instrument panel was covered with sponge rubber in American automobiles, by Tucker in 1948 and Chrysler in 1949.

Also, there was considerable growth in the organizations interested in automotive research, some shortly after World War II, others later. The earliest one was British Motor Industry Research Association (MIRA) (UK), founded in 1946.

The following sections briefly describe automotive human factors studies and their output (mainly standards) and outcomes (products) from 1950 to 1989 by decade. Table 1 summarizes the major developments for each decade.

3.2. Human Factors Research Activities in 1950s: First Decade of Human Factors Research. A survey of the literature on human engineering in the 1950s, conducted by the U.S. Army Human Engineering Laboratory [12], indicated that studies at that time focused on driving visibility (including glare), cab layout based on anthropometric data, and the design of controls.

With regard to anthropometry, in 1955, for the first time, the SAE published data that included 5th- and 95th-percentile values for use in cab layout (Figure 7) [13]. During this decade, research was also conducted on human-body

TABLE 1: Overview of history of human factors researches, their outputs and outcomes.

		Empirical human factors design					WWII						
		1886–1899	1900–1909	1910–1919	1920–1929	1930–1939	1940–1949	1950–1959	1960–1969	1970–1979	1980–1989	1990–1999	2000–2009
Controls	Human factors research and output	Empirical: control by human muscle power			Empirical: access controls while holding steering	Empirical: access controls while holding steering		Symbols to indicate function of controls		Anthropometrical data for hand reach (SAE J287, 1976), Standardized direction of movement of control (SAE J1139, 1977)			
	Outcome	Steering wheel, Foot pedal,			Horn button and timing lever around steering wheel Labels indicate its function	Shift lever, turn signal lever around the steering column		Comprehension of function for people with different languages		Designing location of controls Common design of control direction	Steering wheel switch	Integrated joystick (Toyota, 1998)	i-Drive with controller nob in the center console (BMW, 2001), integrated center control (Nissan 2001)
Displays	Human factors research and output		Empirical: obtain information about vehicle condition	Empirical: visible information	Empirical: decrease amount of eye movements to check meters and gauges	Empirical: smaller eye shift to access the meter cluster		Empirical: smaller eye shift to access the meter cluster Avoid influence of sunlight		Symbols for Motor Vehicle Control, indicators, and tell tales (SAE J1048, 1974, ISO2575)	Investigation of advantage of HUD for vehicle display (1970s)	Measurement of visual accommodation (Toyota 1998)	
	Outcome		Installing gauges and speedometer	Instrument panel with meters and gauges	Clustered meters in instrument panel	Meter cluster in front of driver		Meter cluster in high position with sunshade		Commonly used symbols	Introduction of HUD for vehicle display (GM, Nissan 1988)	Introduction of center on-dash meter (Toyota)	
Visibility to road scene	Human factors research and output		Empirical: perceive approaching vehicles from behind	Empirical: road scene visibility in rain condition					Motor vehicle drivers' eye locations (SAE J941, 1965), Eyellipse (SAE Paper 680105, 1968) Passenger car rear vision (SAE J834, 1967)	Regulation for rear view mirrors (Directive 71/127/EEC, 1971), Field of view from automotive vehicles (SAE SP-381, 1973)	Backing sensor	Investigating visibility using digital human model Rear view monitor	Night Vision System
	Outcome		Rear view mirror	Windshield screen wiper					Define range of drivers' eye positions Examine direct visibility	Specifying visible area in rear view mirrors Investigating location of traffic signals, traffic signs, pedestrians in the driver's view		Time and cost effective design of visibility using CAD	

Primary driving task

TABLE 1: Continued.

		Empirical human factors design						WWII							
		1886–1899	1900–1909	1910–1919	1920–1929	1930–1939	1940–1949	1950–1959	1960–1969	1970–1979	1980–1989	1990–1999	2000–2009
Seating and packaging	Human factors research and output				Empirical: adapt to women drivers	Design drawing		Anthropometrical data for human body (SAE SP142A, 1955), SAE Manikin Subcommittee (1959)	Defining and measuring H-point (SAE J826, 1962), 2DM, 3DM	Chrysler's Digital Human Model CYBERMAN (1974), Measurement of pressure distribution of seat	SAMMIE, CAD with digital human model, is in the market	Commercial digital human model, Ramsis, Jack Body movement analysis using motion capture system	Combining digital human model with CATIA and ALIAS CAD, estimating muscle load using DHM
	Outcome				Seat adjuster	Cabin space design			Cabin space and seat layout design Precise design of seating configuration based on H-point,	Designing seat back angle		Time and cost effective design of packaging	Evaluation of ingress/egress motion
Vibration and comfort	Human factors research and output		Frederick Lanchester (UK) proposed the cabin movement is to be the same as that of human body while walking				Seat cushion and comfort (SAE J940')	Motor Vehicle Seating Manual (SAE J782, 1954)		Relationship between mechanical vibration and discomfort (ISO 2631, 1974)	Analysis of resonance frequency of body parts	Evaluation of vibration discomfort in multiaxis environment (ISO 2631-1, 1997)	Temporal factor in vibration discomfort
	Outcome		Peak frequency of cabin vibration was about 2.0 Hz							Establish evaluation method for vibratory comfort	Designing cushion of seat	Evaluation to integrate vibration in multiaxis	
Climate	Human factors research and output			Empirical: comfort in winter time							Thermal manikin	Equivalent temperature (SAE J2234, 1993)	Ergonomics of thermal environment of vehicle (ISO14505 series)
	Outcome			Cabin heater							Evaluation of vehicle climate	Evaluating thermal comfort using Equivalent Homogeneous Temperature	Evaluation of cabin thermal comfort, combining thermal manikin, calculation of EHT, and subjective evaluation

Driver Workspace

Table 1: Continued.

Category		Empirical human factors design					WWII 1940–1949	1950–1959	1960–1969	1970–1979	1980–1989	1990–1999	2000–2009
		1886–1899	1900–1909	1910–1919	1920–1929	1930–1939							
Driver's Fatigue	Human factors research and output								Physiological measure for fatigue (HR, GSR, BP) CFF study Recommendation of having rest Evaluation of vehicle vibration		HRV as a measure of fatigue		
	Outcome												
Driver's condition — Impairment	Human factors research and output									Physiological measure for drowsiness, EEG, EOG, GSR	Partial eye closure as a measure of drowsiness (Dingus, 1986)	Percentage of eye closure as measure of drowsiness (PERCLOS)	
	Outcome										Algorism to detect drowsiness (NHTSA DOTHS808247, 1994)	Drowsiness detection system	
Crash injury	Human factors research and output						Instrument panel covered by sponge rubber (Tucker 1948, Chrysler 1949)	Investigation of body damage by accident Crash test using high speed camera Crash dummy (GM, Ford)	Crash dummy, FMVSS208 frontal crash test in 30 mph (1966)	Crash dummy, Hybrid II (1974), Hybrid III (1978) Injury index AIS-1971, AIS-1976	Dummy, Euro-SID-1 for side impact (1989)	Dummy for side impact, Bio-SID, more sensors and more biofidelity, (1990) and SID-II (1994)	
	Outcome							Seat belt (Nash, 1949) Nondeformable passenger cell (Daimler-Benz 1952) 3-point seatbelt (Volvo, 1959) Head restraint (AM, 1959)	Collapsible steering column (GM, 1967) Mandatory belt use in front seat (1967, USA; 1969, Japan)	Standardized assessment method	Airbag	Side impact bar, Side airbag	
Interaction with driver information system	Human factors research and output								There was tolerable visual sampling duration while driving (Senders, 1967)	Measurement of eye movement on in-car displays (Mourant, 1978)	Visual sampling model and glance time study (Wierwille)	Visual behavior and driving performance as a measure of distraction (Zwahlen, 1986), mental workload measures	Workload measurement for distraction: occlusion method (ISOI6673, 2007) and LCT (ISO26002, 2011)
	Outcome									Single glance time was 0.5–1.5 seconds. Several glances for radio task	Assessment of IVIS using glance time	Discussion of map display JAMA guideline ver. 1.0 (1990), ver. 2.0 (1999) HARDIE guideline (1995)	Guidelines (JAMA ver. 3.0 (2004), AAM etc.) for information device

TABLE 1: Continued.

		Empirical human factors design				WWII							
		1886–1899	1900–1909	1910–1919	1920–1929	1930–1939	1940–1949	1950–1959	1960–1969	1970–1979	1980–1989	1990–1999	2000–2009
Interaction with advanced driver-assistance system	Human factors research and output												Overreliance, overtrust, and situation awareness with ADAS Analysis of driving behavior
	Outcome							Conventional cruise control			Antilock Braking System (ABS)	Electronic stability control (ESC), vehicle stability control (VSC) Adaptive cruise control (ACC) in the market, night vision	Lane-keep assist Full-range ACC Collision mitigation braking system
External communication access	Human factors research and output								Visual demand while driving (Senders, 1967)			Accident statistics analysis for risk of mobile phone use while driving	Naturalistic driving study Regulations for use of cellular phone (USA, 2001 (NY), Germany, 2001; France, 2003; UK, 2003)
	Outcome									Cellular automobile phone service (ARP Finland, 1971; NTT, Japan, 1979)	Analog cellular service (US 1984)	Prohibition of hand held use of cellular phone (Swiss, 1996; Japan, 1999)	
Driving behavior	Human factors research and output								Driver vehicle control model Development of early DS (GM)	Visual behavior while driving using eye tracker	Researches on UFOV	Visual attention measured by peripheral detection task	Naturalistic Driving study Driving behavior study using DS
Related technologies and vehicles' environment		Very rough road	Increased speed		Radio tuner was installed		Highway mobile telephone (USA, 1946)		Development of highway		RDS-TMC (EU)	Telematics service (US OnStar 1995, Germany TeleAid 1997 BMW Assist 1999, Japan MONET 1997, Carwings 1998)	Smart phones

TABLE 1: Continued.

		Empirical human factors design				WWII							
	1886–1899	1900–1909	1910–1919	1920–1929	1930–1939	1940–1949	1950–1959	1960–1969	1970–1979	1980–1989	1990–1999	2000–2009	
Organization and academic society		VDMI (currently VDA, Germany, 1901), SAE (USA, 1905)		SIA (France, 1927)		JSAE (Japan, 1947), FISTA (1948)	HEFS (USA, 1956), IEA (1959)	JSAE automotive ergonomics study group (Japan, 1962)		HFES Europe Chapter (1983)			
Public institutions				VTI (Sweden, 1923)	TRL (UK, 1933)	MIRA (UK, 1946), TNO human factors (The Netherland 1949)	TTI (USA, 1950), BASt (Germany, 1951)	ONSER (road safety org France, 1961), UMTRI (USA, 1965), JARI (Japan, 1969), TNO Traffic Behavior Department (The Netherland 1969)	NHTSA (USA, 1970), HUSAT (UK, 1970), IRT (France 1972)	INRETS/LESCO (combining ONSER and IRT, currently INFSTTAR/LESCOT) (France 1986)			
Conference meetings		SAE conference (USA, 1906)		TRB (USA, 1920)		FISITA congress (1947)	JSAE conference (Japan, 1951), HFES meeting (1957)	IEA congress (1961), Stapp (USA, 1962)	ESV conference (1971)	Vision in Vehicle (1985)	AVEC (1992), Driving Simulator Conference (1994), ITS World Congress (1994)	Driver Assessment Conference (USA, 2001), International Conference on Driver Distraction and Inattention (EU, 2009), AutomotiveUI (2009), HUMANIST (EU, 2008)	
Social background		Speed violation penalty		Increase of number of vehicles (USA)			Increase of number of vehicles (Europe)	Media promotion for safety (Chevrolet, Corvair USA)	Media promotion for safety (Ford Pinto, USA)	Media promotion for safety (Jeep CJ05, Suzuki Samurai, Audi 5000, USA)	Media promotion for safety (GM CK pickup, USA)	Media promotion for safety (Ford Ram, Crown Victorias, Explorer, USA)	

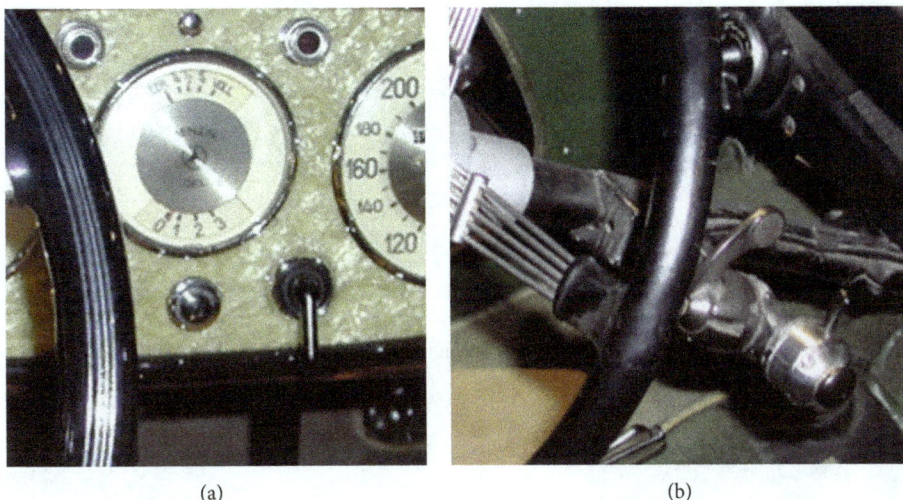

(a)   (b)

FIGURE 5: Turn signal lever in instrument pane (Mercedes-Benz 500K 1935) and that in steering column (Morris Eight Series I 1937) (the author's (MA) photo collection).

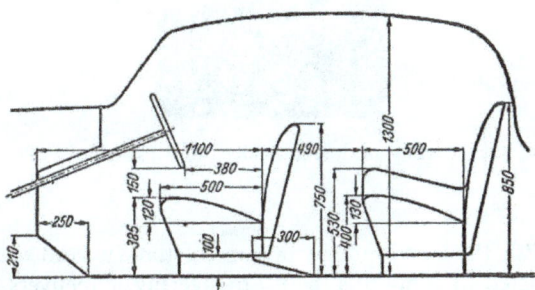

FIGURE 6: Cabin dimensions shown in Kamm's book "*Das Kraftfahrzeug*" (1936).

injuries caused by vehicle crashes [14]. Experimental technologies for crash tests (e.g., dummies, accelerometers, and high-speed cameras) were developed [15].

Following up on some advances in passive safety earlier in the 1940s, Nash Motors installed the first seatbelt in 1949. Other American manufacturers introduced seatbelts in the 1950s. In 1952, Barényi, an engineer at Daimler Benz, invented the nondeformable passenger cell and in later years, the crumple zone and collapsible steering column.

Some European vehicle manufacturers in this period introduced symbols to indicate the functions of controls. The position of the gauge cluster was raised to be closer to the normal line of sight and, therefore, was easier to read.

Subsequent to MIRA's founding in the UK in 1946 was the founding of Texas Transportation Institute (TTI) (US, 1950), German Federal Highway Research Institute (BASt) (Germany, 1951). Also established around this time were organizations specifically focusing on safety and human factors— TNO Human Factors (The Netherlands, 1949), ONSER (road safety organization, currently INFSTTAR, France, 1961) and the automotive ergonomics study group in JSAE (Japan, 1962).

A variety of automotive human factors research efforts began during this period. Methods from psychology, medicine, and anthropology were introduced. An important method involved using statistical distributions of anthropometric dimensions to establish vehicle design standards for those dimensions. This method directly linked human factors research to production of vehicles geometrically adapted to human characteristics, a method that was developed further in the next decade.

3.3. Human Factors Research Activity in 1960s: The Decade of Anthropometry. In the 1950s, automobile manufacturers recognized that anthropometric data could be the basis for laying out the cab to ensure that the driver (1) could see the road, traffic signals, and other vehicles outside of the cab, (2) could see controls and displays inside the cab, and (3) would be able to reach controls. In 1959, the SAE Manikin Subcommittee began developing an easy-to-use tool for ergonomic design based on anthropometric data. The SAE two-dimensional manikin (2DM) and three-dimensional manikin (3DM) were codified in SAE J826, which was published in 1962 [16]. The hip-point (H-point), which was the origin on the human body for automotive cab design, was defined in this standard together with specific measurement procedures. The 2DM was used to design the side view of the vehicle, and the 3DM was used to design cab mockups.

Based on the anthropometric research, the driver's eye position was defined in SAE J941, and the concept of the eyellipse, which specified the range of the driver's eye position, was developed [17–19]. What drivers of widely varying body sizes would be able to see could be examined using the eyellipse. Standards for front-view and rear-view visibility were also published (SAE J834, 1967) [20].

At that time, automobiles were commonly used in the USA and driven by a wide range of people. Therefore, the US car manufacturers were motivated to collect anthropometric data for cab design to accommodate that range

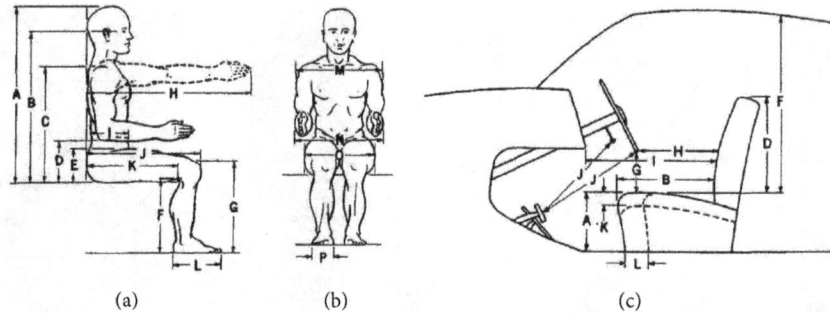

FIGURE 7: Human body measurements and vehicle dimensions shown in SAE SP142 (1955).

FIGURE 8: Field experiment of Critical Fusion of Flicker (CFF) measurement for highway drive in Japan (Brochure of IPRI, AIST, 1969).

FIGURE 9: Helmet for occlusion device (courtesy of J. W. Senders).

of drivers [21]. This data was also helpful to car manufacturers outside the USA who were developing cars for export to the USA and served to further improve various SAE standards that had been developed or were in development.

Frontal-crash test procedures to protect occupants were introduced in FMVSS 208 [22]. In 1959, Volvo was the first manufacturer to provide three-point seatbelts. In the same year, American Motors also equipped their automobiles with head restraints to avoid neck injury in rear-end collisions. In 1967, General Motors conducted pioneering work on collapsible steering columns designed to reduce chest impact injuries [5].

The construction of special-purpose, high-speed roads began with the first autobahn in Germany in the 1930s, followed by construction of interstates (USA), autoroutes (France), motorways (UK), and autostrada (Italy) beginning in the 1950s, and followed by significant highway construction in Japan in the 1960s. Because trips on such roads tended to be long, driver fatigue became a concern. There were many studies done in Japan, mainly by researchers with medical backgrounds, to evaluate driver fatigue using such physiological variables as heart rate, GSR (galvanic skin response), blood pressure [23], and CFF (critical frequency fusion) (Figure 8).

With the development of control theory, studies were conducted to apply this theory to steering maneuvers [24–28]. Studies to measure mental workload, introducing methods from physiology and the cognitive sciences, began in the 1960s. Brown and Poulton assessed drivers' spare mental capacity using auditory subsidiary tasks requiring the driver to identify a digit that differed from the previous one [29]. One pioneering study on driving behavior was Sender's 1967 study to measure visual demand while driving, using an occlusion device with a moving frosted plastic visor on the helmet (Figure 9) [30].

During the 1960s, driving simulators were developed to study vehicle dynamics and to analyze driving behavior. It is not certain when the first simulator was developed, but there were driving simulators in the 1950s. General Motors developed a driving simulator using a gimbal structure to give pitch and roll motion to the driver [31]. The driving simulator developed in 1976 by the Mechanical Engineering Laboratory of AIST (Japan) had a moving cab, and the driving scene was obtained through a movie camera running a miniature diorama of a road in town and in a rural area (Figure 10) [32]. Driving simulators were also developed in US universities. Interestingly, it was not until about 17 years later, with the advent of the Daimler-Benz simulator, that driving simulators received broad attention [33].

In the USA, a major factor in the movement to improve crash safety was the investigative news media. The first vehicle to attract attention was the 1961–1963 Chevrolet Corvair, which in a sharp turn, had a tendency to spin and/or rollover. The Corvair was an unusual rear-engine vehicle, and there

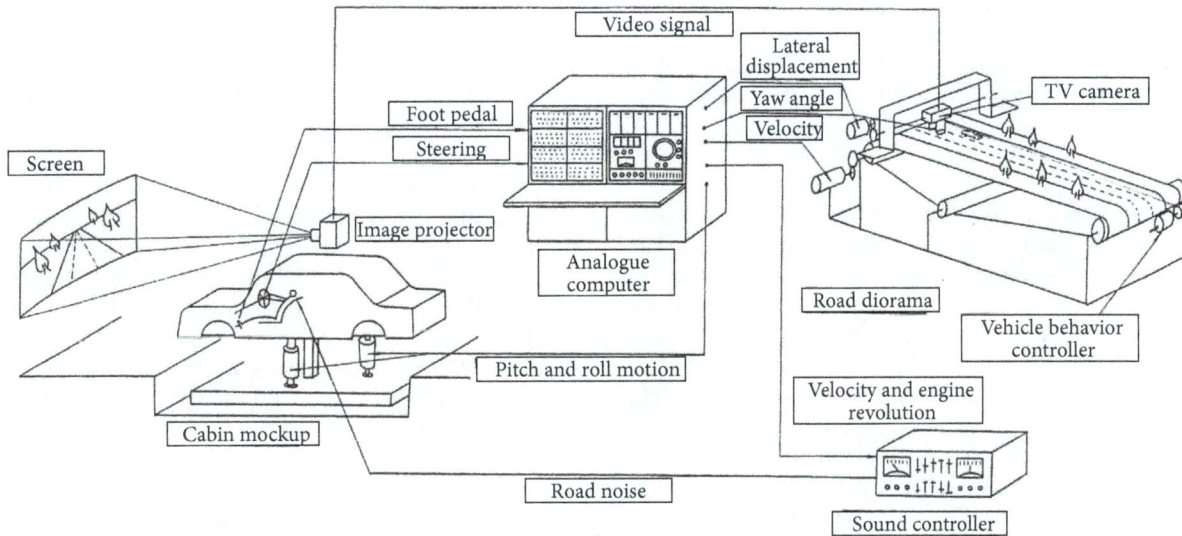

(a)

(b)

FIGURE 10: Driving Simulator of Mechanical Engineering Laboratory of AIST (1968) (Technical Report of MEL, no. 89, 1976).

was considerable discussion of its suspension system in a book by Ralph Nader, a consumer advocate [34]. The book's title, *Unsafe at Any Speed*, captured the way some felt about Corvairs. As a result, there were congressional hearings about vehicle safety (that led to a black eye for General Motors), eventual withdrawal of the Corvair from production, and a significant increase in interest in vehicle safety.

The interest in safety led to the establishment of the Highway Safety Research Institute at the University of Michigan, now the University of Michigan Transportation Research Institute (UMTRI), in 1965 and the National Highway Traffic Safety Administration (NHTSA) in the U.S. Department of

Transportation in 1970. TNO in The Netherlands started a Traffic Behavior Department in 1969, which focused on traffic safety. In the same year, Japan Automobile Research Institute (JARI) was founded. They joined a worldwide collection of organizations (see Table 1).

The growth in the worldwide production of automobiles led to increased interest in designing vehicle cabins suitable for a wide range of people. As the number of traffic accidents rapidly increased with increased production, safety became a major concern for society. Automotive safety technology had evolved since the last decade, but it was facilitated by news media in this decade. Human factors research led first to advances in passive safety and later to advances in active safety. Research on measurement of fatigue, mental workload, and driving-task demand developed in this decade. A shift in human factors research began from a focus on physical characteristics to cognitive characteristics.

3.4. Human Factors Research in the 1970s: Establishing Crash-Safety Assessment and Occupant Comfort. The impact of the US news media in bringing attention to crash safety continued in the early 1970s, focusing on the Ford Pinto. When struck from the rear under certain circumstances, Pintos would dramatically catch fire [35–37], videos of which are still available (http://www.youtube.com/watch?v=rcNeorjXMrE, http://www.youtube.com/watch?v=lgOxWPGsJNY). A critical document in the case was a cost-benefit analysis done by Ford, which compared the cost of making changes to the vehicle to prevent or reduce fires with the cost of injuries and lives lost, an idea that has been the source of numerous ethics discussions over time. However one feels about the Pinto, the case generated an intense focus on vehicle safety, in particular with regard to fires and safety in crashes, especially rear-end crashes. As with the Corvair, the Pinto's poor publicity led to a sharp decline in sales and eventual withdrawal of the Pinto from production. The Pinto case served as the stimulus for further research in the USA.

To help prevent rear-end crashes, Irving and Rutley investigated staged signaling concepts for different braking levels, conveying more information to following vehicles, concepts that led to improved braking over those in use [38]. Also the number and position of brake lights varied, leading to the idea of center, high-mounted stoplights. The effectiveness of the high-mounted stoplight was studied in the 1980s [39, 40]. During this decade, there were also studies of nighttime visibility distance of different headlight beam patterns and technologies (conventional tungsten, sealed beam, and halogen), as well as their effects on glare [41].

Improved understanding of what happened in crashes was also a research focus. Crash dummies were developed by several different organizations. They were integrated into Hybrid I in 1971 and Hybrid II in 1974. Sensors in Hybrid II were located in the head, chest, and femur. To make the dummy more realistic, Hybrid III was developed in 1976 [42]. Ten sensors were located in the head, neck, upper body, femur, knee, and leg, where injury might occur in the event of a crash. The severity of injury of each part of the body could be assessed based on the acceleration of each location. Head Injury Criteria (HIC) were defined by NHTSA in 1971 to assess the severity of head injury using the dummy. The Abbreviated Injury Scales (AIS-1971 and AIS-1976) for determining the level of injury produced by actual accidents were also established during this decade. The assessment method was standardized during this period [43].

However, crash safety was not the only topic of interest during the 1970s. Based on anthropometric research, an SAE standard for hand reach was published in 1976 (SAE J287) [44, 45]. To reduce driver confusion when operating controls, the direction of the movement of controls was standardized in SAE J1139 in 1977 [46].

Symbols to indicate control functions were introduced in the 1950s, mainly for European cars, to avoid the need to produce a different instrument panel for each language region in which a vehicle was sold. These symbols did not require reading written words and were intended to be intuitive. However, when different symbols were used to indicate the same function, drivers could become confused. To avoid such confusion, standard SAE J1048 was established in 1974 [47].

Studies on vehicle vibration and comfort have been conducted since the 1940s. Vibration and shock may cause low back pain and performance changes [48]. Vibration of the vehicle's cab occurs along all three axes, both linearly and rotationally. The most important is vertical movement transferred though the vehicle suspension and car seat. A method to estimate the perception of discomfort was standardized in ISO 2631 in 1974 [49].

In addition to specific research topics, research tools were developed and improved in this decade. Eye trackers, devices used to measure eye-gaze location, became available for vehicle and simulator use in the 1970s. For example, Mourant, Rockwell, and others measured glance time to the mirrors, radio, and the road while driving for novice and experienced drivers [50].

The driving simulator became a tool in human factors research. Volkswagen developed a driving simulator with a three-axis gimbal. A CRT display was used to present a road scene that involved a computer-generated line drawing. Various sounds were also presented. This driving simulator was used to investigate the driver's evasive behavior [51]. A driving simulator using a linear rail was developed at Virginia Tech in 1975 [52].

This most noteworthy result of this decade was the translation of human factors research into practice. Various standards were prepared to design controls and to evaluate seating comfort. Crash dummies were established and utilized by the New Car Assessment Program (NCAP), which began in 1979 in the USA.

3.5. Human Factors Research in the 1980s: Computer-Aided Design for Automobiles, Cab Comfort, Rollovers, and Assessment Methods. As with every recent decade in the USA, the 1980s had a particular vehicle that received attention for issues related to crashworthiness. That vehicle was the Jeep CJ-5, whose rollover propensity was the subject of a broadcast by *60 Minutes*, the most-watched investigative news program on US television. The critical episode, broadcast on December 21, 1980, showed Jeep CJ-5s rolling over when making sharp turns. What many fail to recall is that there was supporting statistical data showing that the CJ-5 was much more likely to roll over than other similar vehicles [53, 54]. For an interesting summary, see [55]. The CJ-5 problems served to spark human factors research on vehicle handling.

Another vehicle that received attention in that decade was the Suzuki Samurai, a short wheelbase, four-wheel drive utility vehicle with a propensity to roll over. Suzuki had a very bitter legal battle with the Consumer's Union, which publishes the most popular consumer magazine in the USA,

Consumer Reports. Unusually, the vehicle was rated as "not acceptable." Sales dropped from 77,500 vehicles in one year to 1,400 the next year. Suzuki sued the Consumers Union but lost, and the production of the Samurai ceased. The Suzuki case emboldened safety advocates who had been sometimes reluctant to challenge the auto companies with "deep pockets" to fund protracted legal actions.

Allegations of unintended acceleration of the Audi 5000 were publicized on *60 Minutes* on November 23, 1986 [56]. Again, given the bad publicity, sales of the Audi 5000 plummeted from 74,000 vehicles in 1984 to 12,000 in 1991. Ironically, the final verdict from the U.S. Department of Transportation was that, while there were design aspects that could startle drivers or contribute to a higher incidence of pedal misapplication, there was nothing requiring a defect notification [57]. The important point here is that this is probably the first time that questions raised by the news media about vehicle safety were not supported by further investigations.

Interestingly, in recent years, there again have been questions raised concerning unintended acceleration; this time was for Toyota vehicles. *Dateline NBC* was the program involved, but in some ways the Toyota case is strikingly similar to that of the Audi 5000. There were allegations of trapped floor mats and concerns about failure of the electronic control systems, a claim that was debunked by NASA [58]. Again, Toyota sales suffered as a consequence, but no vehicles were withdrawn from the market.

In 1980, Brown stated that the improvement in the crash statistics "has undoubtedly resulted from technological advances in the design of steering, braking, tires and suspension systems, affording the driver better control of his vehicle" [59, pages 3–14]. He also emphasized the importance of optimizing information presentation in the vehicle and introducing objective evaluation and quantification instead of pure subjective assessment.

New tools for designing cab dimensions and visibility were developed in the previous decade. Chrysler developed CYBERMAN, a digital human model (manikin) in 1974. However, it was simple and its usefulness was limited. The System for Aiding Man-Machine Interaction Evaluation (SAMMIE) was developed in the UK for a consulting service for ergonomic design by SAMMIE CAD, Ltd., in the 1970s. The three-dimensional, digital human model consisted of 21 links and 17 joints. The cab dimensions and layout of controls in the cab could be evaluated by specifying the joint angles of the three-dimensional human model based upon anthropometric data of representative drivers. Various digital human models were developed during this period. Linked with computer-aided design (CAD), digital human models worked effectively. SAMMIE worked with SAMMIE CAD system, but interchangeability with other systems was limited. Jack (USA), RAMSIS (Germany), and other digital human models were developed during this period. RAMSIS could link with the CATIA CAD system, which was and still is the most commonly used CAD system in the automotive industry. Compared with the traditional anthropometric data and hard manikins, digital human models can lead to shorter development times of vehicle cabs, reduce development cost,

and lead to cabs that accommodate a larger fraction of the population [60, 61].

Head-up displays (HUDs) were initially developed for aviation and superimpose information of aircraft air speed, altitude, and angle of attack onto the forward view. As eye transition and accommodation times were reduced, the user could spend more time looking at the forward scene. In motor vehicles, HUDs have been used to show vehicle speed, warnings, turn signals, and more recently, navigation information. The first studies with HUD prototypes were conducted by Rutley [62], who showed that HUDs can have benefits without the negative distracting effects reported in aviation applications [63]. HUDs were introduced in the market at the end of the 1980s (General Motors 1988, Nissan 1988). As the initial application was to present speed, which was not as time-critical as the flight data shown in aircraft, the customer demand for automotive HUDs when introduced was not great.

Also occurring at this time was considerable research to assess human thermal comfort [64], research that has its origins in Willis Carrier's development of the psychrometric chart [65]. The factors contributing to human thermal comfort, air temperature, radiant temperature, air velocity, humidity, metabolic rate, and the distribution and insulating value of clothing were not all easy to measure in a real vehicle cab. To evaluate space-suit thermal comfort, in 1966, NASA developed a thermal manikin that had a three-dimensional human body and simulated the heat transfer between the human body and the thermal environment. By the end of the 1970s, thermal manikins were used to estimate thermal comfort in vehicle cabs [66].

Drowsiness while driving increases crash risk. A driver's drowsiness, arousal level, and fatigue can be measured using such physiological variables as EEG (electroencephalogram), heart rate, respiration rate, and GSR (galvanic skin response) [67]. As was shown in early studies, physiological measures could be reliably measured in experimental conditions and provided useful information. However, it was difficult to convert the research into practice and develop a commercial drowsiness-detection system, primarily because wired sensors were needed. Thus, in the 1980s there was a shift towards noncontact image sensors (video cameras) that looked for slow eyelid closure to detect drowsiness [68]. Studies were conducted to obtain quantitative measures based on video images, and in the next decade PERCLOS (percentage of eyelid closure time) was established as the index of drowsiness [69]. In 2008, Toyota introduced a crash-mitigation system with eye monitor that detected eyelid closure and warned the driver.

Workload-measurement methods were established during the 1970s [70]. These methods used subjective measures (the Cooper-Harper scale, SWAT-the Subjective Workload Assessment Technique, and NASA TLX-the Task Loading Index), primary task performance measures, secondary task measures (from the task loading and subsidiary task methods), and physiological measures (EEG, pupillary response, eye movement, and heart-rate variability). They were used to measure mental workload while driving. Miura collected detection-reaction times to the illumination of small bulbs located around the front window, as the subsidiary task, to measure the useful field of view [71]. Results indicated that the useful field of view became smaller, and the reaction time of a detection task became longer as task demands increased (e.g., driving in crowded traffic).

With increasing computer power, large driving simulators were developed in the 1980s. In the 1970s, VTI of Sweden began developing a driving simulator with a two-axis gimbal and a linear rail. It had a wide screen and was controlled by a detailed vehicle-dynamics model [72]. An example of its use, which began in 1983, was the investigation of driving on slippery roads and the effects of alcohol. The major development was the Daimler-Benz high-fidelity driving simulator with a motion system that combined the hexapod motion platform and two-dimensional linear rails. A full-size vehicle was placed in the dome on the motion platform. It was introduced in 1984 and was used to investigate active-safety systems, vehicle dynamics, and other topics [73]. During the 1980s, various driving simulators were developed in the USA, Europe, and Japan [74]. Common topics in the 1990s included studying driving behavior in risky conditions, the use of driver information systems [75–78] and the use of advanced driver-assistance systems (ADAS) [79, 80] and the effectiveness of warning systems of various types. One example was using the pedals and steering wheel to provide active feedback to facilitate drivers' performance of a recommended action [81].

The end of the 1980s saw the beginning of an era of driver information and driver-assistance systems (see the next section). One early human factors study of driver information systems involved measuring glance time and number of glances for a variety of conventional tasks and navigation tasks using a prototype computer map navigation system [82]. One study indicated that centerline deviation increased when the driver used a CRT touch screen [83].

The 1980s were the decade of the computer. Digital computers and software began to see wide use in human factors research, including digital human models for designing cabin accommodations, thermal manikins for evaluating thermal comfort in the cabin, and video systems for measuring drowsiness. Computer technology reduced design time and made handling complex data easier. The questionnaire and the secondary-task methods were established for mental-workload measurement based on resource models from psychology. These measurement methods and driving-simulator technology would become useful human factors research tools for the intelligent vehicles and connected vehicles in the following decades.

4. Human Factors Research Since 1990s: The Era of Intelligent Vehicles and Connected Vehicles

4.1. Driver Information Systems and Driver Distraction. Research on automotive human factors reached a turning point in 1990 with the introduction of Intelligent Transportation Systems (ITS), previously known as Intelligent Vehicle Highway Systems (IVHS). With the aim of enhancing vehicle mobility and safety using information and communication technologies, government projects began in the USA and Japan. The Electronic Route Guidance System (ERGS) was conducted in the late 1960s in the USA [84]. The Japanese projects included the Comprehensive Automobile Traffic Control System (CACS) (1973), Road/Automobile Communication System (RACS) (1984), Advanced Road Transportation System (ARTS) (1989), and Vehicle Information Control System (VICS) (1990) [85]. Europe's research initiative Programme for European Traffic of Highest Efficiency and Unprecedented Safety (PROMETHEUS) (1987–1995) initiated the research era of driver information and driver-assistance systems [86]. PROMETHEUS was followed by a sequence of projects (e.g., DRIVE, GIDS, EMMIS, HASTE, and AIDE) that focused on the development of integrated HMI concepts [87] and suitable evaluation methods [88].

The automotive industry also promoted ITS technology developments during this period. In 1981, Honda released Gyrocator, the first in-vehicle navigation system with a map using a transparency sheet. At about the same time, Toyota released NAVICOM, which indicated the direction of a destination using a simple arrow. Etak Navigator, the first after-market car navigation system using a digital map, was released in 1985 in the USA. The digital map was stored in cassette tapes and location was determined by a dead-reckoning system using a compass. In 1987, Toyota launched Electro Multi Vision, which was a predecessor of present-day, in-vehicle car navigation systems (Figure 11). An in-vehicle navigation system manufactured by Sumitomo Electric was installed in the Nissan Cima in 1989 [89]. The Bosch Travelpilot was delivered in the same year in Europe. In-vehicle navigation systems spread after GPS became available in 1990 (officially in 1993). The first on-board installed navigation system including a GPS unit and map material in Europe was delivered in 1994 by BMW using a color-TFT display and a button-operated software menu system. Later versions, which supported audio and communication functions, were moved to the center console and/or operated by a touchscreen, depending on the OEMs human-machine interface (HMI) concept. This development steadily led to unique integrated solutions for each brand as well as unique mobile navigation systems.

There were various efforts to design integrated driver interfaces for in-vehicle information and other existing in-vehicle systems (audio and climate) as the number of functions was increased. Toyota developed the integrated joystick (Toyota Ardeo 1998). BMW introduced i-Drive. Mercedes introduced Command. Audi introduced MMI (Multi Media Interface) as well (2001), which similarly included interaction using a rotary control knob in the center console [63]. Nissan introduced its integrated driver interface in the same year (Figure 12). The position of a central information display close to the windscreen became common at the end of the 1900s.

As with other decades in the USA, the 90s was not without its media controversies over crash risk, the most noteworthy of which was the 1977–1983 CK pickup, the most popular vehicle sold by General Motors. In a very dramatic

FIGURE 11: Early car navigation systems (Toyota 1987 (a) and BMW 1994 (b)).

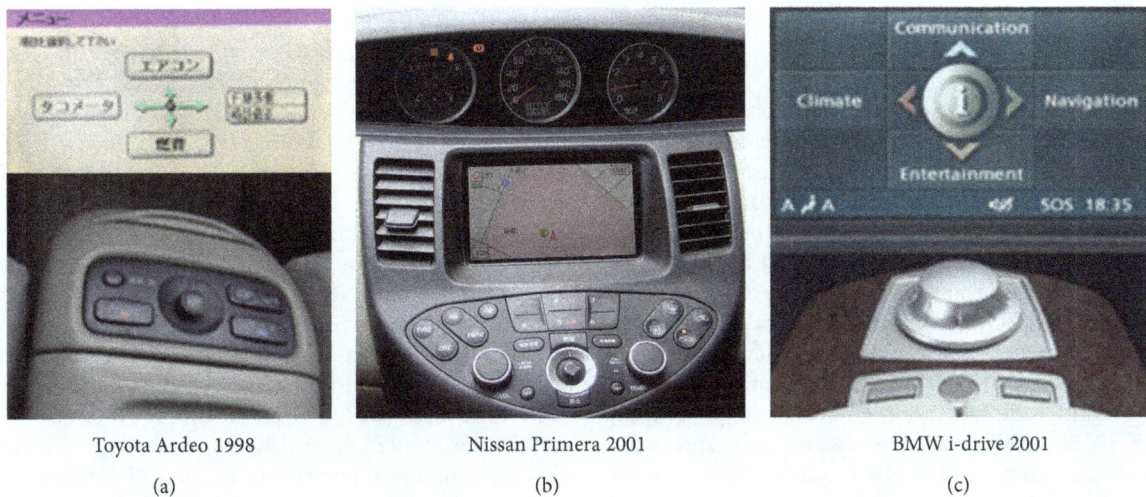

Toyota Ardeo 1998 (a) Nissan Primera 2001 (b) BMW i-drive 2001 (c)

FIGURE 12: Controls for in-vehicle information systems.

presentation on NBC's *Dateline* in 1993, a very popular news investigative show, a CK was shown being struck in the side and bursting into flames. The allegation was that the fuel tanks, mounted outside the frame rails, were vulnerable and could lead to fires if struck. Interestingly, careful investigation by General Motors found that the crashes had been staged, and rocket igniters had started the fires. In response, NBC retracted the story and paid General Motors for the cost of their investigation [90]. This was a huge blow to the news media and reduced its influence in advocating for auto safety.

Until 1990, the driver was regarded as an element of the driver-vehicle system, interacting with the vehicle by operating the steering wheel and pedals to manage the primary driving task. When a navigation system was installed inside the vehicle, the driver had to perform not only vehicle-control tasks by operating the vehicle, but also navigation tasks. When drivers used a paper map, reading the map while the vehicle was in motion was not easy. Often drivers had to stop the vehicle and read a map to find their way to a destination. When a navigation system was installed inside the vehicle, and the system indicated where to turn, the navigation task could easily be performed in parallel with driving tasks (i.e., a dual-task condition).

Concerns about excessive task demands led to studies of mental workload, human cognitive activity, and what is now commonly known as driver distraction. The 1990s saw the delivery of such guidelines as JAMA Guidelines (version 1.0 in 1990, and version 2.0 in 1999), UMTRI Guidelines (1993) [91], TRL Checklist (1999) [92], HARDIE Guidelines (1996) [93, 94], German Code of Practice [95] and other guidelines in Europe [96]. They gave descriptive principles for designing in-vehicle information systems. Also, relevant ISO activities were initiated to develop standardized evaluation methods and formulate minimum standards for in-vehicle HMIs [97].

Studies by Wierwille et al. and Zwahlen et al. in the previous decade suggested that glancing behavior could be an objective measure of driver distraction [98, 99]. Eye-glance evaluations are most readily conducted for information systems that have been developed and are available for on-the-road use. However, the systems must be assessed during the development. The occlusion device developed by Senders in the 1960s (see Figure 9) to measure visual demand in driving was used to simulate glance behavior during driving [100, 101]. Studies using the occlusion method with liquid-crystal shutter goggles were conducted under the aegis of the Alliance of Automotive Manufacturers (AAM) (USA),

FIGURE 13: Occlusion method with the shutter goggles.

ISO/TC22/SC13/WG8, and the Japanese Automobile Manufacturers Association (JAMA) to assess the level of distraction caused by visual-manual tasks and their degree of interruptibility (Figure 13) [102]. This method was internationally standardized as ISO 16673 in 2007 [103] based on input of the Advanced Driver Attention Metrics (ADAM) and Adaptive Integrated Driver-vehicle interface (AIDE) projects among others. In 2004, JAMA delivered JAMA Guideline version 3.0, which prohibited tasks that required a total glance time of more than 8 seconds [102]. SAE Recommended Practices J2364 (15-Second Rule and another occlusion procedure) [104], SAE J2365 [105] (task time estimation), and other procedures were also published as a result. In the search for entry methods that were less demanding than visual-manual interfaces, speech interfaces were examined [106, 107].

Several international design guidelines for in-vehicle information systems have been developed mainly in ISO/TC22/SC13/WG8 since 1994. Published standards were ISO 15005 (dialogue management) and ISO 15007 (measurement of visual behavior) in 2002, ISO 15008 (visual presentation) in 2003, ISO 17287 (suitability of TICS while driving) in 2003, ISO 15006 (auditory information) and ISO TR 16951 (criteria for determining priority of messages) in 2004 [108–112]. ISO 26002 (simulated lane change test, LCT) was published in 2011 for assessing driver distraction based on research from the ADAM project in Europe [113]. LCT was developed to evaluate visual manual secondary tasks but also cognitive loading tasks that used speech interfaces or involved phone conversations [114]. Burns et al. give an overview of the relevant evaluation methods [115].

Driver information systems have been developed as research projects since the 1970s and yielded commercial products such as car navigation systems in the 1990s. During their development, researchers were aware of the importance of human factors because using driver information systems while driving was quite different from using conventional in-vehicle equipment, with some ideas from studies of human-computer interaction for office work providing useful insights. In contrast to conventional in-vehicle systems, drivers could be confronted with a large amount of real-time

information with which they interacted while driving. Measurement techniques for mental workload, glancing/visual behavior, and task demand developed in the last decade were applied to assess the amount of effort to use these driver information systems. Human factors researchers also played important roles in establishing guidelines and standards that offer principles for designing the systems in advance and evaluation methods accompanying the development process. Having guidelines and standards that were publicized by common agreement facilitated entrenching this technology in society.

4.2. Human Factors Research for Advanced Driver-Assistance Systems. The 2000s were another decade in which crash safety received attention in the news media in the USA. High-profile media stories included (1) rollovers of Ford 15-passenger vans (picked up by several television programs), (2) rear-end crashes and subsequent fires involving 2005–2007 Ford Crown Victorias (commonly used as police cars), picked up by both NBC *Dateline* and CBS, and (3) rollovers of the 1998–2001 Ford Explorer. The Explorer received the most attention, including a segment on 60 Minutes and an entire hour on the PBS Frontline program (http://www.pbs.org/wgbh/pages/frontline/shows/rollover/etc/script.html, http://www.pbs.org/wgbh/pages/frontline/shows/rollover/etc/video.html). The Explorer problem was a combination of a high center of gravity combined with a narrow track width, along with failures of particular Firestone tires, which resulted in rollovers [116, 117]. One of the consequences of this matter was the passage of the US government's Transportation Recall Enhancement, Accountability and Documentation (TREAD) Act, which led to new tire-labeling standards, requirements for tire-pressure monitoring systems, and other changes.

Automobile safety technology began with efforts to reduce the consequences of crashes, by designing vehicles that would be less lethal when struck. Over time, there has been somewhat of a shift in human factors research towards active safety, seeking ways to prevent crashes.

The antilock braking system (ABS), first introduced in 1970, marked the formal beginning of active-safety technology. In 1990, electronic stability control (ESC) and vehicle stability control (VSC) came into widespread use. Adaptive cruise control (ACC) systems, which allow a vehicle to follow the preceding vehicle automatically by maintaining a preset time gap, were introduced by the end of the 1990s [118–120].

In addition, backup monitors utilizing the navigation system's display were introduced in the 1990s to reduce backing crashes. The 2000s saw the introduction of lane-keeping systems, which assist drivers by steering to help them stay in the lane (Nissan 2001), and the collision-mitigation braking systems, which intervene with active braking when distance-sensor data indicates that a collision is unavoidable. These systems are an extension of lane-departure warning systems and blind-spot warning systems. Recent entries into the market are the lane-change decision-aid systems, which provide warnings when the driver begins to change lanes, but another vehicle is in the adjacent lane.

Projects such as INTERACTIVE, SAFE-SPOT, and PREVENT deployed the advances in advanced driver-assistance systems (ADASs) and developed human-machine-interaction approaches for assisted driving. These European initiatives were accompanied by national research programs. In Germany, examples of these include MOTIV (1996–2000), INVENT (2001–2006) AKTIV (2006–2010), SIM^{TD} (2008–2013) and UR:BAN (2012–2016). A major achievement of these projects was intense cooperation between European OEMs, suppliers, and university researchers. Similarly, there has been a series of ASV (Advance Safety Vehicle) projects in Japan (ASV1 (1991–1995), ASV2 (1996–2000), ASV3 (2001–2005), ASV4 (2006–2010), ASV5 (2011–)), involving collaboration between the government and car manufacturers.

As part of the research on ADAS, there were a number of new measures of driving performance developed for car following (time headway, THW) [121, 122], lane keeping (time-to-line crossing, TLC) [123, 124], and braking maneuvers (time to collision, TTC) [125–127] over this decade and prior decades.

If several ADASs are installed in a vehicle, various warnings and other information will be given to the driver. In complex driving situations, multiple warning signals may occur simultaneously. In such cases, the driver may become confused and be unable to respond to the warnings or may not react appropriately. Therefore, warning signals should be integrated (ISO TR 12204 [109, 128].

An important human factors element of ADAS as assistive technology is the relationship between the driver and the system and especially the human-machine interface. In many cases, the driver receives feedback on the system state via the speedometer-tachometer cluster, center console displays, or a HUD, supplemented by force feedback from the steering wheel and pedals. If the driver does not comprehend the system's behavior as it actually is, "automation surprise" occurs when the system behaves unexpectedly. Therefore, interaction concepts for these systems have to take into account phenomena such as "over trust" and "over reliance" on the system to avoid serious problems [129]. Currently, numerous ADASs are available as mature products to support longitudinal and lateral vehicle control. Over time, the control authority of these systems has increased, and more complex, cooperative systems have been investigated [130–132]. How to integrate several ADASs and driver information systems has also been the topic of research [133, 134].

By definition ADASs are intended to assist drivers, so these systems must be designed to be compatible with driving behavior. An ADAS that does not consider driver ergonomic requirements may increase the risk of a crash, even though its aim is to enhance safety. Human factors research is necessary to understand how drivers behave with or without the systems in an actual road environment, not in a laboratory experiment. The research methods described in the next section are necessary for such research.

4.3. Naturalistic-Driving Studies and Driving Simulator Studies. One of the research developments of the 1990s has been the completion of several naturalistic-driving studies as

knowing what normally happens on real roads is necessary when developing ADASs. If driving situations are known to be dangerous, then the type of ADAS that should be developed for safety assistance is readily determined. Also, quantitative analysis of driving behavior on actual roads is beneficial for developing vehicle-safety technologies, as well as for developing future driver-assistance systems.

Traditional human factors studies involving controlled experiments are relatively low cost. On the other hand, one cannot conduct a naturalistic-driving experiment of any size for less than $10,000,000, and many cost much more. For that price, one could conduct 20–100 driving simulator experiments, depending upon their complexity. Until the 1990s, there was not sufficient interest in the topics that naturalistic-driving studies address to find funding for them.

Second, naturalistic-driving studies require compact data-collection hardware, low power, a large amount of data-storage capability, and sophisticated wireless communication, so that highly reliable and readily accessed data can be collected. Before the 1990s, that technology did not exist.

In the USA, the National Highway Traffic Safety Administration (NHTSA) conducted the 100-Car Naturalistic-Driving Study in 2001. They collected data on vehicle behavior, road-traffic conditions, and driver behavior in accidents and near-accident incidents, using vehicle-acceleration data as the trigger for recording. This study demonstrated that various distracting situations lead to traffic accidents in the real world [135]. Other relevant studies include the Advanced Collision Avoidance System (ACAS) [136], RDCW [137], and IVBSS [138] projects. Easily installed driving recorders for general-use passenger vehicles became commercially available in Japan in 2003. Detailed causal analysis of accidents and near accidents became possible with this device. JSAE has examined data gathered by driving recorders installed in taxis [139]. They conducted statistical analyses to classify the causes of accidents and also identified specific situations in which drivers committed behavioral errors.

The New Energy and Industrial Technology Development Organization (NEDO) of Japan conducted a three-year project beginning in 2001 to collect driving-behavior data under normal conditions, with no accidents, using instrumented vehicles in real road environments, and compiled the results in a database. This driving-behavior database has been publicly available since 2004 and has been used by universities, research institutions, and industry in research and development activities [140].

In Europe, the EURO-FOT (field operational test) study and the Promoting real Life Observations for Gaining Understanding of road user behaviour in Europe (PROLOGUE) project gathered remarkable naturalistic-driving datasets. EURO-FOT focused on the usage patterns considering ADAS and driver information system applications. An important output from EURO-FOT was the so-called FESTA handbook [141], which provides good practice recommendations for conducting naturalistic-driving studies.

For ADASs that assist steering and pedal control, a control algorithm should be developed to match the driver expectations of the system's behavior. If the control algorithm of ADAS is different from what the driver expects, the driver may feel uneasy and may not use the system. Traditional research methods, designed to repeat a controlled set of conditions so that they can be examined in a cost efficient manner, are an imperfect representation of the real world. For ADAS design, addition information was obtained from naturalistic-driving-behavior studies [142, 143]. To analyze the large data sets from naturalistic-driving studies, specialized statistical-modeling techniques are used [144, 145]. However, prior to applying these methods, the situations and conditions in which the targeted driving behavior occurs must be identified to create the subset of the data desired for analysis.

Improvements in computer-graphics technology and computing performance enabled detailed representation of road structures and traffic-participant behavior. As a result, simulators could be used as tools in driver-behavior research. Using a driving simulator, experiments can be conducted repeatedly, controlling such traffic situations as the positions of other vehicles relative to the subject vehicle. Experiments using a driving simulator are time efficient and do not expose subjects to the risk of real injury in a crash. Taking advantage of these capabilities, researchers are able to analyze the effectiveness of systems being developed and can anticipate potential problems by analyzing drivers' responses to the prototypes [146].

Until the 1990s, driving simulators were only found in a limited number of laboratories, primarily because of their cost. In part this was because rendering of scenes required high-performance graphic processors, and prior to the 1990s systems with adequate performance were specialized and costly. Second, projectors that had adequate resolution and brightness were also quite costly. After the '90s, the simulator hardware components became much less expensive.

Simulators are useful tools for investigating driver behavior. Driving simulators range from those resembling PC games to full-scale driving simulators such as the National Advanced Driving Simulator (NADS) and Toyota driving simulators. Although driving simulators are now commonly used for automotive human factors research, the research must be conducted with a clear understanding of what each simulator is capable of reproducing and to what degree, and with sufficient assessment or validation of the appropriateness of use for the experiment's purpose [147]. New researchers often do not spend enough time to make sure the values of the dependent measures collected are reasonable for real vehicles. A high-quality research program will likely include a balance of simulator experiments and actual road experiments or naturalistic-driving data [148].

Naturalistic-driving studies and driving simulator studies have proven to be powerful tools for analyzing driving behavior, assessing effectiveness, and identifying problems not only of driver information but also of driver-assistance systems. In the past, automotive human factors research typically focused on the relationship between drivers and vehicles. Now, research has gone beyond the human factors laboratories and extended to human behavioral research in the real world.

4.4. Driver Communications External to the Vehicle—Network Service, Mobile Phones, and Internet Access While Driving. The introduction of information-communication technology has been particularly important for driver information systems. The Vehicle Information and Communication System (VICS), which transmits real-time traffic conditions for specific driving regions through FM radio signals and radio/optical beacons, began operating in 1996 in Japan. In Europe, the Radio Data System (RDS), introduced in the 1980s, later became the Traffic Message Channel (RDS-TMC); it conveys traffic information and messages via the FM signal. OnStar, a network service for GM, was started in 1995 in the USA. This was followed by TeleAid in 1997 in Germany, Toyota's MONET in 1997, Nissan Carwings in 1998 in Japan and BMW Assist and Mercedes MBrace in the late '90s. When the driver accesses the remote operations center of one of these systems, the operator assists with the trip according to the driver's request. Analysis of verbal communication between the driver and the operator, such as phrases used, the timing of utterances and pauses, and the number of turns, will provide insights into designing interactive speech systems for driver information systems.

Mobile radio phones installed in vehicles were first developed in 1947 by AT&T, but the service area was very limited and the phone itself was bulky. The A-Netz mobile-phone network started in Germany in 1952. The first cellular network began operating in 1979 in Japan. In the mid-1990s, cellular phones spread rapidly based on the Global System for Mobile communications (GSM) standard, and, not surprisingly, people used the phone while driving. The use of cellular phones while driving soon became a public-safety concern, and using a hand-held cellular phone while driving was forbidden in many European member states in the 1990s, in Switzerland in 1996, and in Japan in 1999 [102]. Use of cellular phones for conversation is also illegal in some states in the USA [http://www.ncsl.org/issues-research/transport/cellular-phone-use-and-texting-while-driving-laws.aspx] and in a number of countries [149]. Hands-free systems for vehicles have since been introduced and have been shown to be less distracting [150]. The nature and extent of the interference of phone conversations while driving continues to be an important research topic [135, 151–154]. Of increasing importance is the effect of using cell phones on situation awareness [155]. Nonetheless, people use phones while driving for many reasons: they may feel that they do not have too much to do, believe driving is wasted time, feel a need to be connected, are bored, or for many other reasons. Use of phones while driving is widespread [156].

Voice communication by phone is one of many ways for people to communicate and interact with each other and with information systems. However, if the in-vehicle system restricts the access to information strictly for safety purposes, drivers might not connect the device to the in-vehicle system, bypassing the restrictions imposed by the vehicle. How to support interaction with data in these devices that drivers need and want while driving without relying on a visual-manual interface needs further human factors research. Interestingly, relative to the amount of research on phone use in conversation, relatively little research has been done on interaction with the Internet and intelligent systems while driving [157].

Some thought should also be given to what drivers really want or need to know. Qualitative methods for recording and analyzing human behavior in daily life are being developed in the field of sociology [158–160]. Such methods include ethnography, which describes detailed human behavior, and action research, in which the researcher explores problems of a society while acting as a member of the targeted society (See also [161, 162]).

Communication with those outside the vehicle that is not relevant to the driving task can cause driver distraction. Compared with interactions with driver information systems or ADASs, communication through mobile phones and the Internet is independent of the driving itself. Incoming alerts for phone calls, e-mail, and Short Message Service are external system-initiated interactions that occur regardless of the driving situation. There is a basic potential of driver distraction. To avoid this, there is a big potential if communication devices (nomadic devices) are connected to an in-vehicle information system that can control interaction with the driver to support the driver in the management of his workload. Discussions of possible mechanisms and interfaces for managing information to reduce workload and enhance situation awareness of the drivers were reported in the ITU-T FG Distraction activity [163–165]. However, the nomadic device should first be connected to the in-vehicle system, but should not bother the user. Connectivity technologies such as Bluetooth are important enablers here. Human factors research must design the in-vehicle system to give the driver an incentive to connect the device. Targets of human factors research are not only reducing workload and improving ease of use, but also designing system to induce safe driving.

4.5. Vehicle Communications with Other Vehicles and the Infrastructure. At first thought, these communications would appear to have nothing to do with human factors, which would be incorrect. The purpose of these communications is ultimately to deliver information to the driver. A major part of the cost of building systems to warn of and avoid crashes is the sensing systems, the radar, LIDAR, video, and sonar technologies to provide 360 degree coverage to support the driver. These sensors provide information to determine where all the threats are to the vehicle. This requires identifying each target from the background, identifying the type of target it is, and then developing a prediction of its path, which is used to determine if the target will collide with the driven vehicle. For locations where crashes are frequent, embedding sensors into the infrastructure is cost effective. Infrastructure-based cooperative systems were developed in Automated Highway System (AHS) projects (1996–2007) and the Driving Safety Support System (DSSS) project in Japan [166]. DSSS became operational in 2011 as a pilot study [167]. The system detects vehicles that are hidden by road structures at intersections, merging zones, and curves and informs the driver using an in-vehicle display and by voice [168]. An alternative approach is being examined under the UMTRI-led Safety Pilot program [169] and in other connected-vehicle activities. In a connected-vehicle approach, every vehicle, every pedestrian, and some key fixed objects that are part of the road infrastructure continuously transmit radio signals that communicate what they are, where they are, and, if they are capable of moving, how fast and in what direction they are moving. This, when fully fielded, could simplify the collision detection problem and lead to a potentially significant reduction in crashes, if the response to potential collisions is automatic.

What remains unknown is how to get drivers to respond to hazards they cannot see and may not become an imminent threat for some time [170]. How drivers should be warned if some of the broad array of information is unavailable, and when vehicles should take over the primary driving task will be a focus of future human factors research.

4.6. Autonomous Vehicles—Removing the Driver from Control. Until recently, self-driving cars seemed like a futuristic concept. However, with DARPA's Grand Challenge program [171], Google's demonstrations (http://spectrum.ieee.org/automaton/robotics/artificial-intelligence/how-google-self-driving-car-works), and other activities such as Stadtpilot in Germany [172], advances in autonomous vehicles are occurring quickly.

Questions of concern to human factors researchers include the following: When can automation do a better job of driving than a human being? How can drivers be kept informed of the driving situation? How does the hand-over (driver to vehicle, vehicle to driver) occur? How do drivers of nonautonomous vehicles negotiate with the behavior of autonomous vehicles?

5. What Can Be Learned from History?

In general, the introduction of the automobile and the related achievements in human factors can be called a success story, having served as a stimulus for other research domains.

(1) Over time, the human factors focus has shifted from relying on personal experience to relying on research data that eventually led to standards from SAE, ISO, and others. However, as vehicles evolve, there will continue to be a need to conduct research to develop new standards, and to support the design of vehicles. Relative to other fields of engineering, the use of models to predict human performance while driving (except for control theory and workspace layout) has been limited [173]. Research on computational models of the heterogeneous group of drivers as information processors in very different traffic situations is needed as well as a significant effort to build practical tools engineers can use [174, 175]. Given what has occurred in the past, an important step would be incorporating those models in SAE and/or ISO standards.

(2) Over time, the primary problems that human factors experts address have increasingly shifted from physical to cognitive, but the original problems never go away. Early human factors efforts concerned making sure that drivers

could operate controls while providing adequate force to steer and brake. Although power-assist systems have assured braking and steering can be accomplished, questions about the optional human-device transfer function remain, as well as where to place controls so they can be comfortably operated. There are still issues of field of view, seating comfort, and thermal comfort, especially in connection with electric vehicles. Designers still wrestle with these issues and continue to request better data, better models, and better tools.

(3) Over time, there has been a shift in what the driver does. Initially, the driver just steered the vehicle, sometimes assisted by the codriver. Now, the driver controls an array of information and communication systems being assisted by the vehicle. Driver distraction and overload are major concerns. Research on how to coordinate performing the primary driving task and communicate with those outside of the vehicle, or both people and vehicles, are needed. The need for driver assistance is continuously increasing, especially in urban settings.

(4) Over time, developments in the automotive industry related to human factors mirror technology developments in general with a shift from providing basic mobility to concerns about crash protection and fuel efficiency. The early developments were related to the physical structure of the vehicle, the province of the mechanical engineer. More recent developments are the province of electrical and computer engineers. The most recent efforts, such as the nomadic device forum of the AIDE project, have involved engineers who develop nomadic and mobile devices brought into the vehicle. The next phase of vehicle evolution may center on the motor vehicle as a social mechanism, thus involving urban planners, sociologists, anthropologists, and others. One example of this concerns how to support the use of social networks (and what should be supported) while driving.

(5) Over time, evaluation methods have changed. The original human factors work was based strictly on intuition. That was followed by decades of research involving single test vehicles in scripted on-road experiments along with the analysis of crash data, almost exclusively from the USA. In the last few decades, the use of driving simulators in combination with eye tracking, but also laboratory evaluation of interaction concepts, has become much more widespread. The major recent development in methods has been naturalistic-driving studies and field operational tests, providing extensive real-world driving data. What remains unknown is at what point these studies transition from independent evaluations to a continuing data collection effort analogous to crash evaluations. Also unknown is when some country other than the USA will make its crash data publically available on the web. Without such information, research and design solutions will invariably focus on American problems, which may not match the driving situation in other countries.

(6) Over time, the way in which designers and researchers interact has changed. Initially, that occurred though major, large conferences such as the SAE Annual Congress, the TRB Annual Meeting, and others. Increasingly, however, the preferred venues are smaller, more focused meetings concerning automotive human factors in general, or specific aspects of that topic such as Driving Assessment and AutoUI.

In addition, an important degree of informal interaction occurs at standardization meetings of various types.

(7) The news media have been a significant factor in bringing issues of crash safety to light, at least in the USA. Fires, crashes in which children are killed, and rollovers invariably get the most attention. At least once every decade there are major questions raised about the safety of at least one vehicle—Chevrolet Corvair, Ford Pinto, GM CK pickup trucks, Jeep CJ-5, Audi 5000, Ford Crown Victoria, Ford Explorer, and so forth. As a result, auto sales plummet for these models, and the manufacturers respond. Not all of the problems receiving attention from them have been genuine. However, at least in the USA, laws have been passed, research funded, and organizations created because of these media investigations.

The role of the news media in the future is uncertain. The USA was traditionally dominated by three television networks—NBC, ABC, and CBS. However, in recent years there has been competition from other networks in the USA, and foreign networks will soon have a greater presence in the USA. The competition has reduced funding for investigative journalism, but in its place, Internet journalism has arisen.

(8) Until now, automotive research and design have been dominated by the USA, Europe, and Japan. However, with China being the largest market for motor vehicles, and a growing market in India, there is the potential for them to be leading contributors to the automotive human factors research and design in the future.

Thus, although many may view traditional motor vehicles as part of an outdated industry, in fact, the industry has continued to evolve, with continuing pressure to introduce new technology into vehicles to increase safety and comfort and to develop cleaner, more fuel-efficient vehicles. However, the challenge the motor vehicle industry faces that the consumer products industry does not face is the high level of reliability and durability required, a concern that dates back decades ago as described in the literature.

As one can tell from the references provided, there has been an abundant and almost overwhelming amount of research conducted on automotive human factors. Those wishing to delve more deeply into the field may wish to begin by considering other overviews of automotive human factors, such as [176–182]. As the field of automotive human factors continues to evolve, it is important for designers, engineers, researchers, and others working on this topic to continue to learn about it. Reading a few papers or taking a human factors class is not enough. To keep informed, one needs to continue reading about the field, attend conferences, and participate in professional activities.

Acknowledgment

Copyright (c) 1956 SAE International. Reprinted with permission from SAE 560061/SP-142A.

References

[1] F. W. Wells, *Occupant Protection and Automobile Safety in the U.S. Since 1900*, SAE International, Warrendale, Pa, USA, 2012.

[2] A. J. Yanik, "The first 100 years of transportation safety: part 1," in *The Automobile: A Century of Progress*, pp. 121–132, Society of Automotive Engineers, Warrendale, Pa, USA, 1997.

[3] "25 Years Progress of SAE," A leaflet of the 25th Anniversary Celebration of the Society of Automotive Engineers, Society of Automotive Engineers, Warrendale, Pa, USA, May 1930.

[4] W. Kamm, *Das Kraftfahrzeug*, Springer, Berlin, Germany, 1936.

[5] A. J. Yanik, "The first 100 years of transportation safety: part 2," in *The Automobile: A Century of Progress*, pp. 133–149, Society of Automotive Engineers, Warrendale, Pa, USA, 1997.

[6] N. Ach, "Psychologie und technik bei bekämpfung von auto-unfällen," *Industrielle Psychotechnik*, vol. 6, no. 3, pp. 87–105, 1929.

[7] T. W. Forbes, "The normal automobile driver as a traffic problem," *The Journal of General Psychology*, vol. 20, pp. 471–474, 1939.

[8] D. Meister, *The History of Human Factors and Ergonomics*, Lawrence Erlbaum Associates, Mahwah, NJ, USA, 1999.

[9] H. W. Sinaiko, *Selected Papers on Human Factors in the Design and Use of Control Systems*, Dover, Mineola, NY, USA, 2000.

[10] A. Chapanis, W. R. Garner, and C. T. Morgan, *Applied Experimental Psychology*, John Wiley & Sons, New York, NY, USA, 1949.

[11] National Research Council, *Human Factors in Undersea Warfare*, National Research Council, Committee on Undersea Warfare, Panel on Psychology and Physiology, Washington, DC, USA, 1949.

[12] P. G. Ronco, "Human factors engineering, bibliographic series volume 1 1940–1959 literature," Technical Report AD 639806, Tufts University, Medford, Mass, USA, 1966.

[13] R. A. McFarland and H. W. Stoudt, "Human body size and passenger vehicle design," SAE Special Publication 142, Society of Automotive Engineers, Warrendale, Pa, USA, 1955.

[14] J. Kulowski, "Orthopedic aspects of automobile crash injuries and deaths," *Journal of the American Medical Association*, vol. 163, no. 4, pp. 230–233, 1957.

[15] E. R. Dye, "Kinematics of the human body under crash conditions," *Clinical Orthopaedics*, vol. 8, pp. 305–309, 1956.

[16] "Manikins for use in defining vehicle seating accommodation," SAE Recommended Practice J826, 1962.

[17] "Motor vehicle driver's eye range," SAE Recommended Practice J941, 1965.

[18] D. Hammond and R. Roe, "Driver head and eye positions," SAE Technical Paper 720200, Society of Automotive Engineers, Warrendale, Pa, USA, 1972.

[19] J. F. Meldrum, "Automobile driver eye position," SAE Technical Paper 650464, Society of Automotive Engineers, Warrendale, Pa, USA, 1972.

[20] "Passenger car rear vision," SAE Recommended Practice J834, 1967.

[21] R. Roe and P. Kyropoulos, "The application of anthropometry to automotive design," SAE Technical Paper 70053, Society of Automotive Engineers, Warrendale, Pa, USA, 1970.

[22] Occupant Crash Protection, "Federal Motor Vehicle Safety Standard (FMVSS) 208, 49 CFR 571.208," Standard 208, 1959.

[23] E. Simonson, C. Baker, N. Burns, C. Keiper, O. H. Schmitt, and S. Stackhouse, "Cardiovascular stress (electrocardiographic changes) produced by driving an automobile," *American Heart Journal*, vol. 75, no. 1, pp. 125–135, 1968.

[24] R. C. Jagacinski and J. M. Flach, *Control Theory For Humans*, Lawrence Erlbaum Associates, Mahwah, NJ, USA, 2003.

[25] D. McRuer and D. H. Weir, "Theory of manual vehicular control," *Ergonomics*, vol. 12, no. 4, pp. 599–633, 1969.

[26] T. B. Sheridan, Ed., *Mathematical Models and Simulation of Automobile Driving*, Conference Proceedings, Massachusetts Institute of Technology, Cambridge, Mass, USA, 1967.

[27] N. Rashevsky, "Man-machine interaction in automobile driving," *Progress in Biocybernetics*, vol. 42, pp. 188–200, 1964.

[28] C. C. MacAdam, "Application of an optimal preview control for simulation of closed-loop automobile driving," *IEEE Transactions on Systems, Man and Cybernetics*, vol. 11, no. 6, pp. 393–399, 1981.

[29] I. D. Brown and E. C. Poulton, "Measuring the spare "mental capacity" of car drivers by a subsidiary task," *Ergonomics*, vol. 4, no. 1, pp. 35–40, 1961.

[30] J. W. Senders, A. B. Kristofferson, W. H. Levison, C. W. Dietrich, and J. L. Ward, "The attentional demand of automobile driving," Highway Research Record 195, 1967.

[31] R. E. Beinke and J. K. Williams, "Driving simulator," in *Proceedings of Automotive Safety Seminar*, vol. 24, General Motors, Warren, Mich, USA, 1968.

[32] E. Kikuchi, T. Matsumoto, S. Inomata, M. Masaki, T. Yatabe, and T. Hirose, "Development and application of high speed automobile driving simulator," Technical Report of Mechanical Engineering Laboratory 89, 1976 (Japanese).

[33] J. Drosdol and F. Panik, "The Daimler-Benz driving simulator: a tool for vehicle development," SAE Technical Paper 850334, Society of Automotive Engineers, Warrendale, Pa, USA, 1985.

[34] R. Nader, *Unsafe at any Speed*, Grossman, New York, NY, USA, 1965.

[35] G. T. Schwartz, "The myth of the Ford Pinto case," *Rutgers Law Review*, vol. 43, pp. 1013–1068, 1991.

[36] M. Dowie, "Pinto madness," *Mother Jones*, September-October 1977.

[37] W. M. Hoffman, "Case study—the Ford Pinto," *Corporate Obligations and Responsibilities: Everything Old is New Again*, 222–229, 1966.

[38] A. Irving and K. S. Rutley, "Some driving aids and their assessments," in *Proceedings of the Symposium on Psychological Aspects of Driver Behavior*, Institute for Road Safety Research, 1971.

[39] R. E. Reilly, D. S. Kurke, and C. C. Buckenmaier, "Validation of the reduction of rear-end collisions by a high-mounted auxiliary stop lamp," Technical Report DOT HS 805 360, U.S. Department of Transportation, National Highway Traffic Safety Administration, Washington, DC, USA, 1980.

[40] P. L. Olson, "Evaluation of a new LED high-mounted stop lamp, in vehicle lighting trends," Special Publication SP-692, Society of Automotive Engineers, Warrendale, Pa, USA, 1987.

[41] K. Rumar, G. Helmers, and M. Thorell, "Obstacle visibility with European Halogen H4 and American sealed beam headlights," Tech. Rep. 133, University of Uppsala, Department of Psychology, Uppsala, Sweden, 1973.

[42] J. K. Foster, J. D. Kortge, and M. J. Wolanin, "Hybrid III-A biomechanically-based crash test dummy," SAE Technical Paper 770938, Society of Automotive Engineers, Warrendale, Pa, USA, 1977.

[43] J. Versace, "A review of the severity index," SAE Technical Paper 710881, Society of Automotive Engineers, Warrendale, Pa, USA, 1971.

[44] D. C. Hammond and R. W. Roe, "SAE controls reach study," SAE Technical Paper 720199, Society of Automotive Engineers, Warrendale, Pa, USA, 1972.

[45] "Driver hand control reach," SAE Recommended Practice J287, 1976.

[46] "Direction-of-motion stereotypes for automotive hand controls," SAE Recommended Practice J1139, 1977.

[47] "Symbols for motor vehicle controls," SAE Standard J1048, 1974.

[48] M. J. Griffin, Handbook of Human Vibration, Elsevier, London, UK, 1996.

[49] "Mechanical vibration and shock—guide for the evaluation of human exposure to whole-body vibration," ISO 2631, 1974.

[50] R. R. Mourant and T. H. Rockwell, "Mapping eye-movement patterns to the visual scene in driving: an exploratory study," Human Factors, vol. 12, no. 1, pp. 81–87, 1970.

[51] B. Richter, "Driving simulator studies: the influence of vehicle parameters on safety in critical situations," SAE Technical Paper 741105, Society of Automotive Engineers, Warrendale, Pa, USA, 1974.

[52] R. C. McLane and W. W. Wierwille, "The influence of motion and audio cues on driver performance in an automobile simulator," Human Factors, vol. 17, no. 5, pp. 488–501, 1975.

[53] R. G. Snyder, T. L. McDole, W. M. Ladd, and D. J. Minahan, "On-road crash experience of utility vehicles," Tech. Rep. UM-HSRI-80-14, Highway Safety Research Institute, Ann Arbor, Mich, USA, 1980.

[54] R. G. Snyder, T. L. McDole, W. M. Ladd, and D. J. Minahan, "An overview of the on-road crash experience of utility vehicles (highlights of the technical report)," Tech. Rep. UM-HSRI-80-15, Highway Safety Research Institute, Ann Arbor, Mich, USA, 1980.

[55] S. Franklin and M. Stepanek, Trouble in Jeep Country, AMC Claima It's Safe, Detroit Free Press, 1983.

[56] P. Niedermeyer, "The Best of TTAC: The Audi 5000 Intended Unintended Acceleration Debacle," http://www.thetruthaboutcars.com/2010/03/the-best-of-ttac-the-audi-5000-intended-unintended-acceleration-debacle/.

[57] R. Walter, G. Carr, H. Weinstock, D. Sussman, and J. Pollard, "Study of mechanical and driver-related systems of the Audi 5000 capable of producing uncontrolled sudden acceleration incidents," Tech. Rep. DOT-TSC-NHTSA-88-4, U.S. Department of Transportation, National Highway Traffic Safety Administration, Washington, DC, USA, 1988.

[58] National Aeronautics and Space Administration, Technical Support to the National Highway Traffic Safety Administration (NHTSA) on the Reported Toyota Motor Corporation (TMC) Unintended Acceleration (UA) Investigation (NESC Assessment TI-10-00618), National Aeronautics and Space Administration, NASA Safety and Engineering Center, January 2011.

[59] J. D. Brown, "The opportunities of ergonomics," in Human Factors in Transport Research. Vehicle Factors: Transport Systems, Workspace, Information and Safety, D. J. Oborne and J. A. Lewis, Eds., vol. 1, Academic Press, New York, NY, USA, 1980.

[60] H. Bubb, F. Engstler, F. Fritzsche et al., "The development of RAMSIS in past and future as an example for the cooperation between industry and university," International Journal of Human Factors Modelling and Simulation, vol. 1, no. 1, pp. 140–157, 2006.

[61] P. Blanchonette, "Jack human modelling tool: a review," Tech. Rep. DSTO-TR-2364, Defense Science and Technology Organization Victoria (Australia) Air Operations Division, Fishermans Bend, Victoria, Australia, 2010, document ADA 518132.

[62] K. S. Rutley, "Control of drivers' speed by means other than enforcement," Ergonomics, vol. 18, no. 1, pp. 89–100, 1975.

[63] K. Bengler, H. Bubb, I. Totzke, J. Schumann, and F. Flemisch, "Automotive," in Information Ergonomics—A Theoretical Approach and Practical Experience in Transportation, P. Sandle and M. Stein, Eds., Springer, Heidelberg, Germany, 2012.

[64] K. Parsons, Human Thermal Environments, Taylor & Francis, London, UK, 2nd edition, 2003.

[65] D. P. Gatley, "Psychrometric chart celebrate 100th anniversary," ASHRAE Journal, vol. 46, no. 11, pp. 16–20, 2004.

[66] D. P. Wyon, C. Tennstedt, I. Lundgren, and S. Larsson, "A new method for the detailed assessment of human heat balance in vehicles, Volvo's thermal manikin, VOLTMAN," SAE Technical Paper 850042, Society of Automotive Engineers, Warrendale, Pa, USA, 1985.

[67] C. W. Erwin, J. W. Hartwell, M. R. Volow, and G. S. Alberti, "Electrodermal change as a predictor of sleep," in Studies of Drowsiness (Final Report), C. W. Erwin, Ed., The National Driving Center, Durham, North Carolina, 1976.

[68] T. A. Dingus, L. H. Hardee, and W. W. Wierwille, "Detection of drowsy and intoxicated drivers based on highway driving performance measures," IEOR Department Report #8402, Virginia Tech, Department of Industrial Engineering and Operations Research, Blacksburg, Va, USA, 1985.

[69] W. W. Wierwille, "Research on vehicle-based driver status/performance monitoring, development, validation, and refinement of algorithms for detection of driver drowsiness," Technical Report DOT HS 808 247, U.S. Department of Transportation, Washington, DC, USA, 1994.

[70] D. A. Spyker, S. P. Stackhouse, S. Khalalfalla, and R. C. McLane, "Development of techniques for measuring pilot workload," Contractor report NASA CR 1888, NASA, Washington, DC, USA, 1971.

[71] T. Miura, "Coping with situational demands: a study of eye movements and peripheral vision performance," in Vision in Vehicles 1, A. G. Gale, M. H. Freeman, C. M. Haslegrave, P. Smith, and S. P. Taylor, Eds., pp. 205–216, North-Holland, Amsterdam, The Netherlands, 1986.

[72] S. Nordomack, H. Jansson, M. Lidstrom, and G. Palmkvist, "A moving base driving simulator with wide angle visual system," VTI Technical Report 106A, Swedish Road and Traffic Research Institute, Linkoping, Sweden, 1986.

[73] S. Hahn and W. Kaeding, "The Daimler-Benz driving simulator-presentation of selected experiments," SAE Technical Paper 880058, Society of Automotive Engineers, Warrendale, Pa, USA, 1988.

[74] E. Blana, "A survey of driving research simulators around the world," ITS Working Paper 481, University of Leeds, Institute for Transport Studies, Leeds, UK, 1996.

[75] C. Y. D. Yang, J. D. Fricker, and T. Kuczek, "Designing advanced traveler information systems from a driver's perspective: Results of a driving simulation study," Transportation Research Record, no. 1621, pp. 20–26, 1998.

[76] W. Janssen and R. van der Horst, "Presenting descriptive information in variable message signing," Transportation Research Record, no. 1403, pp. 83–87, 1993.

[77] M. P. Reed and P. A. Green, "Comparison of driving performance on-road and in a low-cost simulator using a concurrent telephone dialling task," Ergonomics, vol. 42, no. 8, pp. 1015–1037, 1999.

[78] A. Steinfeld and P. Green, "Driver responses to navigation information on full-windshield, head-up displays," *International Journal of Vehicle Design*, vol. 19, no. 2, pp. 135–149, 1998.

[79] J. R. Bloomfield, J. R. Buck, J. M. Christensen, and A. Yenamandra, "Human factors aspects of the transfer of control from the driver to the automated highway system," Tech. Rep. FHWA-RD-94-173, U.S. Department of Transportation, Federal Highway Administration, Washington, DC, USA, 1994.

[80] T. Suetomi and K. Kido, "Driver behavior under a collision warning system: a driving simulator study," SAE Technical Paper 970279, Society of Automotive Engineers, Warrendale, Pa, USA, 1997.

[81] J. Godthelp and J. Schumann, "The use of an intelligent accelerator as an element of a driver support system," in *Proceedings of the 24th ISATA International Symposium on Automotive Technology and Automation*, 1991.

[82] W. W. Wierwille, M. C. Hulse, T. J. Fischer, and T. A. Dingus, "Strategic use of visual resources by the driver while navigating with an in-car navigation display system," SAE Technical Paper 885180, Society of Automotive Engineers, Warrendale, Pa, USA, 1988.

[83] H. K. Zwahlen, in *Information Processing, Driver Performance Data Book, Technical Report DOT HS 807 121*, R. L. Henderson, Ed., U.S. Department of Transportation, National Highway Traffic Safety Administration, Washington, DC, USA, 1987.

[84] F. J. Mammano and R. Favout, "An electronic route-guidance system for highway vehicles," *IEEE Transactions on Vehicular Technology*, vol. 19, no. 1, pp. 143–152, 1999.

[85] H. Ito, "Integrated development of automotive navigation and route guidance system—product development for realization of dreams and standardization," *Synthesiology (English Edition)*, vol. 4, no. 3, pp. 162–171, 2012.

[86] E. J. Blum, R. Haller, and G. Nirschl, "Driver-copilot interaction: modelling aspects and techniques," in *Proceedings of the 2nd Prometheus Workshop*, FHG-IITB, Stockholm, Sweden, 1989.

[87] J. A. Michon, *Generic Intelligent Driver Support*, Taylor & Francis, London, UK, 1993.

[88] A. M. Parkes and S. Franzen, *Driving Future Vehicles*, Taylor & Francis, London, UK, 1993.

[89] H. Ikeda, Y. Kobayashi, and K. Hirano, "How car navigation systems have been put into practical use: development management and commercialization process," *Synthesiology (English Edition)*, vol. 3, no. 4, pp. 280–289, 2011.

[90] P. Frame, "How GM trumped Dateline, maker used vast resources to dismantle pickup story," *Automotive News*, pp. 1, 1993.

[91] P. Green, W. Levison, G. Paelke, and C. Serafin, "Preliminary human factors guidelines for driver information systems," Tech. Rep. FHWA-RD-94-087, U.S. Department of Transportation, Federal Highway Administration, McLean, Va, USA, 1995.

[92] A. Stevens, A. C. Board, P. Allen, and A. Quimby, "A safety checklist of the assessment of in-vehicle information systems: scoring proforma," Project Report PA3536-A/99, Transport Research Laboratory, Crowthorne, UK, 1999.

[93] T. Ross, K. Midtland, M. Fuchs et al., *HARDIE Design Guidelines Handbook: Human Factors Guidelines for Information Presentation by ATT Systems*, Commission of the European Communities, Brussels, Luxembourg, 1996.

[94] P. Green, "Driver interface safety and usability standards: an overview," in *Driver Distraction Theory, Effects, and Mitigation*, M. Regan, J. Lee, and K. Young, Eds., pp. 445–464, CRC Press, Boca Raton, Fla, USA, 2008.

[95] "Wirtschaftsforum VerkehrstelematikVereinbarung zu Leitlinien für die Gestaltung und Installation von Informations- und Kommunikationssystemen in Kraftfahrzeugen," (English translation: Steering Group on the Economic Forum on Telematics in Transport, Agreement on guidelines for the design and installation of information and communication systems in motor vehicles), Bonn, Germany, 1996.

[96] O. M. J. Carsten and L. Nilsson, "Safety assessment of driver assistance systems," *European Journal of Transport and Infrastructure Research*, vol. 1, no. 3, pp. 225–243, 2001.

[97] C. Heinrich, "Automotive HMI International Standards," in *Proceedings 4th International Conference on Applied Human Factors and Ergonomics (AHFE '12)*, 2012.

[98] W. W. Wierwille, J. F. Antin, T. A. Dingus, and M. C. Hulse, "Visual attentional demand of a in-car navigational display system," in *Vision in Vehicles*, A. G. Gale, M. H. Freeman, C. M. Haslegrave, P. Smith, and S. P. Taylor, Eds., vol. 2, pp. 307–316, Elsevier, Amsterdam, The Netherlands, 1988.

[99] H. T. Zwahlen, C. C. Adams Jr., and D. P. DeBald, "Safety aspects of CRT touch panel controls in automobiles," in *Vision in Vehicles*, A. G. Gale, M. H. Freeman, C. M. Haslegrave, P. Smith, and S. P. Taylor, Eds., vol. 2, pp. 335–344, Elsevier, Amsterdam, The Netherlands, 1988.

[100] A. R. A. van der Horst, "Occlusion as a measure for visual workload: an overview of TNO occlusion research in car driving," *Applied Ergonomics*, vol. 35, no. 3, pp. 189–196, 2004.

[101] J. F. Krems, A. Keinath, M. Baumann, C. Gelau, and K. Bengler, "Evaluating visual display designs in vehicles: advantages and disadvantages of the occlusion technique," in *Advances in Network Enterprises, Virtual Organizations, Balanced Automation, and Systems Integration*, L. M. Camarinha-Matos, H. Afsarmanesh, and H. H. Erbe, Eds., pp. 361–368, Kluwer Academic, Norwell, Mass, USA, 2000.

[102] M. Akamatsu, "Japanese approaches to principles, codes, guidelines and checklists for in-vehicle HMI," in *Driver Distraction Theory, Effects, and Mitigation*, M. Regan, J. Lee, and K. Young, Eds., pp. 425–444, CRC Press, Boca Raton, Fla, USA, 2008.

[103] "Ergonomic aspects of transport information and control systems—occlusion method to assess visual distraction," ISO 16673, 2007.

[104] P. Green, "The 15-second rule for driver information systems," in *Proceedings of the ITS America 9th Annual Meeting*, ITS America, Washington, DC, USA, 1999.

[105] P. Green, "Estimating compliance with the 15-second rule for driver-interface bility and safety," in *Proceedings of the Human Factors and Ergonomics Society 43rd Annual Meeting*, Santa Monica, Calif, USA, 1999.

[106] A. Baron and P. Green, "Safety and usability of speech interfaces for in-vehicle tasks while driving: a brief literature review," Technical Report UMTRI 2006-5, University of Michigan Transportation Research Institute, Ann Arbor, Mich, USA, 2006.

[107] V. E. Lo and P. Green, "Development and evaluation of automotive speech interfaces: useful information from the human factors and related literature," *International Journal of Vehicular Technology*, vol. 2013, Article ID 924170, 13 pages, 2013.

[108] "Ergonomic aspects of transport information and control systems—dialogue management principles," ISO 15005, 2002.

[109] "Ergonomic aspects of transport information and control systems—criteria for determining priority of messages," ISO TR16951, 2004.

[110] "Ergonomic aspects of transport information and control systems—visual presentation of information," ISO 15008, 2009.

[111] "Ergonomic aspects of transport information and control systems—suitability of TICS while driving," ISO 17287, 2003.

[112] "Ergonomic aspects of transport information and control systems—auditory information presentation," ISO 15006, 2004.

[113] "Ergonomic aspects of transport information and control systems—simulated lane change test to assess in-vehicle secondary task demand," ISO 26022, 2010.

[114] S. Mattes, "The lane-change-task as a tool for driver distraction evaluation," in *Quality of Work and Products in Enterprises of the Future*, H. Strasser, K. Kluth, H. Rausch, and H. Bubb, Eds., Erognomia, 2003.

[115] P. C. Burns, K. Bengler, and D. H. Weir, "Driver metrics and an overview of user needs and uses," in *Performance Metrics for Assessing Driver Distraction: The Quest for Improved Road Safety*, G. L. Rupp, Ed., chapter 1, pp. 24–30, SAE International, Warrendale, Pa, USA, 2010.

[116] A. Kumar, *Deadly Combination: Ford, Firestone and Florida*, Saint Petersburg Times, Saint Petersburg, Fla, USA, 2001.

[117] A. Kumar, *Attention Shi.s from Fires Tone to Ford Explorer*, Saint Petersburg Times, Saint Petersburg, Fla, USA, 2001.

[118] K. Naab and G. Reichart, "Driver assistance system for lateral and longitudinal vehicle guidance—heading control and active cruise support," in *Proceedings of International Symposium on Advanced Vehicle Control (AVEC '94)*, pp. S449–S454, Tsukuba, Japan.

[119] W. Prestl, T. Sauer, J. Steinle, and O. Tschernoster, "The BMW active cruise control ACC," SAE Technical Paper 2000-01-0344, Society of Automotive Engineers, Warrendale, Pa, USA.

[120] D. M. Hoedemaeker, *Driving with intelligent vehicles. Driving behaviour with adaptive cruise control and the acceptance by individual drivers [Ph.D. thesis]*, Delft University Press, Delft, The Netherlands, 1999.

[121] M. P. Heyes and R. Ashworth, "Further research on car-following models," *Transportation Research*, vol. 6, no. 3, pp. 287–291, 1972.

[122] P. Wasielewski, "Car following headways on freeways interpreted by the semi-Poisson headway distribution model," *Transportation Science*, vol. 13, no. 1, pp. 36–55, 1979.

[123] H. Godthelp, P. Milgram, and G. J. Blaauw, "The development of a time-related measure to describe driving strategy," *Human Factors*, vol. 26, no. 3, pp. 257–268, 1984.

[124] W. van Winsum, K. A. Brookhuis, and D. de Waard, "A comparison of different ways to approximate time-to-line crossing (TLC) during car driving," *Accident Analysis and Prevention*, vol. 32, no. 1, pp. 47–56, 2000.

[125] J. C. Hayward, "Near miss determination through use of a scale of danger," *Highway Research Record*, vol. 384, pp. 24–34, 1972.

[126] W. van Winsum and A. Heino, "Choice of time-headway in car-following and the role of time-to-collision information in braking," *Ergonomics*, vol. 39, no. 4, pp. 579–592, 1996.

[127] K. Vogel, "A comparison of headway and time to collision as safety indicators," *Accident Analysis and Prevention*, vol. 35, no. 3, pp. 427–433, 2003.

[128] "Ergonomic aspects of transport information and control systems—introduction to integrating safety critical and time critical warning signals," ISO TR 12204, 2012.

[129] I. Totzke, S. Jessberger, D. Mühlbacher, and H. P. Krüger, "Semi-autonomous advanced parking assists: do they really have to be learned if steering is automated?" in *Proceedings of European Conference on Human Centered Design for Intelligent Transport Systems*, pp. 123–132, Berlin, Germany, 2010.

[130] M. Kienle, D. Damböck, J. Kelsch, F. Flemisch, and K. Bengler, "Towards an H-Mode for highly automated vehicles: driving with side sticks," in *Proceedings of the 1st International Conference on Automotive User Interfaces and Interactive Vehicular Applications (Automotive UI '09)*, pp. 19–23, September 2009.

[131] K. Bengler, M. Zimmermann, D. Bortot, M. Kienle, and D. Dambock, "Interaction principles for cooperative human-machine systems," *Information Technology*, vol. 54, no. 4, pp. 157–164, 2012.

[132] T. Inagaki and M. Itoh, "Human's overtrust in and overreliance on Advanced Driver Assistance Systems: a theoretical framework," *International Journal of Vehicular Technology*, vol. 2013, Article ID 951762, 8 pages, 2013.

[133] D. Popiv, C. Rommerskirchen, M. Rakic, M. Duschl, and K. Bengler, "Effects of assistance of anticipatory driving on driver's behaviour during deceleration situations," in *Proceedings of the 2nd European Conference on Human Centred Design of Intelligent Transport Systems (HUMANIST '10)*, Berlin, Germany, April 2010.

[134] D. Popiv, M. Rakic, F. Laquai, M. Duschl, and K. Bengler, "Reduction of fuel consumption by early anticipation and assistance of deceleration phases," in *Proceedings of the World Automotive Congress of International Federation of Automotive Engineering Societies (FISITA '10)*, Budapest, Hungary, June 2010.

[135] S. G. Klauer, T. A. Dingus, V. L. Neale, J. Sudweeks, and D. Ramsey, "The impact of driver inattention on near-crash/crash risk: an analysis using the 100-car naturalistic driving study data," Technical Report DOT, HS 810 594, U.S. Department of Transportation, National Highway Traffic Safety Administration, Washington, DC, USA, 2006.

[136] R. Ervin, J. Sayer, D. LeBlanc et al., "Automotive collision avoidance system field operational test report: methodology and results," Technical Report HS 809 900, US Department of Transportation, National Highway Traffic Safety Administration, Washington, DC, USA, 2005.

[137] J. Sayer, C. Winkler, R. Ervin et al., "Road departure crash warning system field operational test: methodology and results. Volume 1: technical report," Tech. Rep. UMTRI-2006-9-1, U.S. Department of Transportation, National Highway Traffic Safety Administration, Washington, DC, USA, 2006.

[138] J. Sayer, D. LeBlanc, S. Bogard et al., "Integrated vehicle-based safety systems field operational test final program report," Technical Report HS 811 482, U.S. Department of Transportation, National Highway Traffic Safety Administration, Washington, DC, USA, 2011.

[139] M. Nagai, "Enhancing safety and security by incident analysis using drive recorders," *Review of Automotive Engineering*, vol. 27, no. 1, pp. 9–15, 2006.

[140] T. Sato and M. Akamatsu, "Influence of traffic conditions on driver behavior before making a right turn at an intersection: analysis of driver behavior based on measured data on an actual road," *Transportation Research F*, vol. 10, no. 5, pp. 397–413, 2007.

[141] "Field opErational teSt supporT Action (FESTA)," in *FESTA Handbook*, European Commission, Brussels, Belgium, 2013, http://www.its.leeds.ac.uk/festa/downloads/FESTA%20Handbook%20v2.pdf.

[142] T. Sato, M. Akamatsu, A. Takahashi et al., "Analysis of driver behaviour when overtaking with adaptive cruise control,"

Review of Automotive Engineering, vol. 26, no. 4, pp. 481–488, 2005.

[143] S. B. McLaughlin, J. M. Hankey, and T. A. Dingus, "A method for evaluating collision avoidance systems using naturalistic driving data," *Accident Analysis and Prevention*, vol. 40, no. 1, pp. 8–16, 2008.

[144] M. Akamatsu, Y. Sakaguchi, and M. Okuwa, "Modeling of driving behavior when approaching intersection based on measured behavioral data on an actual road," in *Proceedings of the Human Factors and Ergonomics Society 47th Annual Meeting*, pp. 1895–1899, 2003.

[145] T. Sato and M. Akamatsu, "Modeling and prediction of driver preparations for making a right turn based on vehicle velocity and traffic conditions while approaching an intersection," *Transportation Research F*, vol. 11, no. 4, pp. 242–258, 2008.

[146] J. D. Lee, D. V. McGehee, T. L. Brown, and M. L. Reyes, "Collision warning timing, driver distraction, and driver response to imminent rear-end collisions in a high-fidelity driving simulator," *Human Factors*, vol. 44, no. 2, pp. 314–334, 2002.

[147] D. L. Fisher, M. Rizzo, J. Caird, and J. D. Lee, *Handbook of Driving Simulation*, CRC Press, Boca Raton, Fla, USA, 2011.

[148] T. Sato, M. Akamatsu, T. Shibata, S. Matsumoto, N. Hatakeyama, and K. Hayama, "Predicting driver behavior using field experiment data and driving simulator experiment data: Assessing impact of elimination of stop regulation at railway crossings," *International Journal of Vehicular Technology*, vol. 2013, Article ID 912860, 9 pages, 2013.

[149] K. L. Young and M. A. Regan, "Driver distraction exposure research: a summary of findings," in *Driver Distraction Theory, Effects, and Mitigation*, M. A. Regan, J. D. Lee, and K. L. Young, Eds., pp. 327–328, 2008.

[150] K. A. Brookhuis, G. de Vries, and D. de Waard, "The effects of mobile telephoning on driving performance," *Accident Analysis and Prevention*, vol. 23, no. 4, pp. 309–316, 1991.

[151] H. Alm and L. Nilsson, "Changes in driver behaviour as a function of handsfree mobile phones—a simulator study," *Accident Analysis and Prevention*, vol. 26, no. 4, pp. 441–451, 1994.

[152] W. J. Horrey and C. D. Wickens, "Examining the impact of cell phone conversations on driving using meta-analytic techniques," *Human Factors*, vol. 48, no. 1, pp. 196–205, 2006.

[153] D. Lamble, T. Kauranen, M. Laakso, and H. Summala, "Cognitive load and detection thresholds in car following situations: safety implications for using mobile (cellular) telephones while driving," *Accident Analysis and Prevention*, vol. 31, no. 6, pp. 617–623, 1999.

[154] D. L. Strayer and W. A. Johnston, "Driven to distraction: dual-task studies of simulated driving and conversing on a cellular telephone," *Psychological Science*, vol. 12, no. 6, pp. 462–466, 2001.

[155] B. Metz, N. Schömig, H. P. Krüger, and K. Bengler, "Situation awareness in driving with in-vehicle information systems," in *Performance Metrics for Assessing Driver Distraction: The Quest For Improved Road Safety*, G. L. Rupp, Ed., chapter 12, SAE International, Warrendale, Pa, USA, 2010.

[156] F. A. Drews and D. L. Strayer, "Cellular phone and driver distraction," in *Driver Distraction Theory, Effects, and Mitigation*, M. Regan, J. Lee, and K. Young, Eds., pp. 169–190, CRC Press, Boca Raton, Fla, USA, 2008.

[157] M. Vollrath, T. Meilinger, and H. P. Krüger, "How the presence of passengers influences the risk of a collision with another vehicle," *Accident Analysis and Prevention*, vol. 34, no. 5, pp. 649–654, 2002.

[158] J. R. Davis and C. M. Schmandt, "The back seat driver: real time spoken driving instructions," in *Proceedings of the IEEE Vehicle Navigation and Information Systems Conference (VNIS '89)*, pp. 146–150, September 1989.

[159] M. Akamatsu and M. Kitajima, "Designing products and services based on understanding human cognitive behavior—development of cognitive chrono-ethnography for synthesiological research," *Synthesiology (English Edition)*, vol. 4, no. 3, pp. 144–155, 2012.

[160] B. Brown, E. Laurier, H. Lorimer et al., "Driving and "passengering": Notes on the ordinary organization of car travel," *Mobilities*, vol. 3, no. 1, pp. 1–23, 2008.

[161] K. Bengler, J. F. Coughlin, B. Reimer, and B. Niedermaier, "A new method to investigate cognitive structures of user's on automotive functionalities," in *Proceedings of the 3rd International Conference on Applied Human Factors and Ergonomics (AHFE '10)*, Miami, Fla, USA, July 2010.

[162] B. Brown and E. Laurier, "The normal, natural troubles of driving with GPS," in *Proceedings of the SIGCHI Conference on Human Factors in Computing Systems (CHI '12)*, pp. 1621–1630, New York, NY, USA, 2012.

[163] ITU-T FG Distraction P.UIA report, FG Distraction report on User Interface requirements for Automotive applications (P.UIA).

[164] ITU-T FG Distraction G.V2A report, FG Distraction report on communications interface between external applications and a Vehicle Gateway Platform (G.V2A).

[165] ITU-T FG Distraction G.SAM report, FG Distraction report on Situational Awareness Management (G.SAM).

[166] H. Hatakenaka, H. Kanoshima, T. Aya, S. Nishi, H. Mizutani, and K. Nagano, "Development and verification of effectiveness of an AHS," in *Proceedings of ITS World Congress*, New York, NY, USA, 2008.

[167] M. Hatakeyama and S. Nakayama, "Progress toward the practical use of vehicle infrastructure cooperation system," in *Proceedings of ITS World Congress*, Stochholm, Sweden, 2009.

[168] K. Daimon, H. Makino, H. Mizutani, and Y. Munehiro, "Study on safety assist information of Advanced Cruise-Assist Highway Systems (AHS) using VICS in blind curve section of urban expressway," *Journal of Mechanical Systems For Transportation and Logistics*, vol. 1, no. 2, pp. 192–202, 2006.

[169] U. S. Department of Transportation, "Safety Pilot Program Overview," http://www.its.dot.gov/safety_pilot/index.htm.

[170] D. Popiv, C. Rommerskirchen, M. Rakic, M. Duschl, and K. Bengler, "Effects of assistance of anticipatory driving on driver's behaviour during deceleration situations," in *Proceedings of the 2nd European Conference on Human Centered Design of Intelligent Transport Systems (HUMANIST '10)*, Berlin, Germany, April 2010.

[171] S. Thrun, M. Monemerlo, H. Dahlkamp et al., "Stanley the robot that won the DARPA grand challenge," in *DARPA Grand Challenge: The Great Robot Race*, M. Buehler, K. Iagnemma, and S. Singh, Eds., vol. 36 of *Springer Tracts in Advanced Robotics*, pp. 1–43, 2007.

[172] T. Nothdurft, P. Hecker, S. Ohl et al., "Stadtpilot: first fully autonomous test drives in urban traffic," in *Proceedings of the 14th International IEEE Annual Conference on Intelligent Transportation Systems*, Washington, DC, USA, 2011.

[173] T. A. Ranney, "Models of driving behavior: a review of their evolution," *Accident Analysis and Prevention*, vol. 26, no. 6, pp. 733–750, 1994.

[174] D. D. Salvucci, "Modeling driver behavior in a cognitive architecture," *Human Factors*, vol. 48, no. 2, pp. 362–380, 2006.

[175] C. P. Cacciabue, Ed., *Modelling Driver Behaviour in Automotive Environment: Critical Issues in Driver Interactions with Intelligent Transport Systems*, Springer, London, UK, 2007.

[176] L. Evans, *Traffic Safety and the Driver*, Van Nostrand Reinhold, New York, NY, USA, 1991.

[177] B. Peacock and W. Karwowski, Eds., *Automotive Ergonomics*, Taylor & Francis, London, UK, 1993.

[178] M. Sivak, M. J. Flannagan, and B. Schoettle, "Driver assessment and training in the 1980s and 1990s: an analysis of the most-cited publications," in *Proceedings of the Driving Assessment Conference*, 2001.

[179] J. D. Lee, "Fifty years of driving safety research," *Human Factors*, vol. 50, no. 3, pp. 521–528, 2008.

[180] H. H. Braess and U. Seiffert, Eds., *Handbuch Kraftfahrzeugtechnik. 6. Auflage*, Vieweg and Teubner, Wiesbaden, Germany, 2011.

[181] N. Gkikas, *Automotive Ergonomics: Driver-Vehicle Interaction*, CRC Press, Boca Raton, Fla, USA, 2012.

[182] V. D. Bhise, *Ergonomics in the Automotive Design Process*, CRC Press, Boca Raton, Fla, USA, 2012.

15

StopWatcher: A Mobile Application to Improve Stop Sign Awareness for Driving Safety

Carl Tucker, Rachel Tucker, and Jun Zheng

Department of Computer Science and Engineering, New Mexico Institute of Mining and Technology, Socorro, NM 87801, USA

Correspondence should be addressed to Jun Zheng, junzheng@ieee.org

Academic Editor: Nandana Rajatheva

Stop signs are the primary form of traffic control in the United States. However, they have a tendency to be much less effective than other forms of traffic control like traffic lights. This is due to their smaller size, lack of lighting, and the fact that they may become visually obscured from the road. In this paper, we offer a solution to this problem in the form of a mobile application implemented in the Android platform: StopWatcher. It is designed to alert a driver when they are approaching a stop sign using a voice notification system (VNS). A field test was performed in a snowy environment. The test results demonstrate that the application can detect all of the stop signs correctly, even when some of them were obstructed by the snow, which in turn greatly improves the user awareness of stop signs.

1. Introduction

Stop signs are the primary form of traffic control in the United States [1] because they are inexpensive and easy to maintain compared with other forms of traffic control like traffic lights. Approximately one third of all vehicular accidents occur in stop sign controlled intersections which seems reasonable considering that a majority of intersections use them. However, these accidents account for over 40% of all fatal vehicular accidents [1]. Research performed by the United States Department of Transportation (USDOT) has shown that the primary reason for this incredibly high number is due to driver inability or failure to see the stop sign, resulting in a collision.

This inability to see the stop sign is generally caused by visual obstructions. Some examples of visually obstructed stop signs are shown in Figure 1. These obstructions may be caused by plant overgrowth, frost, graffiti, parked cars, hills, and many other reasons [2–4]. For example, there is a particular intersection in Santa Fe, New Mexico, where it was reported that roughly half of all approaching vehicles missed the stop sign and continued through the intersection [5]. It was determined that the cause of such a high frequency was because of poor placement of the stop sign, which made it difficult to see from the road. If the primary reason that drivers accidentally run stop signs is because they simply do not see them, a simple solution is to use another method to convey the presence of an upcoming stop sign.

In this paper, we present a novel solution, StopWatcher, for safe drivers who want to be sure that they do not accidentally miss a stop sign. StopWatcher is a mobile application designed for the Android platform. It uses GPS, compass, and web capabilities which are available in a majority of smartphones. The primary design goal is to warn the user about upcoming stop signs on the road in real-time to improve the driving safety.

The rest of this paper is organized as follows. In Section 2, we provide a brief overview of the related work. The design of StopWatcher application is presented in Section 3. In Section 4, we describe the field test and show the test results. Finally, we conclude this paper and point out our future works in Section 5.

2. Related Work

Automatic traffic sign detection and recognition has been a popular research topic in the field of applied computer vision. It uses the traffic scene images acquired by the imaging devices mounted in the car and then applies image processing and pattern recognition techniques to detect

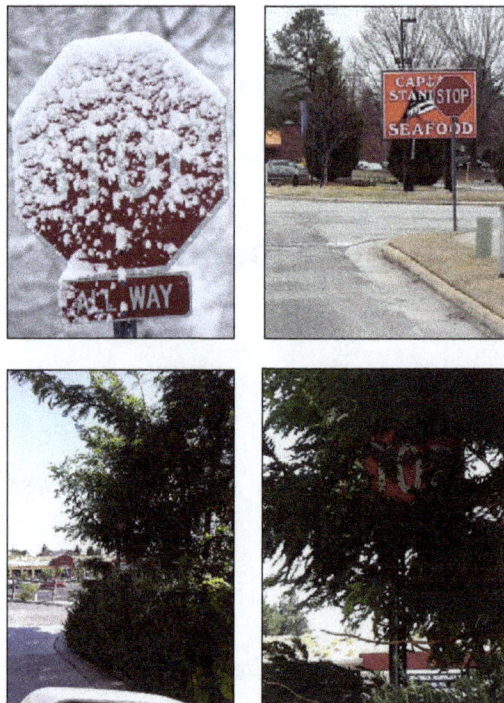

FIGURE 1: Examples of visually obstructed stop signs.

traffic signs. For example, different color information such as the hue, saturation, and HSI color space [6–8], color ranges [9], look-up tables [10] were used to segment the traffic signs in images acquired under various lighting conditions. Shape information of different traffic signs were also used in various works [6, 7, 10, 11]. However, due to the challenging factors of outdoor environment such as various illumination and weather conditions, different view angles, and so forth, those proposed traffic sign detection algorithms and systems cannot achieve perfect performance. Especially when signs are clustered or partially obstructed, the performance is generally very poor. These problems can be easily solved with our StopWatcher application as it is designed to be robust to those challenging environment factors.

Today's smartphones are equipped with numerous sensors such as GPS, accelerometers, compass, microphones and camera, and so forth, which make them useful in some innovative automobile applications. In [12], a system called Nericell was developed by Microsoft which uses accelerometer, microphone, GSM radio, and/or GPS equipped in smartphones to detect road and traffic conditions such as potholes, bumps, braking, and honking. Dai et al. utilized the accelerometer and orientation sensors in a smartphone to detect drunk driving for early alert or calling the police for help [13]. Fazeen et al. proposed to use the accelerometer and GPS of an Android-based smartphone to analyze the driver's behavior and road conditions [14]. The aim of the application is to identify road anomalies and sudden driving maneuvers for driving safety. StopWatcher has a different design goal than these applications to concentrate on automatic traffic sign detection.

3. StopWatcher: A Mobile Application for Stop Sign Detection

StopWatcher is a real-time Android application developed to make drivers more aware of stop signs, even if they are hidden, obscured, or damaged. It is also helpful when general visibility conditions are poor due to darkness or inclement weather. It uses the GPS and Compass features provided by many smartphones. The system architecture of the Stop-Watcher application is shown in Figure 2. The application has a sensor manager that retrieves the sensor data from the hardware and makes it available to the application. The user's current GPS data is used to retrieve the appropriate map from Google Maps for visual interpretation by the user. The application makes use of a MySQL database to store and retrieve information about intersections with stop signs. The MySQL Query Manager coordinates the retrieval of nearby stop sign data when given a user's GPS coordinates. The application then uses an Approach Classifier to interpret the sensor data, as well as the sign location data, in order to determine whether or not a user is approaching a stop sign. The programming of the application was done with Java, the programming language for Android application development.

3.1. Google Maps. While StopWatcher does not technically require any mapping software for its core functionality, we decided that it would be a nice feature for the user, and the debuggers, to be able to see which roads have stop signs ahead of time when planning a route. In this way the driver could be more prepared when actually driving to their destination. Note that the map feature is not intended to be used while the user is driving and might be disabled when the user is driving in future versions.

We use the Google Maps package in Android to handle the maps and road matching functions because it is free and has a very user friendly API. When StopWatcher is launched, it accesses the GPS on the phone and passes the coordinate data to Google Maps. It then queries the MySQL database for the positions of local stop signs. Finally it draws the standard map and centers it on the user's location. In this step it also overlays custom icons which depict the stop signs and the directions that they are facing (Figure 3). Unlike many applications that utilize Google Maps, we have forced the zoom setting to stay at 18 because we found that this setting provides the best view of all of the local streets. Other settings may cause some of the streets to disappear and the overlaid icons lose their meaning or may be misinterpreted as being on an adjacent street. Note that this would only affect the visual functionality of the program and would not affect the VNS since it is controlled by the GPS coordinate data.

There are other minor issues associated with the visual implementation of StopWatcher. While most city roads run either North-to-South or East-to-West, some do not. Since the current system uses a static set of images to denote the road directions that are required to stop at a stop sign (see Figure 3), the system could provide data that might be misinterpreted. For example, take an intersection where

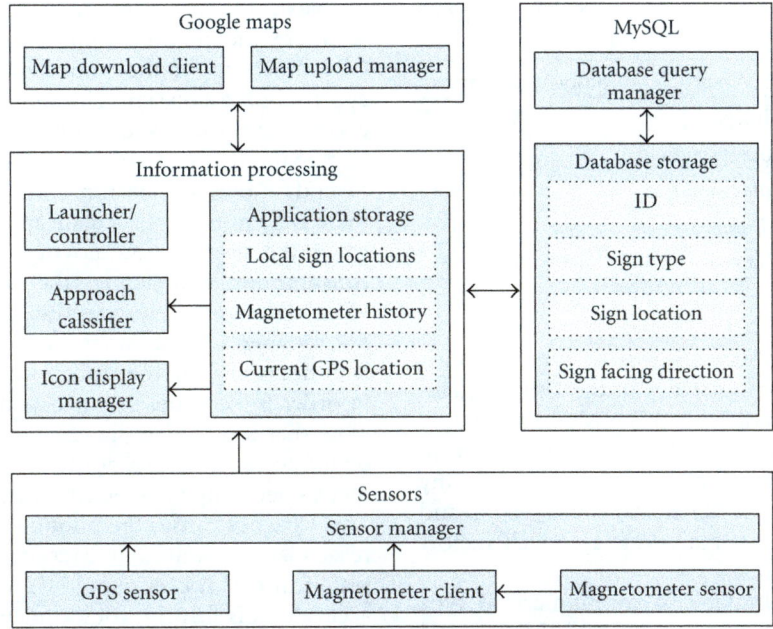

FIGURE 2: System architecture of StopWatcher application.

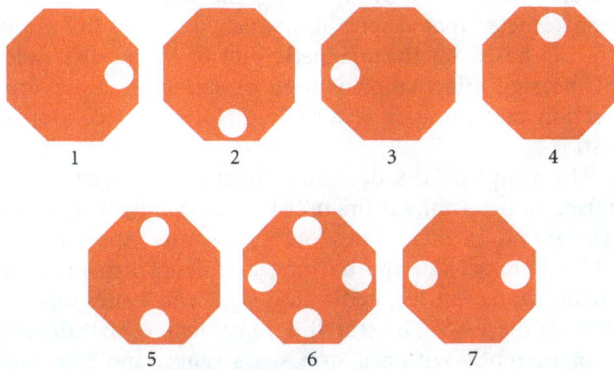

FIGURE 3: Symbols that StopWatcher places in Google Maps denote the directions that are required to stop at an intersection. The white dots indicate the rough directions that the stop signs are facing. For example, symbol 6 represents an all-way stop, while symbol 5 represents a North and South only stop at a 3- or 4-way intersection.

the roads run SouthWest-to-NorthEast and NorthWest-to-SouthEast and assume it contains a 2-way stop. The current icon set could cause confusion as the white dots are only on the North, South, East, and West sections of the image. We tested another method involving the use of the Google Maps snap-to-road functions. We found that in most cases it was not able to accurately place the icons due to the fact that they were being placed on either side of an intersection of two streets. This was only further compounded by roads that had curves in them, causing them to hug another street before making a sharp turn to form an intersection with it. Another problem with the snap-to-road functions is that it does not work well in small towns. This is due to Google making less effort to accurately map these places. As such we decided to

stay with the current method since most streets do in fact run in a standard grid pattern and the graphical output is an accessory function, not a core function of the program.

3.2. MySQL. StopWatcher requires a database in order to hold the GPS positions of stop signs. We decided to use MySQL because it is easy to implement and effective for our proof of concept trials. Our design goal is to keep processing time and memory usage to a minimum. This is because the application is designed to be run on a smartphone which has very limited resources. In keeping with this design goal, we decided to not have entries for every stop sign, but rather have entries that contained all of the information about a given intersection. This greatly reduces the amount of storage space required for intersections containing more than one stop sign. This method offers considerable improvements to the efficiency of the application because most intersections in a city are 4-way intersections that require at least 2 stop signs to be effective [15]. A 2-way intersection occurs when a minor street intersects with a major street, while a one-way intersection only occurs when a street terminates at an intersection, which is usually located at the outer edge of the street grid. This method also reduces the number of necessary distance and approach vector computations, since there are far fewer local nodes than if we had made an entry for every sign.

Each tuple in the database table contains the following intersection data: *ID*, latitude and longitude coordinates, the direction the signs are facing, and the type of road sign at the intersection. The storage cost for each element of the tuple is show in Table 1 based on the specification of MySQL [16]. The latitude and longitude coordinates are taken from the center of the intersection and extend to six places past the decimal in order to ensure a high level of accuracy. This

Table 1: Storage costs for the elements of an entry in MySQL database.

Field name	Field type	Storage size (bytes)
ID	Integer	4
Latitude	Single-precision float	4
Longitude	Single-precision float	4
Sign facing direction	Enum	1
Sign type	Enum	1

is important because the distance across an intersection is typically around 36 feet in most cities of the United States [17]. The program compensates for this position by adding 18 feet, half the distance across a standard intersection, to the distance calculations. Note that less accurate positioning data could place the sign off the center of the intersection and cause StopWatcher to calculate inaccurate approach vectors and distances.

The "facing" field of the table is an enumerated type containing all unique combinations of "N," "S," "E," and "W" denoting North, South, East and West respectively. The 2-way stop sign combinations of NE, NW, SE and SW were excluded as they do not exist in normal traffic conditions due to their inherent ineffectiveness. The values in this field are additive, that is, an intersection with the "facing" field entry of "NS" means that the North and South streets contain a stop sign. It should be noted that the directional terms, like North, do not necessarily mean that the road at the intersection truly runs North, but rather that it is in a more Northern position than the other streets at the intersection. In the case of an intersection that runs in a diagonal to the cardinal directions, we use a 45 degree right turn to determine assignment. That is to say the street that is in the North-East would be considered North and the street in the North-West would be considered West. It should be noted that these types of street intersections were not present in our field test which our simple calculating method worked well. In our future work, we may allow for various combinations of road placements including diagonal positioning for non-standard roads and use the center of the intersection to determine the angle at which the stop signs were placed, but the computational complexity must be considered as it will run on the resource-constrained smartphone.

The last field which denotes the sign type is meant for future expansions to include other traffic signs such as yield signs, crosswalks, school zones, and so forth that could also become obstructed or are otherwise hard to see. Currently the only value being used is the Stop Sign which is the default entry.

3.3. Workflow of StopWatcher Application. Figure 4 shows the workflow of StopWatcher application. After the startup of StopWatcher, the program places the stop signs on the map as shown in Figure 5. The stop sign intersections are retrieved from the MySQL database, grouping them by their "facing" values. All stop sign intersections are then displayed as individual icons and overlaid on the Google map.

Next, the program enters the "Update" stage. This is where the phone samples the user's location via the GPS functions of the phone and detect if the vehicle is approaching a stop sign. The application updates the location every time the user's location changes by at least one meter. The main reason that we decided to force the program to update only on startup and location change is because there is no need to recalculate approach and distance vectors when the user is not moving. This helps reduce power consumption and lighten the workload of the processor when waiting at an intersection or any other time the car is not moving. On the other hand, when the user is moving, it is necessary for the update to be triggered every meter in order to keep the user's location as accurate as possible. Note that in practice the calculating of movement is based on the accuracy of GPS positioning which can vary by several meters per sample, especially when the phone is sitting still. This variance causes the phone to get a false notification to recalculate the values and renders the feature ineffective in real life use. However, should GPS calculations become more accurate, this feature would indeed help to conserve power consumption.

In addition to sampling GPS data in the update stage, the phone also samples what direction the user is currently facing. It does this by accessing the compass sensor, a magnetometer that determines which direction the phone is facing based on the magnetic pull of the Earth's poles. This bearing information is used to determine the relative direction that the user is traveling: North, South, East, or West.

The sampled GPS data and direction information will be used to determine if the user is approaching a stop sign. There are two primary states for this purpose: "Approaching" and "Searching". The approaching state denotes that the user is within range of a specific stop sign, and issues distance alerts via the VNS. The searching state denotes that the user is not currently within a stop sign's range, and it actively compares the stop sign intersection lists against current location and direction data to see if the user is approaching a stop sign. The searching state is the state that the phone starts in when the application is first activated. When it finds a stop sign intersection within range, it decides whether or not it should switch to the approaching state. This is necessary because there is the chance that what it actually found is what we call a "false approach".

In the searching state, the program looks at the stop sign intersection locations to see if the user has entered the range of a stop sign. The program does not need to search all intersection entries, because not all stop signs are relevant to the direction the user is traveling. For example, if a user is traveling North bound and they are approaching an intersection that only has stop signs on the East and West sides, then those signs are not relevant to the driver. Therefore, the program does a preliminary paring down of the stop signs to search based on their "facing" values. The program only searches through the lists of stop sign intersections that are relevant to the driver's current route.

While in the searching state, the program searches all of the relevant stop sign intersection lists, checking if the user

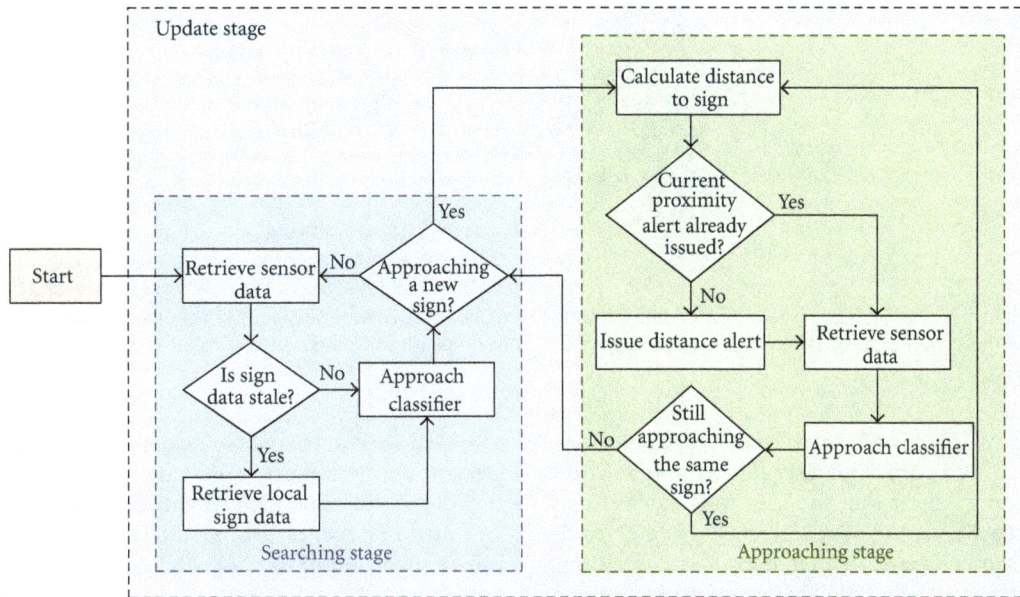

FIGURE 4: Workflow of StopWatcher application.

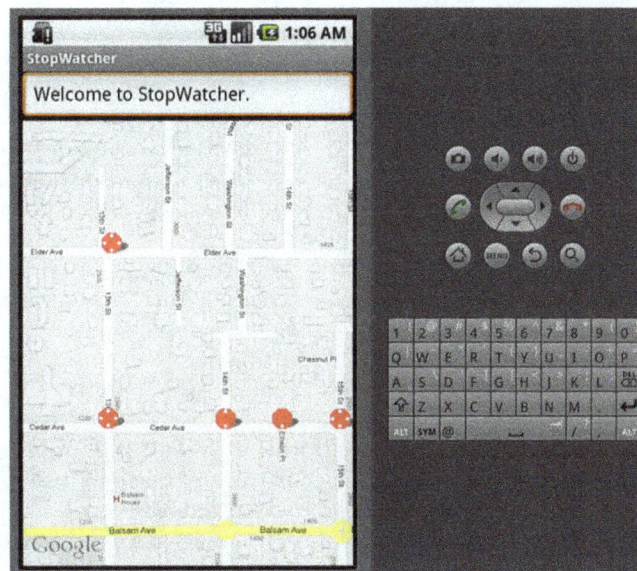

FIGURE 5: Screenshot of StopWatcher application's user interface as seen from the Eclipse Android SDK emulator.

has entered the range of a stop sign. We use a maximum range of 300 feet to determine when the user is considered "on approach" with the stop sign. Using omni-directional range detection can cause errors in the application. This is because a driver could be within 300 feet of a stop sign, but be on a separate road that is running parallel to the one which has the stop sign. To alleviate this type of false detection, we compare the approach angle of the vehicle to the intersection. If the angle between the user's location and the intersection's location is greater than 45 degrees off of the user's traveling direction, then the program does not enter the approaching state. This method shrinks our search area to a cone, searching 300 feet in front of the user, in a 90

degree arc as shown in Figure 6. If the program searches all relevant stop signs and does not enter the approaching state, the program concludes that the user is not approaching a stop sign, and waits for the next location update. On the other hand, if the program detects a stop sign within range that the user is actually approaching, the program enters the approaching state.

It should be noted that we assume that the user will not enter the same stop sign intersection's range twice in a row. That is to say that the driver will not re-enter the same intersection right after he or she just goes through it. This is necessary in order to prevent false detections due to minor inaccuracies and changes in GPS location data.

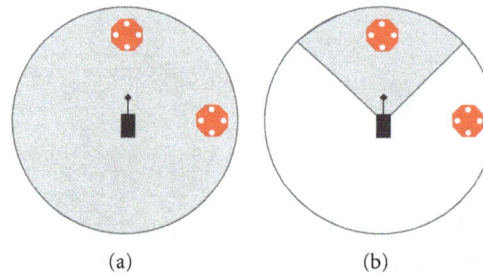

FIGURE 6: The program has a 300 foot search radius when looking for signs. (a) Two separate signs within range that could result in a false positive. (b) Refine the search to only include the intersections that are in line with the current direction of travel.

The approaching state is designed to issue distance alerts, via the VNS, as a user approaches a specific stop sign. There are six possible range alerts that the user may receive: they occur when the user is within 300, 250, 200, 150, 100, or 50 feet from a stop sign. However, if all of these alerts are always triggered, it will be very annoying. Therefore, at a maximum, only four alerts per sign can be triggered in the application. The alerts are designed to skip every other sound byte; that is, if the user hears the 300 foot warning, they will not hear the 250 foot warning, but will hear the 200 foot one. This is to reduce the annoyance factor caused by repeated updates, while still providing the user with the necessary information. The 50 foot warning will always sound no matter the distance that the warnings start at. This is to warn the user that they should be stopping now and allows stop sign detection even in short stretches of road. Once the last alert (50 feet) has been triggered, the program stores the approaching stop sign in a buffer, and returns to the searching state.

One issue to overcome is the scenario when the user is approaching a stop sign, but turns off the road before entering the stop sign intersection. To address this problem, two safeguards are created that continue to check the distance to the intersection and the traveling direction. The first safeguard is that if the user's distance to the stop sign intersection becomes greater than 300 feet, the user has left the stop sign's range. When this happens, the program returns to the searching state. The second safeguard checks the direction the user is traveling while approaching the intersection. If the user's direction is different than the direction of the stop sign approach three rimes in a row, the user has changed direction. There is one other requirement for this that is necessary for it to work correctly. The new direction must be consistent across all 3 samples. This is because there is always the possibility of the phone being bumped or affected in such a way as to give an inaccurate reading. Having three consistent samples helps remove almost all errors that could be caused by the phone mistakenly believing that the user has turned onto a different road. If these criteria are met, then the program will return to the searching state.

4. Field Tests

The StopWatcher application was designed and installed in a Samsung SPH-M920 smartphone equipped with Andriod

OS, GPS, compass, and web capabilities. The field tests were designed to test the StopWatcher application while driving around a typical neighborhood in North America. It should be noted that the test was conducted after a small snow storm which made lots of the stop signs visually obstructed. For the test to be a success it must accurately warn the driver of all intersections where he must stop and not give false warnings when approaching an intersection where the user is not required to stop. It must also give accurate distance estimates to the intersection containing the stop sign when the user is required to stop. We determined that the application must warn the driver of actual stop signs 100% of the time to be considered a success. In addition we decided that it would be allowable to have up to 10% false positives as long as all of the actual stop sign intersections were reported. Figure 7 shows the map of the intersections that were included in the field tests.

During the field test, we set out with the goal to pass through 100 total stop sign intersections, where 50 would require stopping at the sign and the other 50 would not. During the test, all 50 stop sign warnings were triggered correctly which meets our goal of 100% accuracy for hits. However, we did encounter 4 false positive that occurred when turning North from a 4-way stop, which was the only intersection that gave us false positives. However, even with this one intersection creating false positives, we still came out with only 4 false positives out of 50 passes equaling to an 8% false positive rate which is acceptable with our 10% bar. The reason that this 4-way stop sign consistently gave false positives will be in our future investigation.

5. Discussions and Conclusion

Our conclusion is that the primary design goal of Stop-Watcher has been accomplished based on the preliminary results of our field test. With StopWatcher, a driver can be aware of stop signs that may be obscured or otherwise hard to see from the road. As a secondary function it helps to remove the temptation to send text messages or talk on the phone while driving. These in turn, allow drivers to be safer and more responsible when they are behind the wheel.

There are still a few areas of StopWatcher that are in dire need of improving. One important issue is the power consumption. With the StopWatcher application running, the phone only had about 30–45 minutes of battery life due

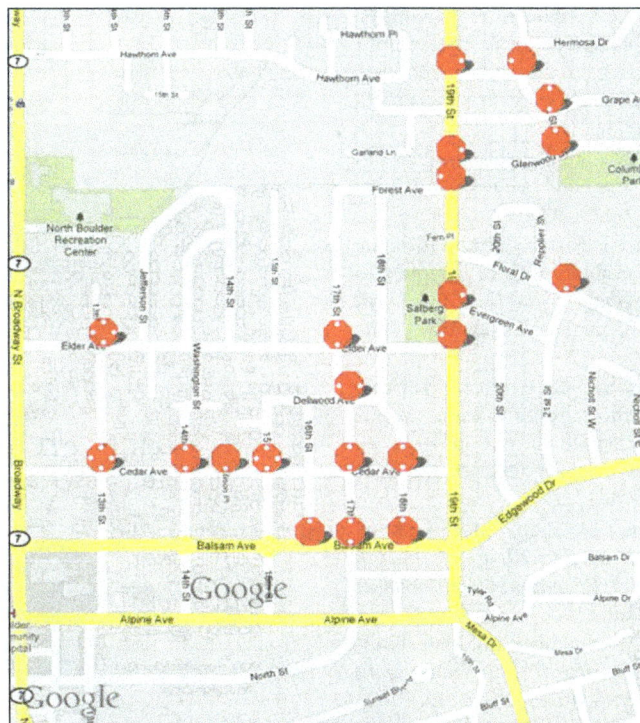

FIGURE 7: Map of all intersections included in the field test.

to the persistent use of GPS, compass, VNS, and graphical user interface. This may limit the usability of our application but can be mitigated by plugging the phone into the power source available in every car. Another improvement we would like to make on StopWatcher in the future is to track more than just stop signs, which has been included in our design but not implemented. In addition, we would like to implement a time system as well for traffic signs that only apply during certain times of the day. These include school zone speed limits, which only apply for about an hour before and after the school session, as well as certain parking signs. This could help people avoid putting children at risk as well as avoid some hefty traffic fines, which are sometimes doubled in school zones.

We would also like to find a way to automatically update the road sign database from the city archives. This could be a bit more problematic as most city governments do not include the exact GPS coordinates of each stop sign that is present in their city. Another possibility would be to parse street names and intersections and compare them to these stop sign maps and find the GPS data from Google Maps directly in order to populate the database.

From the view of proof of concept, StopWatcher is a successful application for driving safety awareness. But there is still a lot of room for expansion in the field of safety through environmental awareness applications. Currently, the majority of applications of location based services are focused on entertainment purposes, such as finding the closest restaurant or night club. Very few works target to improve the safety and well being of the users and those around them. We expect that more applications

like StopWatcher and those mentioned in Section 2 will be developed in future to help reduce automotive accidents and other problems that could be prevented if the user knows more about what is going on around them.

References

[1] U.S. Department of Transportation Federal Highway Administration, "Low-cost safety improvements can improve safety at stop sign controlled intersections," 2002, http://safety.fhwa.dot.gov/intersection/resources/casestudies/fhwasa09010/stop_article.cfm.

[2] Visiting My Councilor, 2008, http://harrumpher.com/?cat=6.

[3] Hidden snow stop sign, 2011, http://peicanada.com/snow_bound_west_prince_pei/image/hidden_snow_stop_sign.

[4] J. Curiel, "Mill Valley: Stop sign obscured by foliage," 2009, http://articles.sfgate.com/2009-07-08/bay-area/17216162_1_sign-stop-intersection.

[5] J. Sena, "Frequently ignored stop sign leads to close calls," 2008, http://www.santafenewmexican.com/Local%20News/Downtown-dilemma.

[6] S. Xu, "Robust traffic sign shape recognition using geometric matching," *IET Intelligent Transport Systems*, vol. 3, no. 1, pp. 10–18, 2009.

[7] A. De la Escalera, J. M. Armingol, and M. Mata, "Traffic sign recognition and analysis for intelligent vehicles," *Image and Vision Computing*, vol. 21, no. 3, pp. 247–258, 2003.

[8] C. Y. Fang, C. S. Fuh, P. S. Yen, S. Cherng, and S. W. Chen, "An automatic road sign recognition system based on a computational model of human recognition processing," *Computer Vision and Image Understanding*, vol. 96, no. 2, pp. 237–268, 2004.

[9] X. Gao, L. Podladchikova, D. Shaposhnikov, K. Hong, and N. Shevtsova, "Recognition of traffic signs based on their colour and shape features extracted using human vision models," *Journal of Visual Communication and Image Representation*, vol. 17, no. 4, pp. 675–685, 2006.

[10] S. H. Hsu and C. L. Huang, "Road sign detection and recognition using matching pursuit method," *Image and Vision Computing*, vol. 19, no. 3, pp. 119–129, 2001.

[11] P. Jimenez, S. Arroyo, H. Moreno, F. Ferreras, and S. Bascon, "Traffic sign shape classification evaluation II: FFT applied to the signature of blobs," in *Proceedings of the IEEE Intelligent Vehicles Symposium*, pp. 607–612, Las Vegas, Nev, USA, June 2005.

[12] P. Mohan, V. N. Padmanabhan, and R. Ramjee, "Nericell: rich monitoring of road and traffic conditions using mobile smartphones," in *Proceedings of the 6th ACM Conference on Embedded Network Sensor Systems (SenSys '08)*, pp. 323–336, Raleigh, NC, USA, November 2008.

[13] J. Dai, J. Teng, X. Bai, Z. Shen, and D. Xuan, "Mobile phone based drunk driving detection," in *Proceedings of the 4th International Conference on Pervasive Computing Technologies for Healthcare*, pp. 1–8, March 2010.

[14] M. Fazeen, B. Gozick, R. Dantu, M. Bhukhiya, and M. C. Gonzalez, "Safe driving using mobile phones," *IEEE Transactions on Intelligent Transportation Systems*, vol. 13, no. 3, pp. 1462–1468, 2012.

[15] "Manual on uniform traffic control devices (MUTCD)," 2003, http://mutcd.fhwa.dot.gov/pdfs/2003/pdf-index.htm.

[16] http://dev.mysql.com/doc/refman/5.1/en/data-types.html.

[17] E. Ben-Joseph, "Residential street standards & neighborhood traffic control: a survey of cities' practices and public officials' attitudes," Institute of Urban and Regional Planning, University of California at Berkeley, 1995, http://web.mit.edu/ebj/www/Official%20final.pdf.

Optimisation of the Nonlinear Suspension Characteristics of a Light Commercial Vehicle

Dinçer Özcan, Ümit Sönmez, and Levent Güvenç

Mekar Mechatronics Research Labs, Department of Mechanical Engineering, İstanbul Okan University, Akfırat-Tuzla, TR-34959 İstanbul, Turkey

Correspondence should be addressed to Levent Güvenç; levent.guvenc@okan.edu.tr

Academic Editor: Shinsuke Hara

The optimum functional characteristics of suspension components, namely, linear/nonlinear spring and nonlinear damper characteristic functions are determined using simple lumped parameter models. A quarter car model is used to represent the front independent suspension, and a half car model is used to represent the rear solid axle suspension of a light commercial vehicle. The functional shapes of the suspension characteristics used in the optimisation process are based on typical shapes supplied by a car manufacturer. The complexity of a nonlinear function optimisation problem is reduced by scaling it up or down from the aforementioned shape in the optimisation process. The nonlinear optimised suspension characteristics are first obtained using lower complexity lumped parameter models. Then, the performance of the optimised suspension units are verified using the higher fidelity and more realistic Carmaker model. An interactive software module is developed to ease the nonlinear suspension optimisation process using the Matlab Graphical User Interface tool.

1. Introduction

Vehicle suspension design and performance problems have been studied extensively using simple car models such as two degrees-of-freedom (d.o.f.) quarter car, four or six d.o.f. half car, or seven d.o.f. full car models. Usually, the suspension design methodologies are based on analytical methods where a linear vehicle model is investigated by solving linear ordinary differential equations. Laplace and Fourier transforms are valuable tools that are used while investigating suspension units with linear characteristics. The performance functions represented by transfer functions in Laplace and/or Fourier domains are considered to be related to ride comfort, tire forces, and handling criteria versus road roughness input to achieve an optimum design. On the other hand, the investigation of nonlinear suspension characteristics must be based more on numerical methods rather than analytical methods due to the more complicated nature of the problem. In this investigation, both linear and nonlinear spring and damper characteristics of a light commercial vehicle are considered and used in an optimisation study.

Lumped parameter suspension models are used in this paper. Mass, stiffness, and damping are distributed spatially throughout a mechanical system like a suspension. Mass, stiffness, and damping are therefore functions of spatial variables (x, y and z coordinates) in a mechanical system, resulting in what is called a distributed parameter system since the mass, stiffness, and damping parameter values are distributed over the mechanical system. An easier approach is to lump the continuous mass, stiffness, and damping characteristics into ideal mass, stiffness, and damping elements. The result is a lumped parameter system, as the mass of the mechanical system is concentrated at the ideal mass element, the stiffness of the mechanical system is concentrated in the ideal stiffness element (the spring), and the damping of the mechanical system is concentrated in the ideal damping element (the damper). The lumped parameter models used in this paper are the quarter car and half car suspension models.

The optimisation requirements of suspension systems and the state-of-the-art of suspension research in the last decade are reviewed first. It should be noted, however, that the available literature is vast and only a small portion of it can be presented here. This paper includes the well-known ride, handling trade-off optimisation, and geometrical optimisation of light commercial vehicle suspension systems. Some heavy vehicle suspension optimisation papers are also reviewed due to their conceptual contribution to the subject.

Vehicle suspensions can be regarded as interconnections of rigid bodies with kinematic joints and compliance elements such as springs, bushings, and stabilizers. Design of a suspension system requires detailed specification of the interconnection points (or so called hard points) and the characteristics of compliance elements. Tak and Chung [1] proposed a systematic approach of achieving optimum geometric design of suspension systems where design variables are determined to meet some prescribed performance targets expressed in terms of suspension design factors, such as toe, and camber, compliance steer. Koulocheris et al. [2] proposed to combine deterministic and stochastic optimisation algorithms for determining optimum vehicle suspension parameters. They showed that such combination yields significantly faster and more reliable convergence to the optimum. Their method combines the advantages of both categories of deterministic and stochastic optimisation. They used a half car model of suspension systems, subject to various road profiles considering the improvement of the passenger ride comfort, leading to minimisation of the maximum acceleration of the sprung mass while paying attention to the geometrical constraints of the suspension as well as the necessary traction of the vehicle.

Maximizing tractive effort is essential to competitive performance in the drag racing environment. Antisquat is a transient vehicle suspension phenomenon which can dramatically affect tractive effort available at the motorcycle drive tire. Wiers and Dhingra [3] addressed the design of a four-link rear suspension of a drag racing motorcycle to provide anti-squat. This design increases rear tire traction, thereby improving vehicle acceleration performance. For the drag racing application considered, any increase in normal forces at the tire patch helps improve race competitiveness. Mitchell et al. [4] used a genetic algorithm for the optimisation of automotive suspension geometries considering the description of a suspension model and a scoring method. Their approach is to design with a unit-free measure of fitness for each test and then to combine these with a weighing function. They showed that the genetic algorithm and the scoring mechanism worked effectively and significantly faster than the more common grid optimisation technique. Raghavan [5] presented an algorithm to determine the attachment point locations of the tie rod of an automotive suspension, in order to achieve linear toe change characteristics with jounce and rebound of the wheel. This linear behaviour is advantageous for achieving good ride and handling. Raghavan's procedure can be applied to all suspension mechanism types such as short-long arm, McPherson struts, and five-link front and five-link rear suspensions.

The design of suspension systems generally demands a compromise solution to the conflicting requirements of handling and ride comfort. The following examples demonstrate this compromise.

(i) For example, for better comfort a soft suspension and for better handling a stiff suspension is needed.

(ii) A high ground clearance is required on rough terrain, whereas a low centre of gravity height is desired for swift cornering and dynamic stability at high speeds.

(iii) It is advantageous to have low damping for low force transmission to the vehicle frame. On the contrary, high damping is desired for fast decay of oscillations.

Considering the aforementioned requirements, Deo and Suh [6] proposed a design for a customizable automotive suspension system with independent control of stiffness, damping, and ride height. Their design enabled the achievement of desired performance depending on user preference, road conditions, and maneuvering inputs while avoiding the performance trade-offs.

Goncalves and Ambrosio [7] proposed a methodology in order to investigate flexible multibody models for the ride and stability optimisations of vehicles. Their methodology allows the use of complex shaped deformable bodies, represented by finite elements. The ride optimisation is achieved by finding the optimum of a ride index that is the outcome of a metric that accounts for the acceleration measured at several key points of the vehicle, weighed according to their importance for occupant comfort. Duysinx et al. [8] developed a mechatronic approach to model, simulate and optimize a passenger car (Audi A6) incorporating a controlled semiactive suspension. They paid particular attention to the formulation of the mechatronic model of the car and compared two different modelling and optimisation approaches. The first approach is carried out in the Matlab-Simulink environment and the derivation of the equations is based on a symbolic multi-body model. The optimisation procedure has also been investigated in Matlab. Their second approach relies on a multi-body model based on the finite-element method where the optimisation has been realized with an open-source industrial optimisation tool.

Eskandari et al. [9] optimized the handling behaviour of a midsized passenger car by altering its front suspension parameters using Adams/Car software. They utilized an objective function combination of eight criteria indicating handling characteristics of the car and reduced the amount of optimisation parameters, by implementing the design of experiments method capabilities. The amount of the parameters was reduced from fifteen to ten by using a sensitivity analysis. A similar Adams/Car-based study was conducted by Boyalı et al. [10]. More recently, He and McPhee [11] reviewed the state-of-the-art related to modelling approaches, considering vehicle system models, design variable and performance criteria definitions, optimisation problem formulation methods, optimisation search algorithms, sensitivity analysis, computational efficiency, and other related techniques. They applied these techniques to the design synthesis of ground vehicle suspensions and proposed a methodology for automated design synthesis of ground vehicle suspensions.

Li et al. [12] considered a five-link suspension optimisation for improving the ride safety and comfort using

Adams/Insight. They investigated the relations among multi-link suspension structural parameter, wheel location parameter, and wheel track. Uys et al. [13] reported an investigation to determine the spring and damper settings that will ensure optimal ride comfort of an off-road vehicle, on different road profiles and at different speeds. Spring and damper settings can be set either to the ride mode or the handling mode, and therefore a compromise ride-handling suspension is avoided. They found that optimizing for a combined driver plus rear passenger seat weighed root mean square vertical acceleration rather than using driver or passenger values only returns the best results. Their results indicated that optimization of suspension settings using the same road and constant speed will improve ride comfort on the same road at different speeds and these settings will also improve ride comfort for other roads at the optimisation speed and other speeds, although not as much as when optimisation has been done for the particular road. In the present paper, one vehicle speed and one road profile were used for optimization taking this statement from [13] into account. They also concluded in [13] that for improved ride comfort, damping generally has to be lower than the standard (compromised) setting, the rear spring as soft as possible, and the front spring ranging from as soft as possible to stiffer depending on road and speed conditions. Ride comfort is the most sensitive to a change in rear spring stiffness.

The roll steer of a front McPherson suspension system is studied, and the design characteristics of the mechanism are optimized by Habibi et al. [14] using the genetic algorithm method. The roll steer affects handling, and dynamic stability of the vehicle due to variation of the angles of the wheel and the suspension links (i.e., camber, caster, and toe). However, these changes cause other problems. In their paper, Habibi et al. [14] used a genetic algorithm method to determine the optimum length and orientation of the mechanism's members to minimize the variations of the toe, camber, and caster angles. They defined a performance index which expresses the overall variations of the main parameters in the whole range of rolling of the body. A general formulation for multi-body flexible systems, with linear elastic deformations, is considered by Goncalves and Ambrosio [15] whose application involved a road vehicle where flexibility plays an important role in ride and handling dynamic behaviour. Using finite elements to describe the flexibility of the body and the modal superposition method has the advantage of greatly reducing the dimensionality of the system. The results presented in [15] showed that the use of the detailed vehicle model within the framework of ride optimisation leads to a measurable improvement of the comfort conditions for different road profiles and driving conditions.

An optimum concept to design "road-friendly" heavy vehicles with the recognition of pavement loads as a primary objective function of vehicle suspension design was investigated by Sun [16]. A walking-beam suspension system is used as an illustrative example of the vehicle model to demonstrate the concept and process of optimisation. Dynamic response of the walking-beam suspension system was obtained by means of stochastic process theory. Using the direct update method, optimisation is carried out when tire load magnitudes are taken as the objective function of suspension design. Their results showed that tires with high air pressure could lead to more damage in pavement structures, and increasing suspension damping and tire damping can reduce the tire loads and pavement damage.

This paper concentrates on optimisation of the nonlinear shape of the spring and damper of a light commercial vehicle. The work reported in the paper is motivated by the fact that the spring and damping characteristics in an actual road vehicle are designed to be nonlinear on purpose by the automanufacturer. Larger spring forces are used in the rebound motion of the wheel in order to keep forcing an appropriate level of tire-road contact at all times, for example. When the automanufacturer starts working on a new model, a previous, successful suspension design is used as the base characteristic which is modified to fit the characteristics of the new vehicle model. The work presented here follows the same approach as spring, and damper characteristics of an existing base design are used as the starting point in the optimisation. The optimisation procedure is embedded into an interactive Matlab Graphical User Interface to offer ease of use to suspension designers.

The organization of the rest of the paper is as follows. The vehicle models used in this study are introduced in Section 2 along with the scope of this paper. Some of the performance criteria available in the literature are discussed in Section 3. The optimisation procedure used is the topic of Section 4, while the optimisation results obtained using this procedure are treated in Section 5. The paper ends with the usual conclusions and recommendations being presented in Section 6.

2. Vehicle Models Used and Scope of the Investigation

In this section, the scope of the current investigation is summarized by presenting the vehicle models that are used in this study. The use of a complex three-dimensional model of the vehicle, with a detailed description of all suspension systems and road/tire interaction, is necessary to fully investigate the problem. However, such models are computationally expensive especially when used in an iterative optimisation design process. A good alternative which is used here is the optimisation of a subsystem of a complex model. The suspension subsystem is very important in terms of vehicle dynamics. Its spring and damper load deflection characteristics are treated as the basic design variables here.

The ride comfort optimisation is achieved by finding the optimum of a ride comfort index which results from a metric that accounts for the linear and the angular accelerations of the model's suspended mass centre and properly combined in a cost function, considering their importance for the comfort of the occupant. Two lumped parameter models are built in Matlab considering the independent front suspension unit (a quarter car model) and the rear axle suspension unit (a half car model) of a light commercial vehicle. Vertical displacement z_s and acceleration a_s of the suspended mass and the tire force F_{Tire} of the quarter car model are considered as the key variables in ride comfort and handling, respectively. Similarly, vertical and angular displacement z_s and ϕ_s of the mass centre and the tire forces at both of the rear wheels of the half car model are selected as the key variables for the rear suspension half car model.

The dynamics of the quarter car and the half car rear axle suspension models are governed by nonlinear differential equations of motion. The road profile described in Cebon [17] is used in the optimisation study and simulations to determine suitable linear and/or nonlinear spring and nonlinear damper characteristics. Considering a nonlinear model, a suspension characteristic optimisation routine written as an m-file in Matlab gives more insight than using commercial vehicle simulation/analysis software. In some cases, this option (nonlinear suspension characteristics optimisation) is not readily available in commercial vehicle suspension packages. The main objective and the contribution of this investigation are to determine the optimum nonlinear functions of the damper and the spring characteristics for the improvement of the passengers' ride comfort and vehicle handling leading to minimisation of the chosen objective function.

Basic shapes representing the spring and the damper characteristics (force versus deflection for the spring and force versus velocity for the damper) are used according to automotive manufacturer's specifications. Basic functional shapes in each operating mode (extension or compression regions of the spring and the damper) are predetermined and the functional fits to these shapes are obtained. These functions and their linear combinations are then scaled searching for the optimum characteristics. The emphasis of this investigation is placed on finding nonsymmetric optimum nonlinear functions of the spring and the damper force characteristics. Optimised functional relations are then incorporated into a model built in a high fidelity, realistic commercial vehicle dynamics software to evaluate the performance of the vehicle model with the optimised suspension. Carmaker software [18] is used for this purpose to study the handling behaviour of the car in standard tests (double-lane change, fishhook, etc.). The Carmaker vehicle model [18] is a highly realistic one that incorporates engine dynamics, tire dynamics, steering dynamics, suspension dynamics, vehicle sprung body dynamics, longitudinal and lateral dynamics, a driver model, and road and environment models. The investigation also looks at a scenario where ride comfort and handling are simultaneously required. The cornering behaviour of a road vehicle is an important performance mode often equated with handling. In order to analyze both of ride and handling requirements, a double lane change manoeuvre is performed after travelling over an irregular road profile with a disturbance, and then the vehicle comes to a stop. The simulation results of optimised nonlinear damper and nonlinear spring characteristic functions are compared with those of the optimised linear ones in simulations.

2.1. Suspension Models Used.
The mathematical models of the quarter car and the half car representing one of the

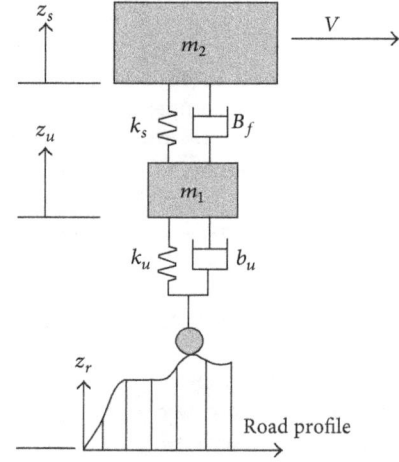

FIGURE 1: Quarter car model.

front quarters and the rear axle suspension unit of a light commercial vehicle are presented in this section.

2.1.1. Quarter Car Model. The quarter car model subject to road disturbances is shown in Figure 1. The equation of motion considering the vertical displacement of the vehicle body with linear spring and nonlinear damper characteristics may be written as

$$m_s\,\ddot{z}_s + B_F\{\dot{z}_s - \dot{z}_u\} + k_s\,(z_s - z_u) = 0. \tag{1}$$

Similarly the equation of motion of the vehicle wheel may be written as

$$m_u + \ddot{z}_u + B_F\{\dot{z}_u - \dot{z}_s\} + k_s\,(z_u - z_s) - b_u\,(\dot{z}_u - \dot{z}_r)$$
$$- k_u\,(z_r - z_u) = 0, \tag{2}$$

where $B_F\{\dot{z}_u - \dot{z}_s\}$ represents the nonlinear functional relation of suspension damper force versus velocity characteristic.

2.1.2. Half Car Model. The rear solid axle suspension unit of the light commercial vehicle considered here is represented with a half car model (see Figure 2) and subjected to road disturbances coming from both sides of the track (the left and the right wheels). The torsional antiroll bar is also considered in the half car model.

The half car model can represent the bounce (z_s, z_u) and roll motions (ϕ_s, ϕ_u) of the car body and solid rear axle. Therefore, it has four d.o.f. The equations of motions of the sprung (car body) and unsprung (rear axle) masses considering the bounce and roll motions may be written as

$$m_s\ddot{z}_s + F_{\text{sleft}} + F_{\text{sright}} = 0 \tag{3}$$

for the bounce motion of the sprung mass and

$$m_u\ddot{z}_u - F_{\text{sleft}} - F_{\text{sright}} + F_{\text{uleft}} + F_{\text{uright}} = 0 \tag{4}$$

for the bounce motion of the rear axle. The roll motion of the sprung inertia is given by

$$I_{xx}\ddot{\phi}_s + F_{\text{sleft}}\frac{L}{2} - F_{\text{sright}}\frac{L}{2} = 0 \tag{5}$$

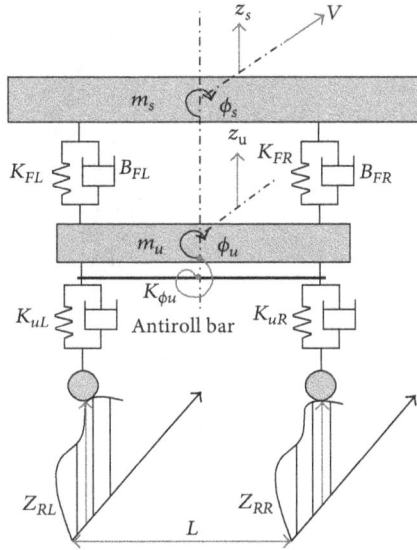

FIGURE 2: Half car model.

and the roll motion of the rear axle (considering rotational unsprung inertia I_{uxx}) is represented by

$$I_{uxx}\ddot{\phi}_u + K_{\phi u}\phi_u - \left(F_{\text{sleft}} + F_{\text{uright}}\right)\frac{L}{2}$$
$$+ \left(F_{\text{sright}} + F_{\text{uleft}}\right)\frac{L}{2} = 0, \tag{6}$$

where the forces F_{sleft}, F_{sright}, F_{uleft}, and F_{uright} acting on the sprung mass are given by

$$F_{\text{sleft}} = B_{FL}\left\{\left(\dot{z}_s + \dot{\phi}_s\frac{L}{2}\right) - \left(\dot{z}_u + \dot{\phi}_u\frac{L}{2}\right)\right\}$$
$$+ K_{FL}\left\{\left(z_s + \phi_s\frac{L}{2}\right) - \left(z_u + \phi_u\frac{L}{2}\right)\right\},$$

$$F_{\text{sright}} = B_{FR}\left\{\left(\dot{z}_s - \dot{\phi}_s\frac{L}{2}\right) - \left(\dot{z}_u - \dot{\phi}_u\frac{L}{2}\right)\right\}$$
$$+ K_{FR}\left\{\left(z_s - \phi_s\frac{L}{2}\right) - \left(z_u - \phi_u\frac{L}{2}\right)\right\}, \tag{7}$$

$$F_{\text{uleft}} = k_{uL}\left(z_u + \phi_u\frac{L}{2} - z_{RL}\right),$$

$$F_{\text{uright}} = k_{uR}\left(z_u - \phi_u\frac{L}{2} - z_{RR}\right),$$

where L stands for the track width, and I_{xx} and I_{uxx} represent the moments of inertia of the sprung mass and axle, respectively. $K_{\phi u}$ is the torsional spring stiffness of the antiroll bar which is calculated as

$$K_{\phi u} = \frac{GJ}{L_b}, \tag{8}$$

where G is the shear modulus, J is the polar moment of inertia, and L_b is the length of the stabilizer bar which is

subjected to torsional stress. In the simulations and in the optimisation process, spring and damper characteristics of the left and right sides are assumed to be identical.

3. Performance Criteria Available in the Literature

3.1. Some Performance Indices Available in the Literature. The choice of subobjective functions and their weights in the combined (main objective) function plays a very critical role in the optimisation process. In this section, the literature review of the vehicle suspension objective functions and the performance indices are summarized, and our approach to the objective function formulation is presented. The nonlinear stiffness and damping characteristics are optimized by Koulocheris et al. [2] considering a half car model subject to different types of road irregularities. As an objective function, the maximum value of vertical acceleration of the vehicle body at the passenger seat is minimized from the view point of ride comfort. The objective function is formed according to the quadratic penalty given by

$$f(z) = \max\left(\ddot{z}_s\right) + M\sum c_i^2(z), \tag{9}$$

where $\ddot{z}_s$ is the vertical acceleration of the vehicle mass, M is the penalty parameter, and c_i are the constraint functions for parameter vector z.

Geometrical parameters of the suspension were considered in Mitchell et al. [4] when determining the fitness of a given suspension design. Since these parameters are not all at the same magnitude or even have the same units, coming up with a single fitness value is difficult. Their basic approach was to carry out the design with a unitless measure of fitness for each test and then to combine these results with a weighing function. Several functions were analyzed and compared while evaluating the speed and the accuracy of the method using the genetic optimisation algorithm. A first-order normal distribution was chosen due to its convergence speed. The score equals to 100 for the ideal score and 28.3 at the bound. Each metric score (score$_i$) is combined by way of a weighing function (W_i). Then, the scoring metric and total score are normalised using

$$\text{Total Score} = \frac{\sum_i W_i\text{Score}_i}{\sum_i W_i}. \tag{10}$$

The coordinates of the front and rear suspension hard points, the stiffness and damping properties of the front and rear suspension springs and damper, sprung mass, gear ratio, the inertia of steering wheel and so forth. were selected by Li et al. [19] as the design variables, considering vehicle handling. The objective evaluation index was adapted to evaluate the performance of vehicle handling. The index included course following indices, driver burden indices, indices for the risks of roll over, index for driver's road feeling, and index for lateral slip. The double lane change maneuver was selected for a virtual test, and the objective evaluation index was calculated.

The objective function in Eskandari et al. [9] was selected to represent several aspects of the handling behaviour of a vehicle and is of the form

$$F = \sum_i W_i X_i, \tag{11}$$

where X_i represent yaw velocity overshoot, yaw velocity rise time, lateral acceleration overshoot, lateral acceleration rise time, roll angle steady-state response, RMS of the understeering coefficient, RMS of the steering torque, and RMS of the steering sensitivity. Determination of the weighing factors W_i was made based on the importance of each quantity and is adjustable.

A global performance index is considered in Tak and Chung [1] as the linear combination of each individual performance index. Through kinematic analysis, toe and camber curves were obtained, and target values of the toe and camber curves were set up. The squared value of the reaction force at each tie rod was also included in the performance index. The global performance index was determined as the weighed linear combination of the wheel angles and reaction forces at the joints as

$$I = W_1 I_{toe} + W_2 I_{camber} + \cdots + W_j I_{roll\ rate} \\ + \cdots + W_k I_{reaction\ forces}. \tag{12}$$

The performance of an active suspension system was evaluated by Jonasson and Roos [20], covering comfort and road holding capabilities as well as the energy consumption of the system. The formulation of three different performance indices was considered: two of them are based on the RMS norm described by

$$|z|_{RMS} = \sqrt{\frac{1}{\tau} \int_0^\tau z^2\, dt}. \tag{13}$$

Comfort is strongly related to the vertical accelerations of the vehicle body. Therefore, the performance index for comfort is formulated considering vertical accelerations. The comfort index for a vehicle with an active suspension system is weighed with respect to the acceleration of the body in a conventional system and is described as a ratio $I_1 = |\ddot{z}_{active}|_{RMS}/|\ddot{z}_{passive}|_{RMS}$. A value of more than one means that the current design is inferior as compared to the passive suspension. Road holding capability is directly related to the variation in vertical tire force, a constant tire force is ideal, and therefore the second index is described by $I_2 = |F_{T,active}|_{RMS}/|F_{T,passive}|_{RMS}$. The third index I_3 of the objective function considers the energy used by the active system as represented with $I_3 = |P_{loss,active}|/|P_{loss,passive}|$. Then, these indices are combined to form an overall objective function for the optimisation algorithm

$$I = W_1 I_1 + W_2 I_2 + W_3 I_3. \tag{14}$$

3.2. Use of Frequency Weighing Based on the ISO2631 Standard. In order to optimize ride characteristics, human sensitivity to vibrations needs to be considered. For that purpose,

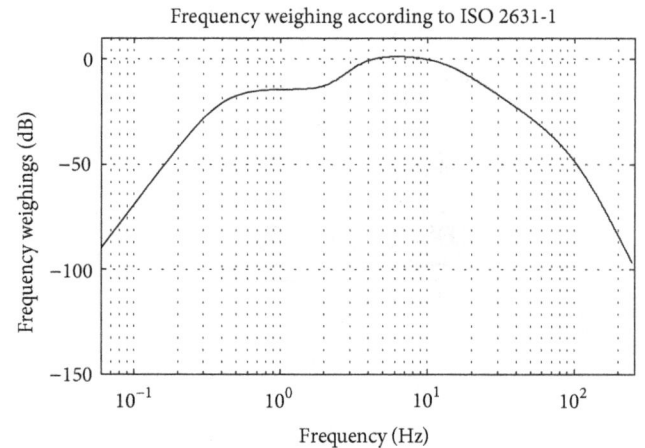

FIGURE 3: Frequency weights as specified in ISO 2631-1 standard.

the vertical motion is weighed according to the ISO 2631 [21] standard. The different characteristics of the excitation, including magnitude, frequency, axis, and duration based on the human tolerance for vibrations should be considered. As suggested by the ISO 2631 standard, the complete acceleration time histories for each of the target points are measured. Then, each Cartesian component of the acceleration history is decomposed into a Fourier series. After that, a frequency weight given in ISO 2631 standards is multiplied by each term of the Fourier series. The single objective function value is determined as the sum of the weighed terms of the Fourier series previously obtained in the decomposition process. Frequency weights of acceleration as specified in ISO 2631-1 standards are shown in Figure 3.

Figure 4 shows the body acceleration response of a quarter car model to a chirp (swept sine) signal, its power spectral density (PSD), and their weighed counterparts obtained after using the ISO 2631 standard. The original and the weighed signals are presented in time domain (top plot) for comparing their magnitudes and in the frequency domain (bottom plot) for comparing their power spectral densities (see Figure 4).

The target accelerations of the vehicle models are weighed according to the ISO 2631 standards in this investigation [21, 22]. The cases belonging to each vehicle model (total of two cases) are investigated. The performance indices of these cases are presented in the simulation section (Section 6).

4. The Optimisation Procedure

In this section, the optimisation procedure is explained in detail. In the current investigation, the complexity of the optimisation problem is reduced by deciding on the basic shape/behaviour of the force versus displacement and force versus velocity characteristics and then scaling them up or down during optimisation. This is motivated by the suspension design procedure used in the automotive industry where a previous, successful design is used as the base design, modified for the vehicle model being developed. The procedure used here has two steps.

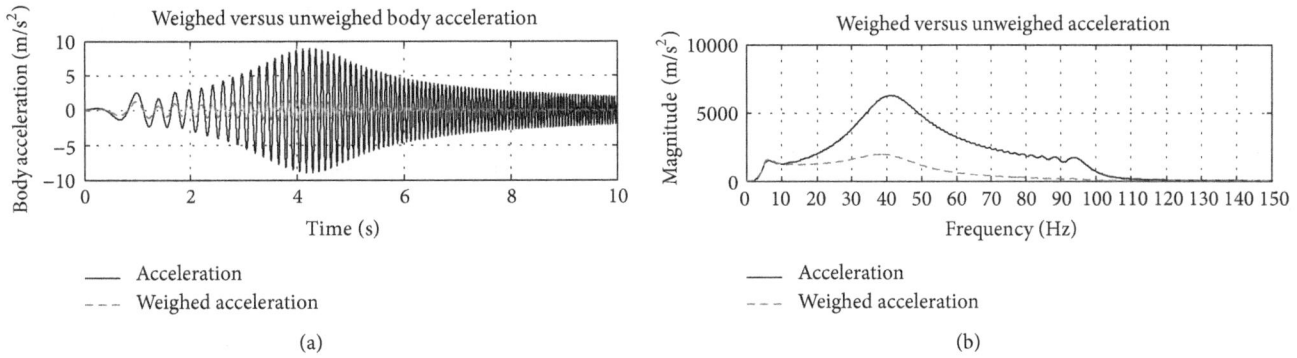

(a)

(b)

FIGURE 4: Body acceleration signal is weighed according to ISO2631-1 standards.

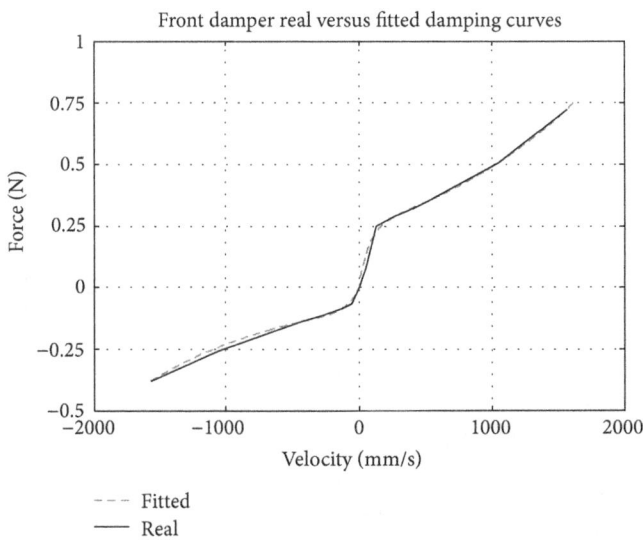

FIGURE 5: Normalized damper characteristic and its functional fit (front).

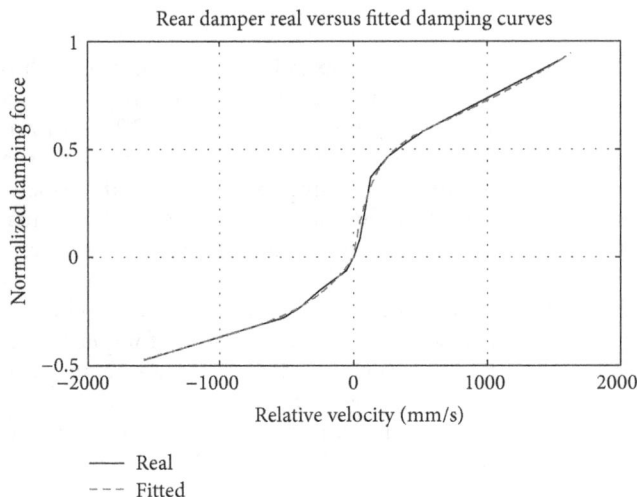

FIGURE 6: Normalized damper characteristic and its functional fit (rear).

(1) A function like a polynomial, rational, or an exponential function that can fit the basic initial data of force versus displacement/velocity profile is chosen first.

(2) Then, a scaling factor C_{opt} is used for the function, such that

$$F_{opt}(x) = C_{opt}F(x).\tag{15}$$

The optimisation procedure requires computations of the objective function at each iteration step. The performance index used here includes both time and frequency domain analyses. The performance index evaluation can be summarized with the following steps.

(1) First, the simulations of the quarter and the half car models subjected to a road excitation are carried out in the time domain. Note that while running the optimisation routines of the vehicle suspension models, two aspects could significantly affect results and might cause errors in the optimisation process as follows:

  (a) it is preferable to consider the steady state response of the vehicle run. Since a constant vehicle speed is assumed, it takes some time for the vehicle to reach the steady state conditions. Therefore, the beginning part of the time domain simulation containing the transient response is omitted,

  (b) attention should be paid to the static deflections (due to weights) and the initial conditions considering the static equilibrium points for the springs.

(2) Then, the target point's accelerations are weighed according to ISO-2631 standards in the frequency domain and used as part of the objective function.

(3) A suitable global objective function is established according to the needs of automotive manufacturers. This is the most subjective step of the methodology. Since the choice of the objective function and weighing of the particular objectives will result in different

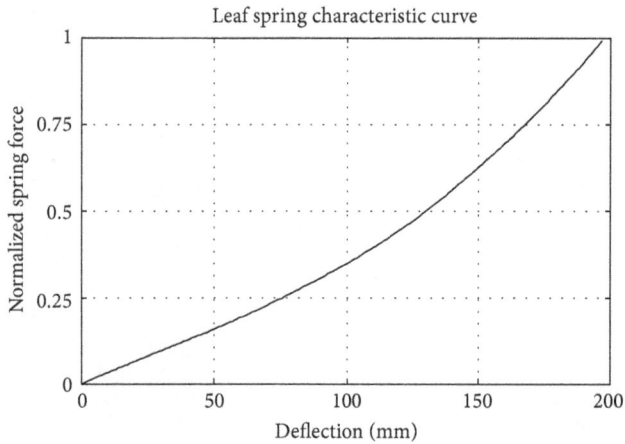

FIGURE 7: Normalized spring characteristic (rear).

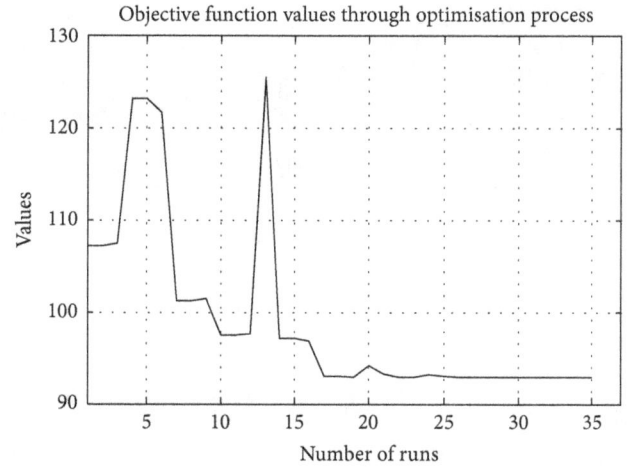

FIGURE 8: Road profile PSD versus wavenumber.

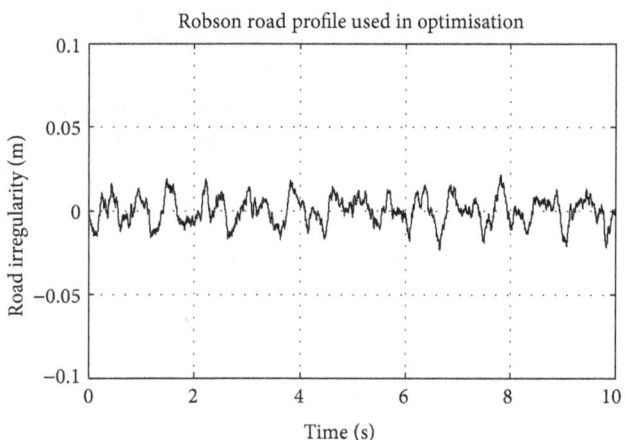

FIGURE 9: Generated road profile versus time.

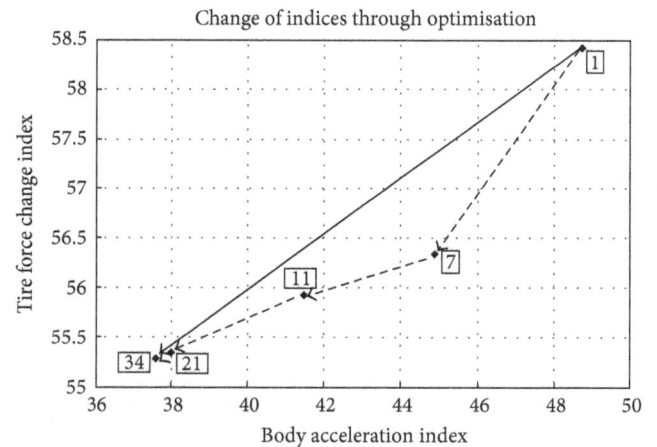

FIGURE 10: Objective function's run history.

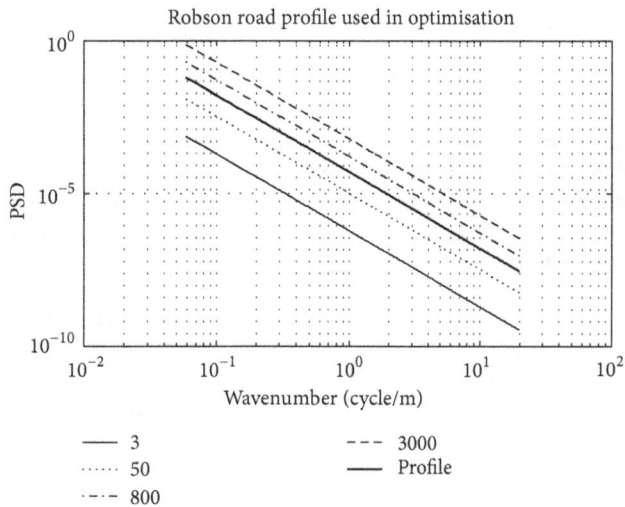

FIGURE 11: Change of indices throughout optimisation process.

optimum outcomes. Our choice of objective functions (performance index) for the current paper is explained in the previous section and in the following section on optimisation results.

(4) As the final step, the optimisation type and algorithm are selected, and the optimisation step is performed in Matlab. The optimisation toolbox SQP algorithm with Quasi-Newton line search is implemented. The SQP algorithm like Simplex, Complex, and Hook-Jeeves belongs to the family of local search algorithms. The local search algorithms converge to the nearest optimum, since they depend upon the starting values of the design variables. Examples in the following section illustrate the simulation result for quarter car and half car vehicle models used here. Numerical simulation results show that the SQP algorithm can efficiently and reliably find the optimum in the neighbourhood of the initial point.

Finally, the optimum spring and damper characteristics obtained should be checked to see if they are manufacturable.

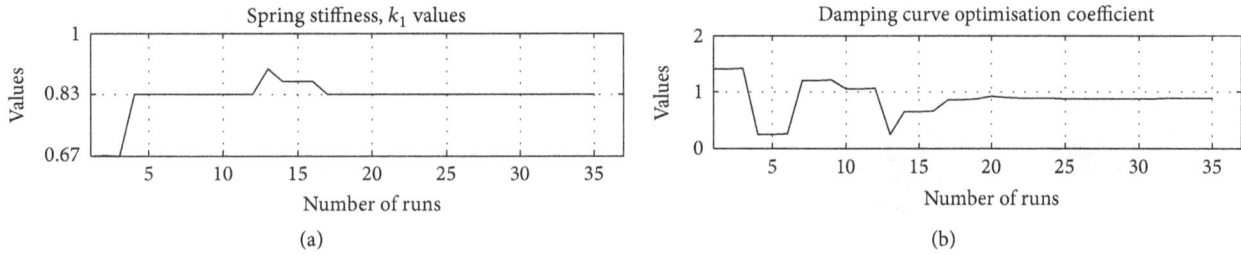

FIGURE 12: Optimum normalised linear spring stiffness and nonlinear damper characteristic scaling coefficients.

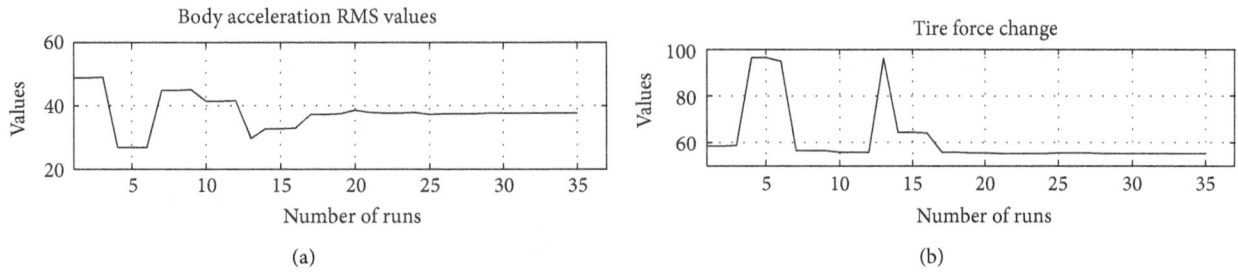

FIGURE 13: Performance indices throughout the optimisation process.

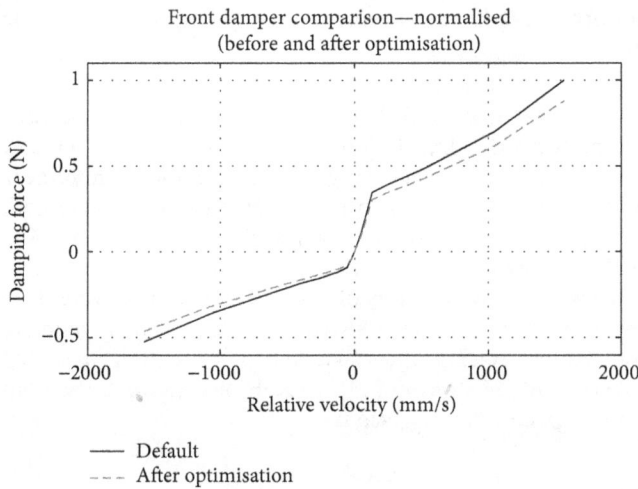

FIGURE 14: Default and optimized damper characteristic curves.

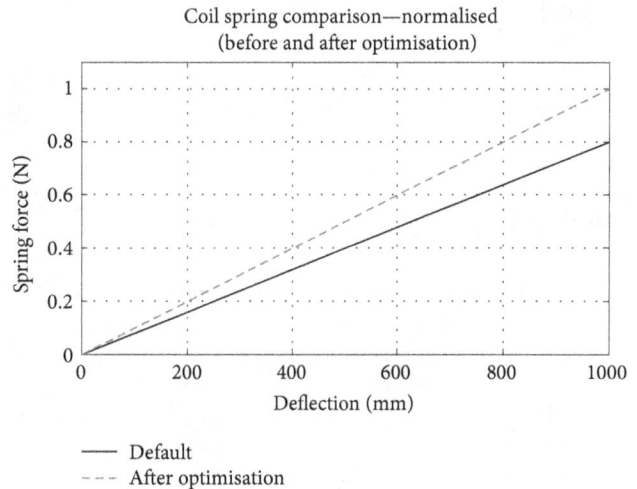

FIGURE 15: Default and optimised spring stiffness.

The nonlinear damper characteristic of the front independent suspension and rear solid axle of a light commercial vehicle are presented in normalized form in Figures 5 and 6, respectively. Since the shape of the curve is essential for manufacturing, an appropriate functional representation should be used in the optimisation process. A function with the following form is suitable for the whole range of the damper data

$$F(v) = Ae^{-kv} + Be^{qv}. \qquad (16)$$

Equation (16) is used as an empirical curve fit to real damper characteristics. The parameters A, B, k and q in (16) are, therefore, determined empirically. v in (16) represents the independent variable, which is suspension velocity here. The real experimentally determined suspension damping characteristics are shown as solid lines in Figures 5 and 6. The empirical data fit obtained using (16) is shown as dashed lines in Figures 5 and 6. The close fit between the real and fit (using (16)) damping characteristics shows that (16) provides a highly accurate representation of the real damping characteristics considered in this paper.

For the nonlinear modelling of the spring of the rear axle, a lookup table which represents the nonlinear characteristic of the spring is used (see Figure 7). The spring characteristic of the front independent suspension is linear in all optimisation processes.

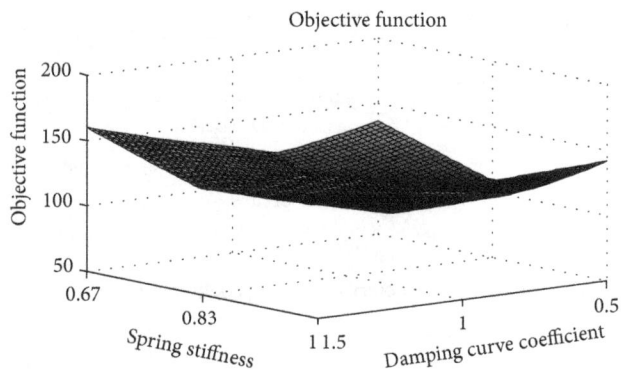

FIGURE 16: Values of the objective throughout two-dimensional search domain.

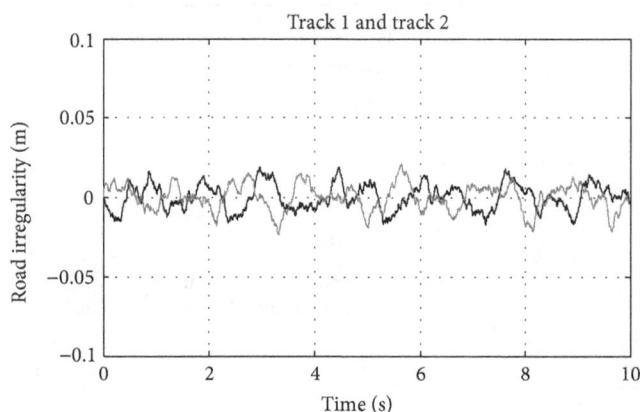

FIGURE 17: Left and right side of the road generated with Robson Method.

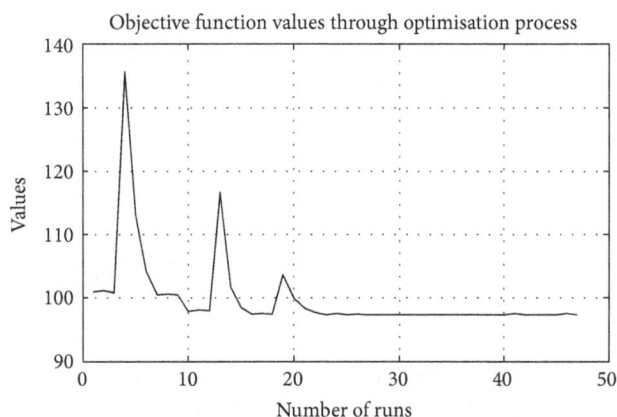

FIGURE 18: Total objective function progress.

5. Optimisation Simulation Results

The optimisations results for the front and rear suspension units of the quarter and the half car models are presented in this section. The performance indices used in the optimisation process are presented in the following Sections (5.1 and 5.2). Finally, Carmaker software is used to check and to

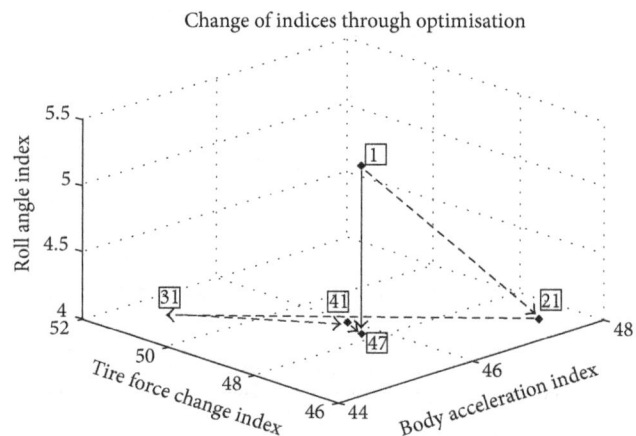

FIGURE 19: Change of indices throughout optimisation process.

TABLE 1: Different road classes.

Road class	$C/10^{-8}$ m$^{0.5}$ cycle$^{1.5}$
Motorway	3–50
Principal road	3–800
Minor road	50–3000

confirm the optimised results using a high fidelity, full vehicle model.

5.1. Quarter Car Optimisation Results. Optimisation runs are performed using the quarter car model in this part of the investigation considering the vehicle with a nonlinear damper and linear spring unit subjected to a generated road. It is required to have road surface input profiles for the realistic response of the vehicle dynamic simulations. These input profiles may come directly from the measurements made by a test vehicle. It is also possible to artificially generate the random road profiles like Robson's Method, presented in Kamash and Robson [23]. In Robson's Method, the road profile spectra can be given by

$$G(n) = Cn^{-w}, \qquad (17)$$

where w is a constant which is equal to 2.5, n is the wavenumber in cycle/m as described in ISO-8608, and $G(n)$ is the displacement spectral density in m^3/cycle. For different classes of roads C takes the values listed in Table 1 [24].

In the conventional spectral analysis the process is simply squaring the spectral coefficients which are determined by using discrete Fourier transform. It is required to have uniformly distributed random phases between 0 and 2π. The inverse Fourier transform is given by:

$$z_{r1} = \sum_{k=0}^{N-1} \sqrt{G_k} e^{-i(\theta_k + (2\pi kr/N))}, \qquad r = 0, 1, 2, \ldots (N-1), \quad (18)$$

where G_k is described by

$$G_k = \left(\frac{2\pi}{N\Delta}\right) G_{11}(\gamma_k) \qquad (19)$$

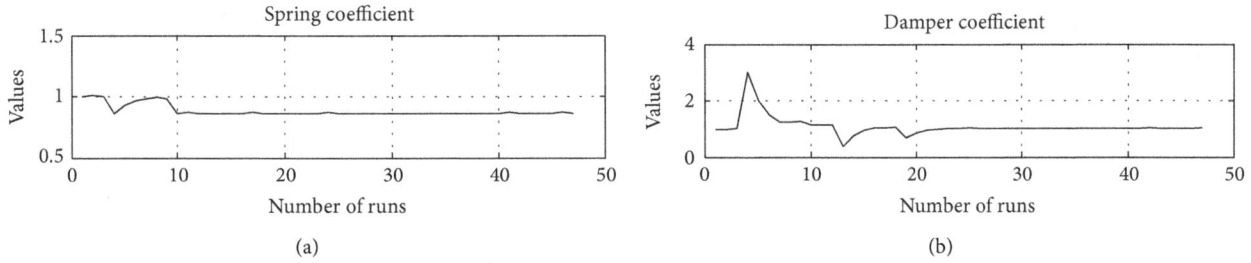

FIGURE 20: Spring and damper characteristics scaling coefficient progress.

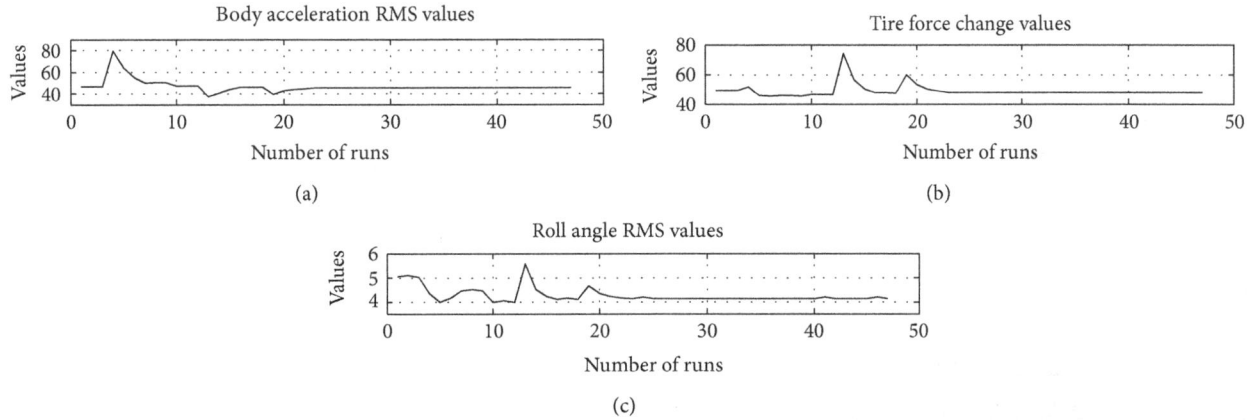

FIGURE 21: Objective function components' progress.

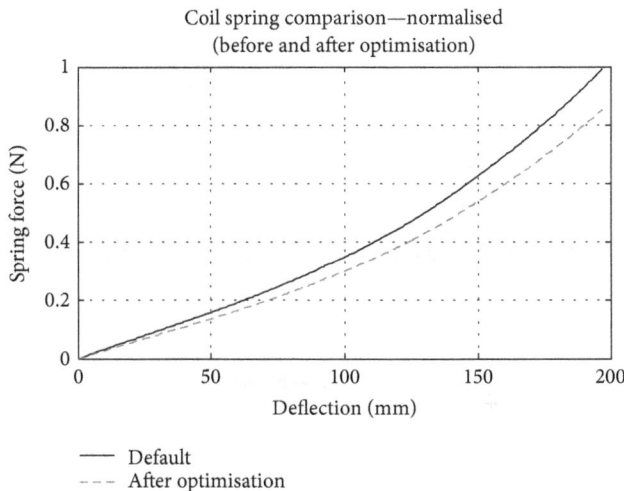

FIGURE 22: Optimized and the default spring characteristics.

and γ_k is given by

$$\gamma_k = \frac{2\pi k}{N\Delta},\tag{20}$$

where γ_k is the wave-number in rad/m, $G_{11}(\gamma_k)$ is the target spectral density, Δ is the distance interval between two spot heights, and θ_k is a set of independent random phase angles uniformly distributed between 0 and 2π. The road profile was generated via Robson's Method as is given in Kamash and

Robson [23] and also in Cebon [17]. The road profile used in the optimisation has $C = 250$ (see Figures 8 and 9).

The objective function has two components which are the weighed body acceleration RMS value and the penalty function of tire force difference. The objective function is defined as

$$I = w_1 \ddot{Z}_s + w_2 \Delta F_{\text{tire}},\tag{21}$$

where $\ddot{Z}_s$ is the RMS value of the weighed body acceleration, and ΔF_{tire} is the difference between the maximum and minimum tire force picked from the tire force history. The total objective function changes as shown in Figure 10 during the optimisation run, and an optimum point is reached after 35 iterations. The minimisation of both indices in the objective function as the process goes on is illustrated in Figure 11 via representative steps. Figure 12 shows the normalised values of the design variables: the spring stiffness and damper characteristic scaling coefficients for the quarter car model. The spring stiffness increases, and the damper characteristic scaling coefficient decreases until an optimum point is reached. How the body acceleration RMS value $\ddot{Z}_s$ and the tire force change ΔF_{tire} change during the optimisation run are displayed in Figure 13. The initial and optimised damper and spring characteristics are shown in Figures 14 and 15, respectively. Using a brute force approach, the objective function is evaluated on a grid of spring stiffness and damping curve scaling coefficient values. The resulting plot in Figure 16 demonstrates that the optimisation results shown in Figure 13 are accurate.

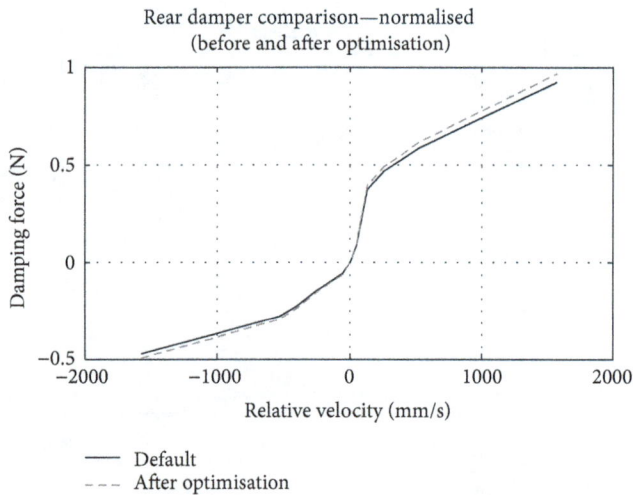

FIGURE 23: Optimised and default damper characteristics.

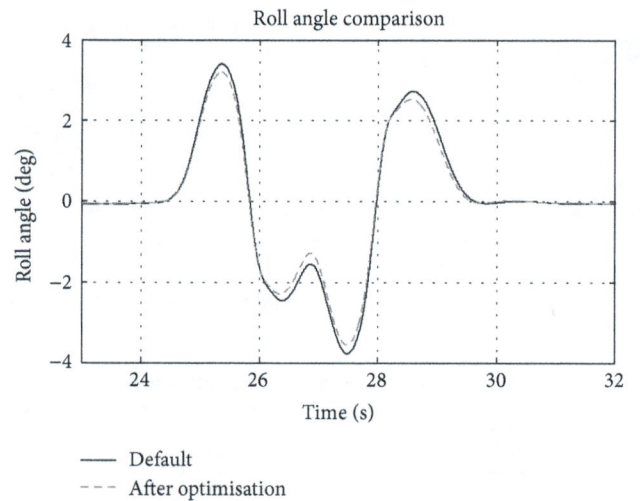

FIGURE 24: Values of the objective throughout two-dimensional search domain.

FIGURE 25: Snapshot of Carmaker animation.

5.2. Half Car Model. The half car model embodying a nonlinear rear suspension unit of the light commercial vehicle incorporating nonlinear dampers and nonlinear springs whose basic characteristic curve shapes are given in the previous sections is considered in this subsection. The performance index used in the previous case (the quarter car) is also considered here with the addition of a function including body roll motion. The objective function (performance index) used

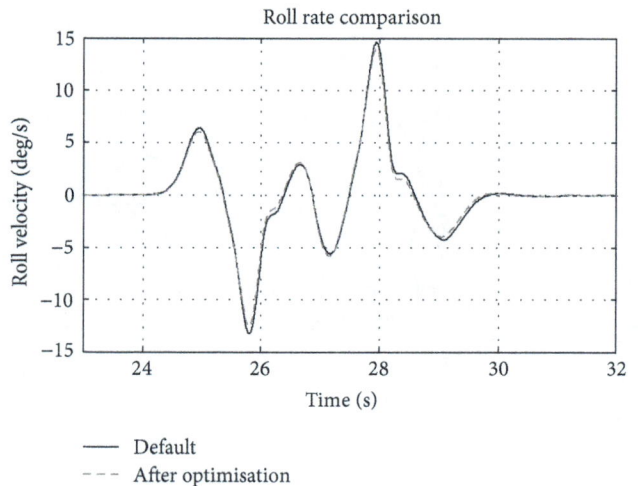

FIGURE 26: Roll angle comparison double lane change.

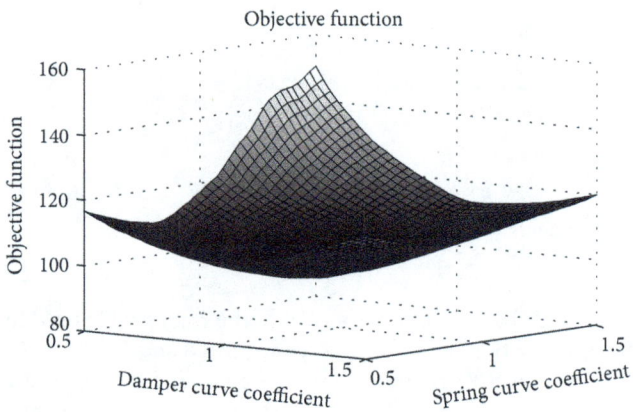

FIGURE 27: Roll rate comparison double-lane change.

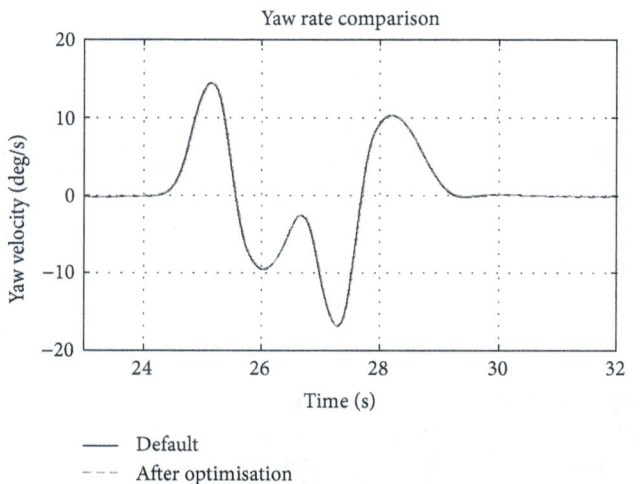

FIGURE 28: Yaw rate change during the manoeuvre.

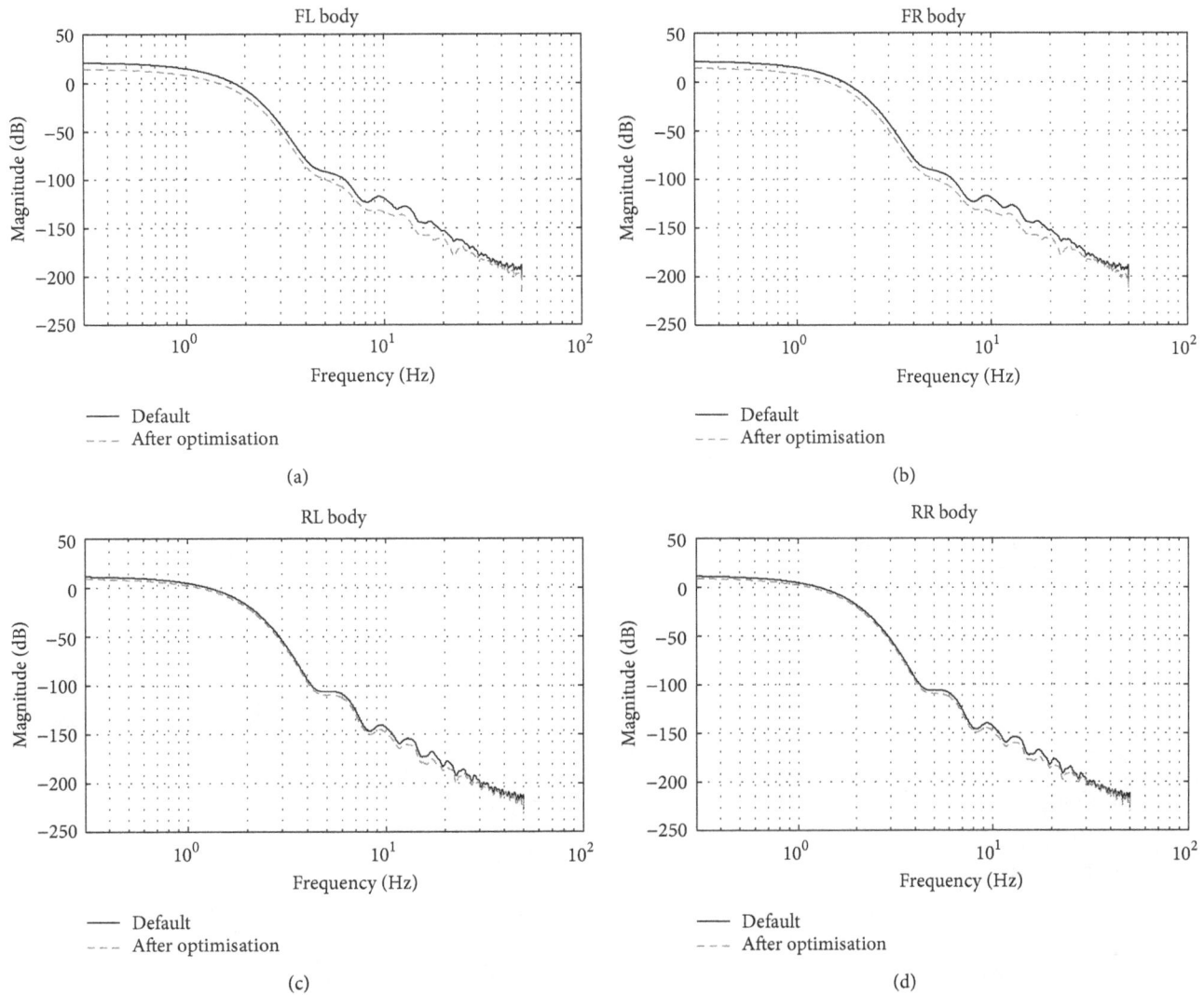

FIGURE 29: Vertical vehicle body acceleration PSD plots at four corners with high order model.

in optimisation is made up from three functions, as described below:

$$I = w_1 \ddot{Z}_s + w_2 \Delta F_{\text{tire}} + w_3 \Phi_s, \qquad (22)$$

where Φ_s represents the RMS value of the roll angle of the vehicle. The difference in the performance index for this case lies on the existence of the roll angle.

The left and the right sides of the road being followed are shown in Figure 17, and they are generated independently using the Robson Method. Since both sides of the road are independent of each other, they have zero correlation between them. Arbitrary body roll motions will be induced as a result of this road profile in the simulation runs. The rear axle nonlinear suspension unit characteristics are optimized considering body roll angle.

The optimisation results for this case are presented in Figures 18–21. The change of the performance index during the course of the optimisation run that is given in Figures 18 and 19 emphasizes the minimisation of all three indices

step by step throughout the process. How the nonlinear spring and nonlinear damper scaling coefficients change during the optimisation is shown in Figure 20. The changes of components of the performance index are shown separately in Figure 21. Figures 22 and 23 show the initial and optimised damper and spring characteristic curves, respectively. Figure 24 demonstrates the search domain of the objective function utilized according to two design parameters changing in a predetermined interval. Based on these figures, it is seen that after the optimisation the spring characteristic tends to be softer, while the damper characteristic tends to be more damped.

5.3. Handling Tests Using Carmaker Software. Finally, the ride handling and comfort performances of the optimized suspension parameters are confirmed by employing the high fidelity Carmaker model of the considered vehicle (see Figure 25). A standard double lane change maneuver is utilized to observe the body roll improvement which mostly indicates better

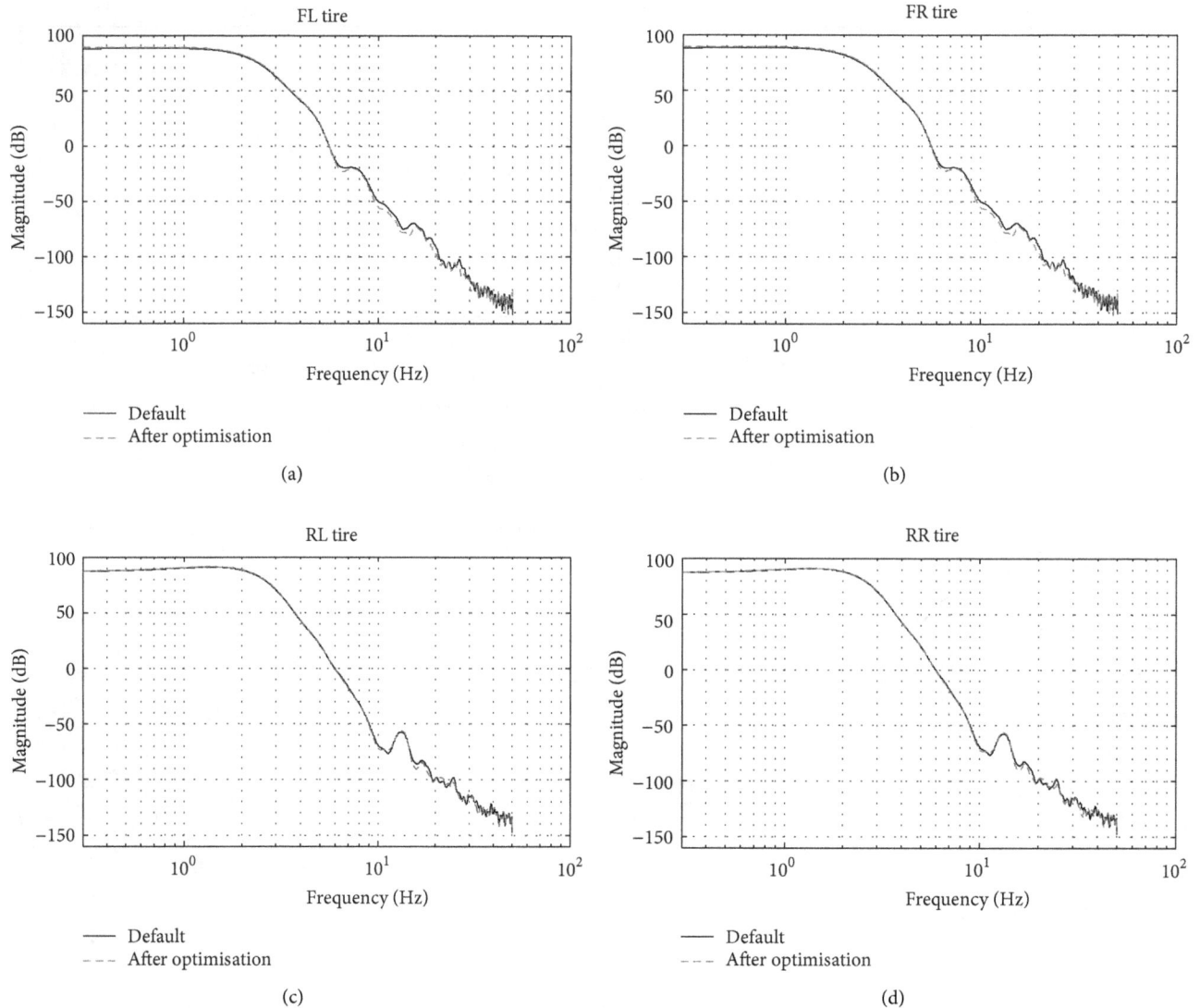

FIGURE 30: Vertical wheel acceleration PSD plots at four corners with high order model.

handling. The simulation results for roll angle and roll rate in Figures 26 and 27 show that the nonlinear optimisation has improved roll handling performance. Figure 28 shows that the yaw rate handling has not changed significantly. A test track with road irregularities which contains various frequencies is employed for the confirmation of comfort performance of the optimized suspension settings. The simulation results in Figure 29 show the ride comfort improvement in the form of lower vehicle body vertical acceleration PSD values at the four corners of the vehicle. The simulation results in Figure 30 show that the tire accelerations have not changed significantly, illustrating the maintenance of the previous road holding characteristic.

6. Conclusions

Light commercial vehicle front and rear suspension units incorporating both linear or nonlinear springs and nonlinear

dampers were optimized to improve vehicle ride and handling. The nonlinear equations of motions of the quarter car and the half car representing the front and rear suspension models were presented and simulated in the Matlab/Simulink environment. Several aspects of performance criteria were considered for ride comfort and handling such as RMS values of weighed body acceleration, the range of tire forces, and RMS values of body roll angle. For each aspect of performance, time-domain performance measures were evaluated after the optimisation run. A simple optimisation methodology of nonlinear suspension unit was presented, incorporating typical data provided by car manufacturers for the initial characteristics. The methodology was based on keeping the shape of the damper and spring properties and curve fitting a proper function to these data and then scaling it throughout the optimisation process. Finally, the advantage of the nonlinear optimised suspension unit compared to a default suspension unit was demonstrated using a double lane change maneuver with a high fidelity full vehicle model in

Carmaker. In order to generalise the nonlinear suspension unit optimisation problem, an interactive Matlab toolbox was constructed and used in obtaining the results presented here.

The fact that the improvement between original performances and optimized ones is not too big in the results presented in the paper is due to the fact that we started with an already optimized suspension which had been designed earlier. We decided against creating a nonideal starting value for the suspension parameters and showing a large improvement in performance. We were indeed able to show that the suspension design provided to us was very close to its optimum configuration, and only small performance achievements were possible. This is also a very useful outcome as the suspension designer can analytically evaluate his design against the optimal one. The method presented in the paper is fully automated in the form of a Matlab program and its interactive graphical user interface. The results are also obtained much faster as compared to the conventional method of suspension design including a lot of trial and error.

Abbreviations

DOF: Degree of freedom
I_i: Performance index, $i = 1, 2, 3 \ldots$
I: Total index
RMS: Root mean square
PSD: Power spectral density.

List of Symbols

Quarter Car Model

m_s: Sprung mass, kg
m_u: Unsprung mass, kg
z_s: Sprung mass vertical displacement, m
z_u: Unsprung mass vertical displacement, m
z_r: Road irregularity, m
b_u: Linear tire damping coefficient
k_s: Linear spring stiffness
k_u: Linear tire stiffness
B_F: Nonlinear damper characteristic function.

Half Car Model

m_s: Sprung mass, kg
m_u: Unsprung mass, kg
z_s: Sprung mass vertical displacement, m
z_u: Unsprung mass vertical displacement, m
z_{RL}, z_{RR}: Left and right road irregularities, m
k_{uL}, k_{uR}: Left and right linear tire stiffness, N/m
B_{FL}, B_{FR}: Left and right nonlinear damper characteristic functions, Ns/m
K_{FL}, K_{FR}: Left and right nonlinear damper characteristic functions, N/m
K_{Θ}:   Antiroll bar torsional stiffness, Nm/rad
L: Track width, m
ϕ_s: Sprung mass roll angle, deg/sec

ϕ_u: Unsprung mass roll angle, deg/sec
I_{xx}:   Sprung mass moment of inertia about x-axis
I_{uxx}: Unsprung mass moment of inertia about x-axis.

Optimisation Process

w_i: Weighing, $i = 1, 2, 3 \ldots$
ΔF_{tire}:   Tire force change, N
$\Delta F_{\text{tire opt}}$: Optimised value of tire force change, N
$\ddot{Z}_s$:   RMS of weighed sprung mass acceleration
$\ddot{Z}_{\text{spot}}$:   Optimised value of the RMS of weighed sprung mass acceleration
Φ_s: RMS of sprung mass roll angle.

Acknowledgments

The authors thank Yıldıray Koray for his help in the road profile used. The authors thank Server Ersolmaz, Erhan Eyol, Mustafa Sinal. and Selçuk Kervancıoğlu of Ford Otosan for helpful discussions on suspension design. The authors also thank Ford Otosan and the EU FP6 project AUTOCOM INCO-16426 for their support. The authors dedicate this paper to the dear memory of their second coauthor Professor Ü. Sönmez who passed away so suddenly and unexpectedly.

References

[1] T. Tak and S. Chung, "An optimal design software for vehicle suspension systems," in *SAE Automotive Dynamics & Stability Conference*, Troy, Mich, USA, May 2000, SAE Paper no: 2000-01-1618.

[2] D. Koulocheris, H. Vrazopoulos, and V. Dertimanis, "Optimisation algorithms for tuning suspension systems used in ground vehicles," in *Proceedings of International Body Engineering Conference & Exhibition and Automotive & Transportation Technology Conference*, Paris, France, July 2002, SAE Paper no: 2002-01-2214.

[3] P. C. Wiers and A. K. Dhingra, "On the beneficial effects of antisquat in rear suspension desion of a drag racing motorcycle," *Journal of Mechanical Design*, vol. 124, no. 1, pp. 98–105, 2002.

[4] S. A. Mitchell, S. Smith, A. Damiano, J. Durgavich, and R. MacCracken, "Use of genetic algorithms with multiple metrics aimed at the optimization of automotive suspension systems," in *Proceedings of Motorsports Engineering Conference and Exhibition*, Dearborn, Mich, USA, November 2004, SAE Paper no: 2004-01-3520.

[5] M. Raghavan, "Suspension design for linear toe curves: a case study in mechanism synthesis," *Journal of Mechanical Design*, vol. 126, no. 2, pp. 278–282, 2004.

[6] H. V. Deo and N. P. Suh, "Axiomatic design of customizable automotive suspension," in *Proceedings of the 3rd International Conference on Axiomatic Design Seoul (ICAD '04)*, 2004.

[7] J. P. C. Goncalves and J. A. C. Ambrosio, "Road vehicle modelling requirements for optimization of ride and handling," *Multibody System Dynamics*, vol. 13, pp. 3–23, 2005.

[8] P. Duysinx, O. Brüls, J. F. Collard, P. Fisette, C. Lauwerys, and J. Swevers, "Optimization of mechatronic systems: application

to a modern car equipped with a semi-active suspension," in *Proceedings of the 6th World Congresses of Structural and Multidisciplinary Optimization*, Rio de Janeiro, Brazil, 2005.

[9] A. Eskandari, O. Mirzadeh, and S. Azadi, "Optimization of a McPherson suspension system using the design of experiments method," in *SAE Automotive Dynamics, Stability & Controls Conference and Exhibition*, Novi, Mich, USA, February 2000.

[10] A. Boyalı, S. Öztürk, T. Yiğit, B. Aksun Güvenç, and L. Güvenç, "Use of computer aided methods in optimisation in vehicle suspension design and in designing active suspensions," in *Proceedings of the 6th Biennial Conference on Engineering Systems Design and Analysis (ESDA '02)*, Istanbul, Turkey, 2002.

[11] Y. He and J. McPhee, A review of automated design synthesis approaches for virtual development of ground vehicle suspensions SAE Paper no: 2007-01-0856, 2007.

[12] L. Li, C. Xia, and W. Qin, "Analysis of kinetic characteristic and structural parameter optimization of multi-link suspension," in *Proceedings of the 14th Asia Pacific Automotive Engineering Conference*, Hollywood, Calif, USA, August 2007, SAE Paper no: 2007-01-3558.

[13] P. E. Uys, P. S. Els, and M. Thoresson, "Suspension settings for optimal ride comfort of off-road vehicles travelling on roads with different roughness and speeds," *Journal of Terramechanics*, vol. 44, no. 2, pp. 163–175, 2007.

[14] H. Habibi, K. H. Shirazi, and M. Shishesaz, "Roll steer minimization of McPherson-strut suspension system using genetic algorithm method," *Mechanism and Machine Theory*, vol. 43, no. 1, pp. 57–67, 2008.

[15] J. P. C. Goncalves and J. A. C. Ambrosio, "Optimization of vehicle suspension systems for improved comfort of road vehicles using flexible multibody dynamics," *Nonlinear Dynamics*, vol. 34, no. 1-2, pp. 113–131, 2003.

[16] L. Sun, "Optimum design of "road-friendly" vehicle suspension systems subjected to rough pavement surfaces," *Applied Mathematical Modelling*, vol. 26, no. 5, pp. 635–652, 2002.

[17] D. Cebon, *Handbook of Vehicle-Road Interaction: Vehicle Dynamics, Suspension Design, and Road Damage*, Taylor & Francis, Boca Raton, Fla, USA, 1999.

[18] CarMaker Reference Manual and User's Guide.

[19] M. Li, Z. Changfu, P. Zhao, and J. Xiangping, "Parameters sensitivity analysis and optimization for the performance of vehicle handling," in *Proceedings of the 14th Asia Pacific Automotive Engineering Conference*, Hollywood, Calif, USA, August 2007, SAE Paper no: 2007-01-3573.

[20] M. Jonasson and F. Roos, "Design and evaluation of an active electromechanical wheel suspension system," *Mechatronics*, vol. 18, no. 4, pp. 218–230, 2008.

[21] BS 6841, "British standard guide to measurement and evaluation of human exposure to whole-body mechanical vibration and repeated shock," 1987.

[22] International Standard ISO-2631-1, *Mechanical Vibration and Shock-Evaluation of Human Exposure To Whole-Body Vibration, Part 1: General Requirements*, 2nd edition, 1997.

[23] K. M. A. Kamash and J. D. Robson, "The application of isotropy in road surface modelling," *Journal of Sound and Vibration*, vol. 57, no. 1, pp. 89–100, 1978.

[24] International Standard ISO-8608, *Mechanical Vibration—Road Surface Profiles—Reporting of Measured Data*, 1st edition, 1995.

17

DSP-Based Sensor Fault Detection and Post Fault-Tolerant Control of an Induction Motor-Based Electric Vehicle

**Bekheïra Tabbache,[1,2] Mohamed Benbouzid,[1]
Abdelaziz Kheloui,[2] and Jean-Matthieu Bourgeot[3]**

[1] University of Brest, EA 4325 LBMS, rue de Kergoat, CS 93837, 29238 Brest Cedex 03, France
[2] Ecole Militaire Polytechnique, UER ELT, Algiers 16111, Algeria
[3] ENIB, EA 4325 LBMS, 945, Avenue Technopole, 29280 Plouzané, France

Correspondence should be addressed to Mohamed Benbouzid, mohamed.benbouzid@univ-brest.fr

Academic Editor: Tee Cheng

This paper deals with sensor fault detection within a reconfigurable direct torque control of an induction motor-based electric vehicle. The proposed strategy concerns current, voltage, and speed sensors faults that are detected and followed by post fault-tolerant control to allow the vehicle continuous operation. The proposed approach is validated through experiments on an induction motor drive and simulations on an electric vehicle using a European urban and extraurban driving cycle.

1. Introduction

Fault tolerance is gaining interest as a means to increase the reliability, the availability, and the continuous operation of electromechanical systems among them automotive ones [1, 2]. In the automotive context, electric vehicle is a key application where the propulsion control depends on the availability and the quality of sensor measurements. Measurements, however, can be corrupted or interrupted due to sensor faults. If some sensors are missing, the controllers cannot provide the correct control actions for the EV propulsion. It is therefore compulsory to have a sensor fault detection and isolation system to improve the reliability of the electric drive. Thereafter, reconfiguration should be achieved with equivalent observed signals. This will allow fault-tolerant operation.

In this context, an FTC approach is proposed for an induction motor-based EV experiencing sensor faults (current, voltage, and speed) [3, 4]. As DTC is recognized as a high-performance control strategy for EVs electric propulsion, it has been adopted [5, 6]. In general, DTC-based induction motor drives use two current sensors, one or two voltage sensors, and a speed sensor.

For sensor fault tolerance purposes, the tendency is to use three currents sensors and introduce observers and estimation techniques to detect and isolate current and speed sensor faults [6–12]. Some proposed FTC approaches use three nonlinear observers to guarantee the information redundancy [9]. Unfortunately, multiple observer schemes cannot be so easily implemented due to the limited sampling period even with recently developed DSPs. In [9], a Luenberger observer is used to estimate the speed and generates residuals. Unfortunately, even if this observer gives good results, the obtained performances in case of induction motors are proven to be limited in particular at low speed. This is mainly due to the induction motor strong nonlinearity.

The proposed FTC approach, which is based on a bank of observers, the Extended Kalman Filter is adopted for the estimation of the stator flux, the speed, and the generation of the current and speed residuals for fault detection and replacement signals for the reconfiguration. To

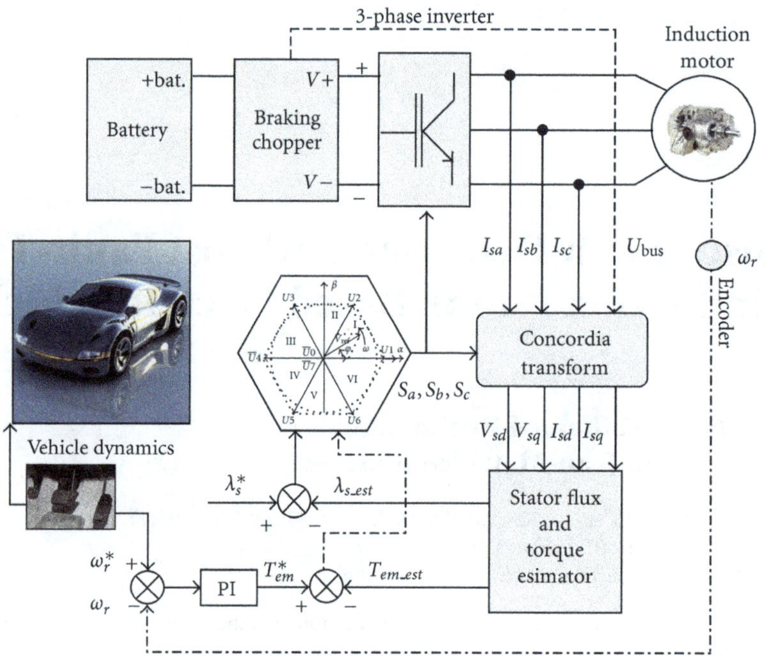

FIGURE 1: DTC block diagram.

detect and isolate sensor faults, nonlinear observers are used to guarantee the redundancy [10].

2. Induction Motor-Based EV DTC Briefly

The basic idea of the method is to calculate flux and torque instantaneous values only from the stator variables. Flux, torque, and speed are estimated. The input of the motor controller is the reference speed, which is directly applied by the pedal of the vehicle. Control is carriedout by hysteresis comparators and a switching logic table selecting the appropriate voltage inverter switching configurations [5].

Figure 1 gives the global configuration of a DTC scheme and also shows how the EV dynamics is taken into account.

3. Sensor Fault Detection and Isolation

Sensor fault detection and isolation (SFDI) is based on two parts. The first one generates sensor residuals. The second part detects and isolates the faulty sensors (current, voltage, or speed sensors).

3.1. EKF for Residual Generation. In our case, only one observer, the EKF for instance, has been adopted.

The Kalman filter is a special class of linear observer (deterministic type), derived to meet a particular optimality stochastic condition.

The Kalman filter has two forms: basic and extended. The EKF can be used for nonlinear systems where the plant model is extended by extravariables, in our case by the mechanical speed [13].

In an induction motor drive, the Kalman filter is used to obtain unmeasured state variables (rotor speed ω_r, rotor flux vector components $\lambda_{r\alpha}$, and $\lambda_{r\beta}$) using the measured state variables (stator current i_s and voltage components V_s in the Concordia frame α-β). Moreover, it takes into account the model and measurement noises.

The induction motor state model used by the EKF is developed in the stationary reference frame and summarized by (1) [5, 14], where R is the resistance, L is the inductance, and L_m is the magnetizing inductance.

The implementation of the Kalman filter is based on a recursive algorithm minimizing the error variance between the real variable and its estimate

$$
\frac{d}{dt}\begin{bmatrix} i_{s\alpha} \\ i_{s\beta} \\ \lambda_{r\alpha} \\ \lambda_{r\beta} \\ \omega_r \end{bmatrix} = \begin{bmatrix} -\dfrac{K_R}{K_L} & 0 & \dfrac{L_m R_r}{L_r^2 K_L} & \dfrac{L_m \omega_r}{L_r K_L} & 0 \\ 0 & -\dfrac{K_R}{K_L} & \dfrac{L_m \omega_r}{L_r K_L} & \dfrac{L_m R_r}{L_r^2 K_L} & 0 \\ \dfrac{L_m}{T_r} & 0 & -\dfrac{1}{T_r} & -\omega_r & 0 \\ 0 & \dfrac{L_m}{T_r} & \omega_r & -\dfrac{1}{T_r} & 0 \\ 0 & 0 & 0 & 0 & 1 \end{bmatrix} \begin{bmatrix} i_{s\alpha} \\ i_{s\beta} \\ \lambda_{r\alpha} \\ \lambda_{r\beta} \\ \omega_r \end{bmatrix}
$$

$$
+ \frac{1}{K_L}\begin{bmatrix} 1 & 0 \\ 0 & 1 \\ 0 & 0 \\ 0 & 0 \\ 0 & 0 \end{bmatrix} \begin{bmatrix} V_{s\alpha} \\ V_{s\beta} \end{bmatrix}, \quad (1)
$$

$$
\begin{bmatrix} i_{s\alpha} \\ i_{s\beta} \end{bmatrix} = \begin{bmatrix} 1 & 0 & 0 & 0 & 0 \\ 0 & 1 & 0 & 0 & 0 \end{bmatrix} \begin{bmatrix} i_{s\alpha} \\ i_{s\beta} \\ \lambda_{r\alpha} \\ \lambda_{r\beta} \\ \omega_r \end{bmatrix},
$$

where $K_L/K_R = R_s/L_s + 1 - \sigma/\sigma T_r$, $T_r = L_r/R_r$, $T_s = L_s/R_s$, and $\sigma = 1 - L_m^2/L_s L_r$.

Let us consider a linear stochastic system whose discrete state model is given by

$$x(k + 1) = Ax(k) + Bu(k) + w(k),$$
$$y(k + 1) = Cx(k) + v(k), \tag{2}$$

where $w(k)$ represents the disturbances vector applied to the system inputs. It also represents the modeling uncertainties; $v(k)$ corresponds to system output measurement noises. It is supposed that the random signals $v(k)$ and $w(k)$ are Gaussian noises not correlated and with null average value. They are characterized by covariance matrixes, Q and R, respectively, which are symmetrical and positive definite. The initial state vector x_0 is also a random variable with covariance matrix P_0 and average value $\overline{x}_0$.

The Kalman filter recursive algorithm is illustrated by Figure 2. For an induction motor, the Kalman filter must be used in its extended version. Therefore, a nonlinear stochastic system discrete state equation is given by

$$x_{k+1} = f(x_k, u_k) + w_k,$$
$$y_k = h(x_k) + v_k, \tag{3}$$

where f and h are vector functions

$$f = \begin{bmatrix} \left(1 - T\frac{K_R}{K_L}\right)i_{s\alpha} + T\frac{L_m R_r}{L_r^2 K_1}\lambda_{r\alpha} + T\frac{L_m \omega_r}{L_r K_L}\lambda_{r\beta} + T\frac{1}{K_L}V_{s\alpha} \\ \left(1 - T\frac{K_R}{K_L}\right)i_{s\beta} - T\frac{L_m R_r}{L_r^2 K_1}\lambda_{r\alpha} + T\frac{L_m \omega_r}{L_r K_L}\lambda_{r\beta} + T\frac{1}{K_L}V_{s\beta} \\ T\frac{L_m}{T_r}i_{s\alpha} + \left(1 - T\frac{1}{T_r}\right)\lambda_{r\alpha} - T\omega_r\lambda_{r\beta} \\ T\frac{L_m}{T_r}i_{s\beta} + T\omega_r\lambda_{r\alpha} + \left(1 - T\frac{1}{T_r}\right)\lambda_{r\beta} \\ \omega_r \end{bmatrix},$$

$$h = C_d x_{k|k+1} = \begin{bmatrix} i_{s\alpha} \\ i_{s\beta} \end{bmatrix}. \tag{4}$$

The notation $k + 1$ is related to predicted values at $(k + 1)$th instant and is based on measurements up to kth instant. T is the sampling period.

The EKF equations are similar to those of the linear Kalman filter with the difference that A and C matrices should be replaced by the Jacobians of the vector functions f and h at every sampling time as follows:

$$A_k[i, j] = \left.\frac{\partial f_i}{\partial x_j}\right|_{x=\hat{x}(k|k)},$$
$$C_k[i, j] = \left.\frac{\partial h_i}{\partial x_j}\right|_{x=\hat{x}(k|k-1)}. \tag{5}$$

The covariance matrices R_k and Q_k are also defined at every sampling time.

For the induction motor control, the EKF is used for the speed real-time estimation. It can also be used to estimate states and parameters using the motor voltages and currents measurements.

3.2. Sensor Fault Detection and Isolation

3.2.1. Current Sensor Faults.
Three sensors are used to measure the motor currents. To detect current sensor faults, the following equation is used:

$$i_{\text{sum}} = i_{as}^m + i_{bs}^m + i_{cs}^m, \tag{6}$$

where the upper script m means a measured quantity.

Indeed, if one or two sensors fail, i_{sum} will be a nonzero sinusoidal signal. Therefore, additional logic and information (redundancy) are required to isolate the failed sensor. The required redundancy can be obtained from the EKF which is driven by the scheduled test input sets

$$\text{CSFI}(1) = \left\{i_{as}^m, i_{bs}^m, i_{cs}^{m1}\right\} \quad \text{when } i_{cs}^{m1} = -\left(i_{as}^m + i_{bs}^m\right),$$
$$\text{CSFI}(2) = \left\{i_{as}^m, i_{bs}^{m2}, i_{cs}^m\right\} \quad \text{when } i_{bs}^{m2} = -\left(i_{as}^m + i_{cs}^m\right), \tag{7}$$

where CSFI is the Current Sensor Fault Isolation input for phase a or b. CSFI(1) is used to isolate a faulty current sensor in phase c and CSFI(2) to isolate a faulty current sensor in phase a or b. It should be noticed that the two current residuals are calculated using Concordia components to isolate the faulty current sensor.

Sensor fault detection is performed using i_{sum}. The first residue is calculated using the Concordia currents provided by CSFI(1)

$$r_1 = \left|i_{s\alpha}^{m1} - \hat{i}_{s\alpha}\right| + \left|i_{s\beta}^{m1} - \hat{i}_{s\beta}\right|. \tag{8}$$

If this residue is lower than a predefined threshold, current sensor (c) should be the failed one. Otherwise, the fault sensors are (a) or (b). In this case, an additional residue is calculated using CSFI(2)

$$r_2 = \left|i_{s\alpha}^{m2} - \hat{i}_{s\alpha}\right| + \left|i_{s\beta}^{m2} - \hat{i}_{s\beta}\right|. \tag{9}$$

If this residue is greater than a predefined threshold, the faulty current sensor is (a). Otherwise, it is (b). The estimated currents ($\hat{}$) are provided by the EKF.

It should be noted that the proposed fault detection method avoid merging CSFI(1) and CSFI(2) information.

3.2.2. Voltages Sensor Faults.
The fault detection may be performed using a simple threshold test on the parity equation (10), which describes the three-phase simple voltage equivalence

$$V_{\text{sum}} = V_{as}^m + V_{bs}^m + V_{cs}^m. \tag{10}$$

It is used to monitor the EKF inputs and therefore ensure the efficient operation of the proposed FTC.

3.2.3. Speed Sensor Faults.
The fault detection is achieved by comparing the measured speed with the estimated one given by the EKF. The encoder fault detection is given by the following residual:

$$r_\omega = \left|\omega_m^m - \hat{\omega}_m^m\right|. \tag{11}$$

Figure 3 illustrates the SFDI principle. It includes the EKF, the residual generation, and the Concordia transforms.

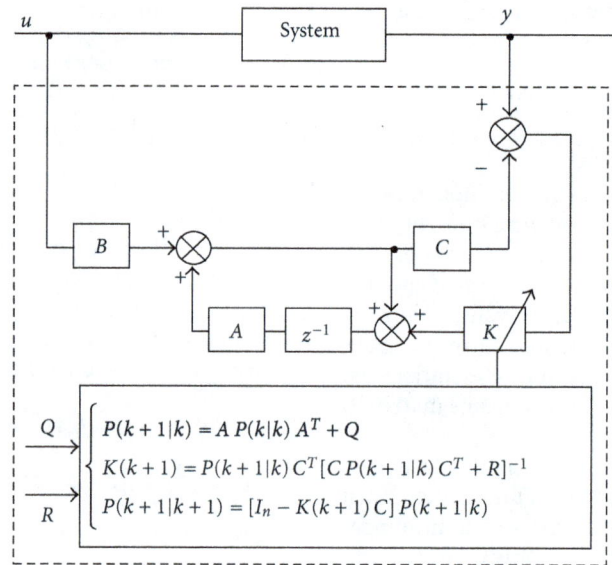

FIGURE 2: The Kalman filter recursive algorithm.

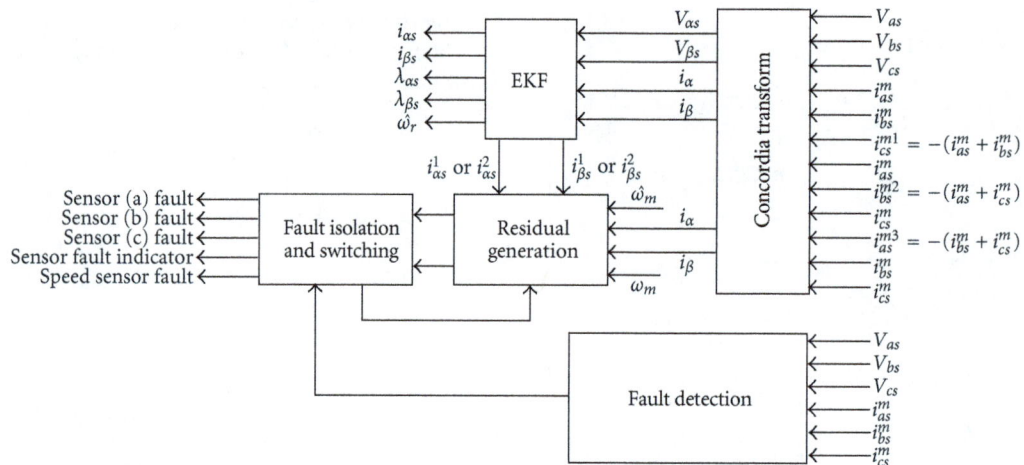

FIGURE 3: Sensor fault detection and isolation scheme.

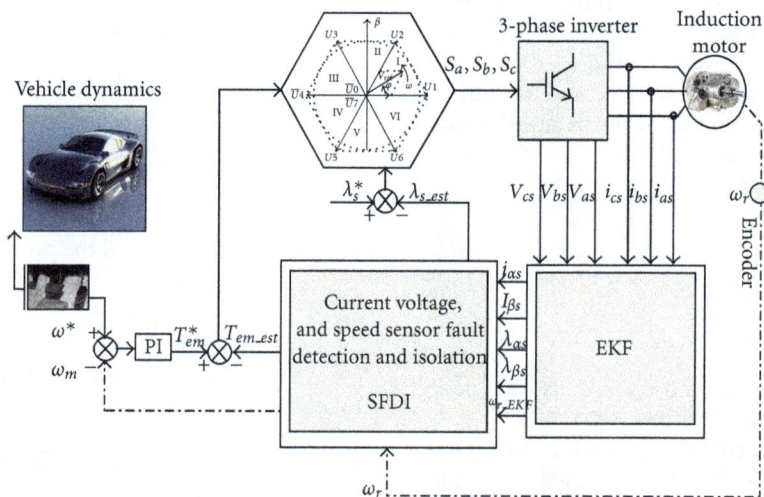

FIGURE 4: The proposed sensor fault-tolerant control scheme.

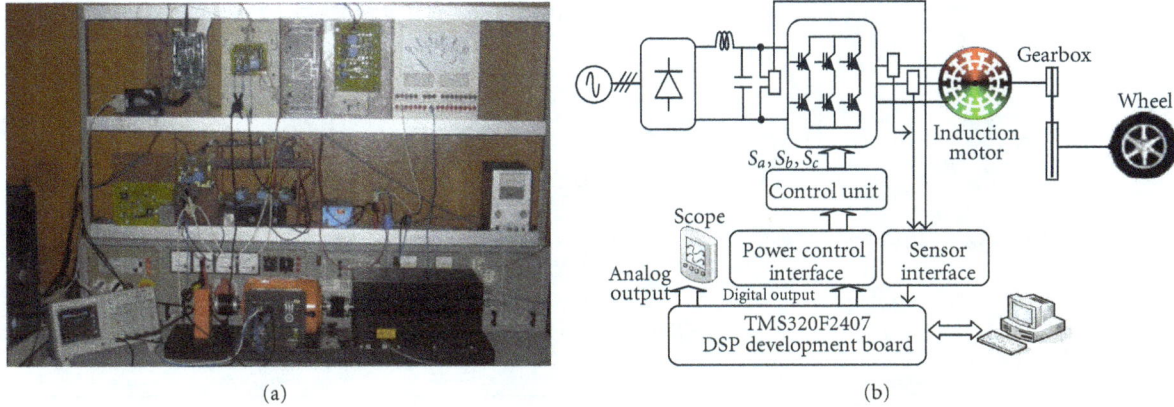

(a)   (b)

FIGURE 5: (a) View and (b) schematic description of the experimental bench.

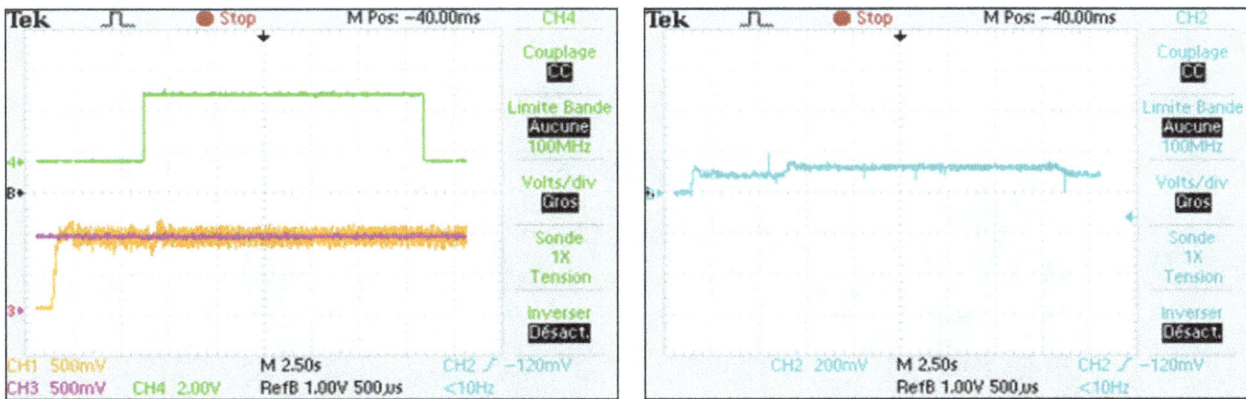

(a) Sensor fault indicator (green), motor speed (orange), reference speed (purple)

(b) Current residual

FIGURE 6: Induction motor drive FTC performance under a current sensor fault.

3.3. Sensor Fault-Tolerant Control Scheme. Figure 4 describes the proposed fault-tolerant control scheme in terms of current, voltage, and speed sensor faults.

4. Experimental and Simulation Tests

4.1. Induction Motor Sensor Fault-Tolerant Control. Experimental tests have been first carriedout to check the sensor fault-tolerant control performances on a 1-kW induction motor drive (Figure 5).

The used cage induction motor rated data are given in the Appendix. This motor is supplied by a 2-level voltage inverter. The setup main components are a DSP system (single fixed-point TMS320LF2407), a speed sensor attached to the motor shaft, current and voltage sensors. The DSP system is interfaced to a standard PC.

Figures 6 and 7 illustrate experimental results for current and speed sensors, respectively. Figure 6 shows the response of the proposed sensor FTC scheme in the event of one current sensor failure (phase b). In this case, the proposed algorithm ensures the control as shown by the speed and its reference in Figure 6(a). Figure 7 shows the sensor FTC

performances in the event of a speed sensor failure. In this case, the sensor speed is replaced by the EKF estimates ensuring then the control as illustrated by Figure 7(a).

The obtained results confirm the effectiveness of the proposed sensor fault detection and post fault-tolerant control approach. Indeed, quiet good speed tracking performances are achieved.

4.2. EV Sensor Fault-Tolerant Control. The proposed sensor fault detection and post fault-tolerant control approach is now evaluated for an electric vehicle using a 37-kW induction motor based powertrain. The EV and the used cage induction motor rated data and parameters are given in the Appendix. Simulations are carriedout using a European urban and extraurban driving cycle as speed reference. For that purpose, two sensor faults are introduced: a current sensor fault in phase a (an offset) at 2-sec and a speed sensor fault (a signal disconnection) at 7-sec.

Figure 8 shows the EV fault-tolerant performances in terms of speed (via a single gear). The obtained results clearly confirm the effectiveness of the proposed post fault-tolerant control approach.

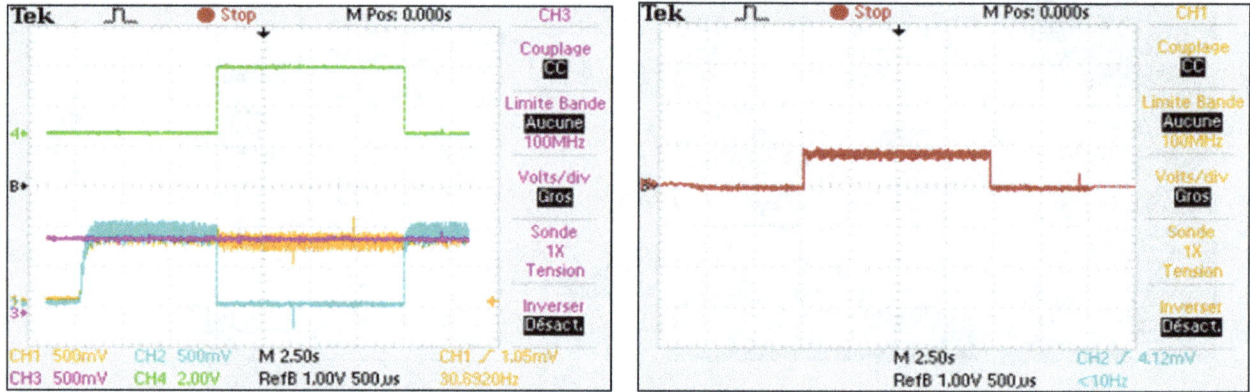

(a) Sensor fault indicator (green), motor speed (orange), speed sensor output (blue), reference speed (purple)

(b) Speed residual

FIGURE 7: Induction motor drive FTC performance under a speed sensor fault.

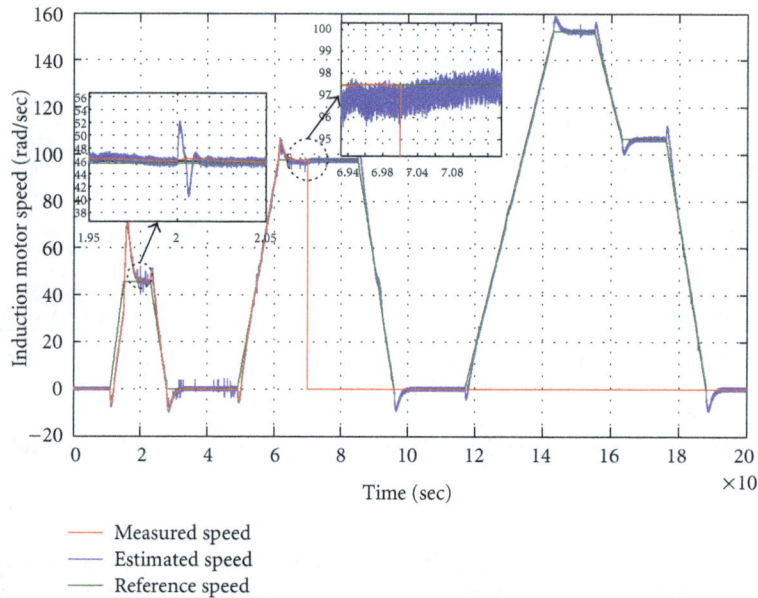

FIGURE 8: EV FTC performances: measured speed (red), estimated speed (blue), and reference speed (green).

5. Conclusion

This paper dealt with fault-tolerant control of an induction motor-based EV experiencing sensor faults (current, voltage, and speed). The carried-out simulations and experiments confirm that the proposed sensor post fault-tolerant control approach seems to be effective and provides a simple configuration with high performance in terms of speed response.

Appendix

EV Mechanical and Aerodynamic Parameters: $m = 1540$ kg (two 70 kg passengers), $A = 1.8$ m^2, $r = 0.3$ m, $\mu_{rr1} = 0.0055$, $\mu_{rr2} = 0.056$, $C_{ad} = 0.19$, $G = 3.29$, $\eta_g = 0.95$, $v_0 = 4.155$ m/sec, $g = 9.81$ m/sec^2, $\rho = 0.23$ kg/m^3.

Rated Data of the Simulated Induction Motor: 37 kW, 1480 rpm, $p = 2$, $R_s = 0.0851\,\Omega$, $R_r = 0.0658\,\Omega$, $L_s = 0.0314$ H, $L_r = 0.0291$ H, $L_m = 0.0291$ H, $J = 0.37$ kg · m^2, $k_f = 0.02791$ Nmsec.

Rated Data of the Tested Induction Motor: 1 kW, 2.5 Nm, 2830 rpm, $p = 1$, $R_s = 4.750\,\Omega$, $R_r = 8.000\,\Omega$, $L_s = 0.375$ H, $L_r = 0.375$ H, $L_m = 0.364$ H, $J = 0.003$ kg · m^2, $k_f = 0.0024$ Nmsec.

References

[1] D. U. Campos-Delgado, D. R. Espinoza-Trejo, and E. Palacios, "Fault-tolerant control in variable speed drives: a survey," *IET Electric Power Applications*, vol. 2, no. 2, pp. 121–134, 2008.

[2] M. E. H. Benbouzid, D. Diallo, and M. Zeraoulia, "Advanced fault-tolerant control of induction-motor drives for EV/HEV

traction applications: from conventional to modern and intelligent control techniques," *IEEE Transactions on Vehicular Technology*, vol. 56, no. 2, pp. 519–528, 2007.

[3] M. Zeraoulia, M. E. H. Benbouzid, and D. Diallo, "Electric motor drive selection issues for HEV propulsion systems: a comparative study," *IEEE Transactions on Vehicular Technology*, vol. 55, no. 6, pp. 1756–1764, 2006.

[4] B. Tabbache, M. E. H. Benbouzid, A. Kheloui, and J. M. Bourgeot, "Sensor fault-tolerant control of an induction motor based electric vehicle," in *Proceedings of the 14th European Conference on Power Electronics and Applications (EPE 2011)*, pp. 1–8, Birmingham, UK, September 2011.

[5] B. Tabbache, A. Kheloui, and M. E. H. Benbouzid, "An adaptive electric differential for electric vehicles motion stabilization," *IEEE Transactions on Vehicular Technology*, vol. 60, no. 1, pp. 104–110, 2011.

[6] K. S. Lee and J. S. Ryu, "Instrument fault detection and compensation scheme for direct torque controlled induction motor drives," *IEE Proceedings*, vol. 150, no. 4, pp. 376–382, 2003.

[7] K. Rothenhagen and F. W. Fuchs, "Doubly fed induction generator model-based sensor fault detection and control loop reconfiguration," *IEEE Transactions on Industrial Electronics*, vol. 56, no. 10, pp. 4229–4238, 2009.

[8] K. Rothenhagen and F. W. Fuchs, "Current sensor fault detection, isolation, and reconfiguration for doubly fed induction generators," *IEEE Transactions on Industrial Electronics*, vol. 56, no. 10, pp. 4239–4245, 2009.

[9] S. M. Bennett, R. J. Patton, and S. Daley, "Sensor fault-tolerant control of a rail traction drive," *Control Engineering Practice*, vol. 7, no. 2, pp. 217–225, 1999.

[10] A. Akrad, M. Hilairet, and D. Diallo, "Design of a fault-tolerant controller based on observers for a PMSM drive," *IEEE Transactions on Industrial Electronics*, vol. 58, no. 4, pp. 1416–1427, 2011.

[11] L. Baghli, P. Poure, and A. Rezzoug, "Sensor fault detection for fault tolerant vector controlled induction machine," in *Proceedings of the European Conference on Power Electronics and Applications*, pp. 1–10, Dresden, Germany, September 2005.

[12] B. Tabbache, M. E. H. Benbouzid, A. Kheloui, and J. M. Bourgeot, "DSP-based sensor fault-tolerant control of electric vehicle powertrains," in *Proceedings of the IEEE International Symposium on Industrial Electronics (ISIE '11)*, pp. 2085–2090, Gdansk, Poland, June 2011.

[13] L. Harnefors, "Instability phenomena and remedies in sensorless indirect field oriented control," *IEEE Transactions on Power Electronics*, vol. 15, no. 4, pp. 733–743, 2000.

[14] B. Akin, U. Orguner, A. Ersak, and M. Ehsani, "Simple derivative-free nonlinear state observer for sensorless AC drives," *IEEE/ASME Transactions on Mechatronics*, vol. 11, no. 5, pp. 634–643, 2006.

Permissions

The contributors of this book come from diverse backgrounds, making this book a truly international effort. This book will bring forth new frontiers with its revolutionizing research information and detailed analysis of the nascent developments around the world.

We would like to thank all the contributing authors for lending their expertise to make the book truly unique. They have played a crucial role in the development of this book. Without their invaluable contributions this book wouldn't have been possible. They have made vital efforts to compile up to date information on the varied aspects of this subject to make this book a valuable addition to the collection of many professionals and students.

This book was conceptualized with the vision of imparting up-to-date information and advanced data in this field. To ensure the same, a matchless editorial board was set up. Every individual on the board went through rigorous rounds of assessment to prove their worth. After which they invested a large part of their time researching and compiling the most relevant data for our readers. Conferences and sessions were held from time to time between the editorial board and the contributing authors to present the data in the most comprehensible form. The editorial team has worked tirelessly to provide valuable and valid information to help people across the globe.

Every chapter published in this book has been scrutinized by our experts. Their significance has been extensively debated. The topics covered herein carry significant findings which will fuel the growth of the discipline. They may even be implemented as practical applications or may be referred to as a beginning point for another development. Chapters in this book were first published by Hindawi Publishing Corporation; hereby published with permission under the Creative Commons Attribution License or equivalent.

The editorial board has been involved in producing this book since its inception. They have spent rigorous hours researching and exploring the diverse topics which have resulted in the successful publishing of this book. They have passed on their knowledge of decades through this book. To expedite this challenging task, the publisher supported the team at every step. A small team of assistant editors was also appointed to further simplify the editing procedure and attain best results for the readers.

Our editorial team has been hand-picked from every corner of the world. Their multi-ethnicity adds dynamic inputs to the discussions which result in innovative outcomes. These outcomes are then further discussed with the researchers and contributors who give their valuable feedback and opinion regarding the same. The feedback is then collaborated with the researches and they are edited in a comprehensive manner to aid the understanding of the subject.

Apart from the editorial board, the designing team has also invested a significant amount of their time in understanding the subject and creating the most relevant covers. They scrutinized every image to scout for the most suitable representation of the subject and create an appropriate cover for the book.

The publishing team has been involved in this book since its early stages. They were actively engaged in every process, be it collecting the data, connecting with the contributors or procuring relevant information. The team has been an ardent support to the editorial, designing and production team. Their endless efforts to recruit the best for this project, has resulted in the accomplishment of this book. They are a veteran in the field of academics and their pool of knowledge is as vast as their experience in printing. Their expertise and guidance has proved useful at every step. Their uncompromising quality standards have made this book an exceptional effort. Their encouragement from time to time has been an inspiration for everyone.

The publisher and the editorial board hope that this book will prove to be a valuable piece of knowledge for researchers, students, practitioners and scholars across the globe.

List of Contributors

Othman Nasri, Hassan Shraim and Phillippe Dague
LRI, University Paris-Sud 11, CNRS & INRIA Saclay Île-de-France, Bât 490, 91405 Orsay Cedex, France

Olivier Heron and Michael Cartron
CEA LIST, Saclay, 91191 Gif-sur-Yvette Cedex, France

Youssef A. Ghoneim
Research and Development Center, General Motors Corporation, 30500 Mound Road, Warren, MI 48090, USA

Kamini Rawat, Vinod Kumar Katiyar and Pratibha Gupta
Department of Mathematics, IIT Roorkee, Roorkee 247667, India

Youssef Tawk, Aleksandar Jovanovic, Phillip Tomé, Jérôme Leclère, Cyril Botteron and Pierre-André Farine
Electronics and Signal Processing Laboratory, Institute of Microengineering (IMT), École Polytechnique Fédérale de Lausanne, Breguet 2, 2000 Neuchâtel, Switzerland

Ruud Riem-Vis and Bertrand Spaeth
Jiiva, Stadtbachstrasse 40, 3012 Bern, Switzerland

Omar Yaqub and Lingxi Li
Department of Electrical and Computer Engineering, Indiana University-Purdue University Indianapolis (IUPUI), Indianapolis, IN 46202, USA
Transportation Active Safety Institute (TASI), Indiana University-Purdue University Indianapolis (IUPUI), Indianapolis, IN 46202, USA

Mohamad-Hoseyn Sigari
Control and Intelligent Processing Center of Excellence (CIPCE), School of Electrical and Computer Engineering, College of Engineering, University of Tehran, Tehran 14399, Iran

Mahmood Fathy and Mohsen Soryani
Computer Engineering Department, Iran University of Science and Technology, Tehran 16846, Iran

Payman Shakouri and Andrzej Ordys
School of Mechanical and Automotive Engineering, Kingston University of London, London SW15 3DW, UK

Paul Darnell and Peter Kavanagh
Jaguar Land Rover, Banbury Road, Gaydon, Warwickshire CV35 ORR, UK

Michael A. Roscher, Roland Michel and Wolfgang Leidholdt
imk automotive GmbH, Annaberger Straße 73, 09126 Chemnitz, Germany

King Hann Lim
Electrical and Computer Department, School of Engineering, Curtin University Sarawak, CDT 250, Sarawak, 98009 Miri, Malaysia

Kah Phooi Seng
School of Computer Technology, Sunway University, No. 5, Jalan Universiti, Bandar Sunway, Selangor, 46150 Petaling Jaya, Malaysia

Li-Minn Ang
Centre for Communications Engineering Research, Edith Cowan University, Joondalup, WA 6027, Australia

Md. Jahidur Rahman
Department of Electrical and Computer Engineering, The University of British Columbia, Vancouver, Canada V6T 1Z4

Jiaxin Yang
Department of Electrical and Computer Engineering, McGill University, Montreal, Canada H3A 0E9

Jonathan Ledy, Anne-Marie Poussard and Rodolphe Vauzelle
Laboratoire XLIM-SIC, UMR CNRS 6172, Université de Poitiers, 86034 Poitiers, France

Hervé Boeglen and Benoît Hilt
Laboratoire MIPS-GRTC, Université de Haute Alsace, 68000 Colmar, France

Daiheng Ni, Jia Li, Steven Andrews, and Haizhong Wang
Department of Civil and Environmental Engineering, University of Massachusetts Amherst, Amherst, MA 01003, USA

Bing-Fei Wu, Chih-Chung Kao, Ying-Feng Li and Min-Yu Tsai
Institute of Electrical and Control Engineering, National Chiao Tung University, Hsinchu 300, Taiwan

Motoyuki Akamatsu
Human Technology Research Institute, AIST, Japan

Paul Green
University of Michigan Transportation Research Institute (UMTRI), USA

Klaus Bengler
Institute of Ergonomics, Technische Universität München, Germany

Carl Tucker, Rachel Tucker and Jun Zheng
Department of Computer Science and Engineering, New Mexico Institute of Mining and Technology, Socorro, NM 87801, USA

Dinçer Özcan, Ümit Sönmez and Levent Güvenç
Mekar Mechatronics Research Labs, Department of Mechanical Engineering, İstanbul Okan University, Akfırat-Tuzla, TR-34959 İstanbul, Turkey

Bekheïra Tabbache and Mohamed Benbouzid
University of Brest, EA 4325 LBMS, rue de Kergoat, CS 93837, 29238 Brest Cedex 03, France

Bekheïra Tabbache and Abdelaziz Kheloui
Ecole Militaire Polytechnique, UER ELT, Algiers 16111, Algeria

Jean-Matthieu Bourgeot
ENIB, EA 4325 LBMS, 945, Avenue Technopole, 29280 Plouzan´e, France

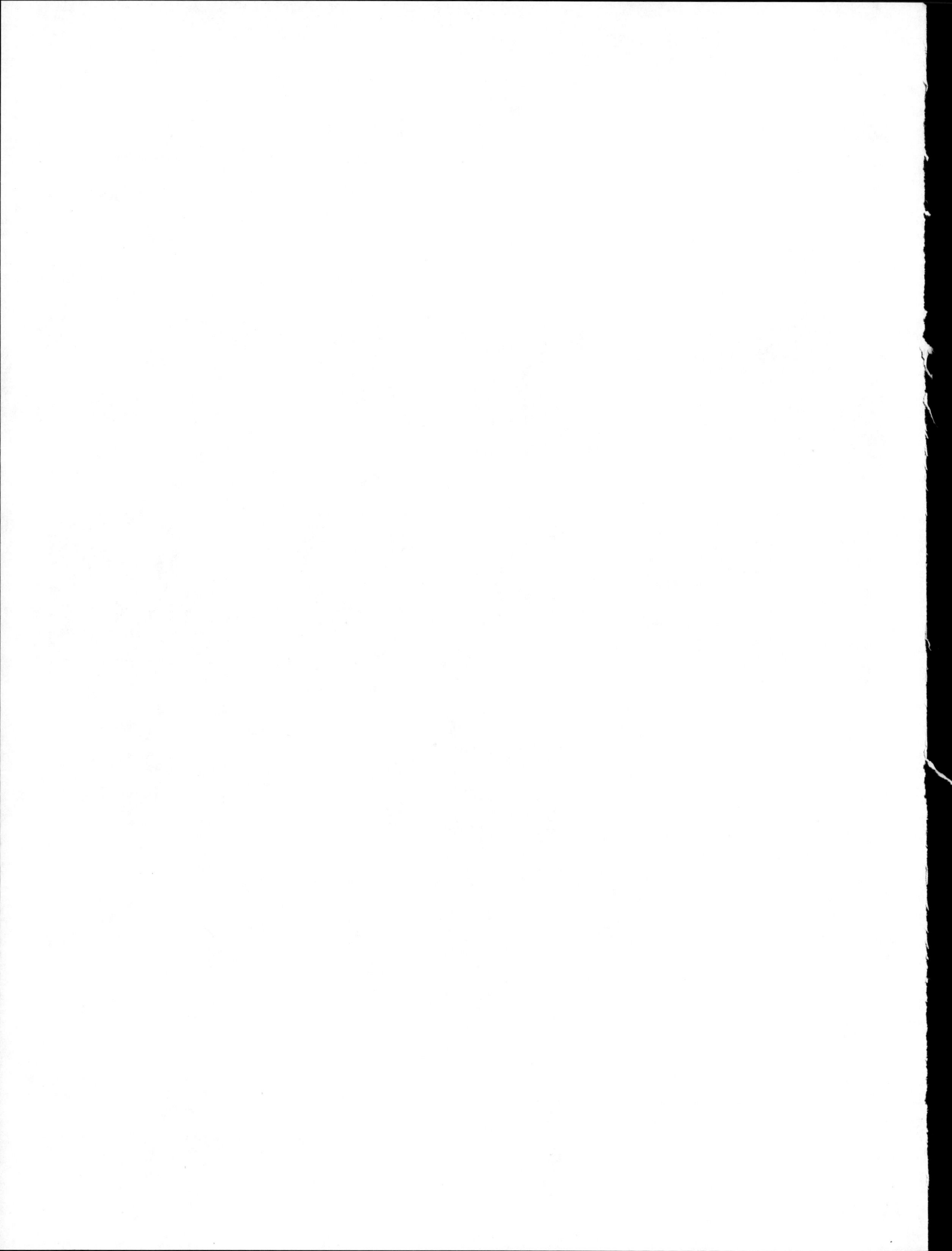